THE BRAIN FROM 25,000 FEET

SYNTHESE LIBRARY

STUDIES IN EPISTEMOLOGY,

LOGIC, METHODOLOGY, AND PHILOSOPHY OF SCIENCE

VOLUME 317

THE BRAIN FROM 25,000 FEET

High Level Explorations of Brain Complexity, Perception, Induction and Vagueness

by

MARK A. CHANGIZI

Sloan-Swartz Center for Theoretical Neurobiology,
Caltech, Pasadena, CA, U.S.A.

KLUWER ACADEMIC PUBLISHERS
DORDRECHT / BOSTON / LONDON

A C.I.P. Catalogue record for this book is available from the Library of Congress.

ISBN 1-4020-1176-8

Published by Kluwer Academic Publishers,
P.O. Box 17, 3300 AA Dordrecht, The Netherlands.

Sold and distributed in North, Central and South America
by Kluwer Academic Publishers,
101 Philip Drive, Norwell, MA 02061, U.S.A.

In all other countries, sold and distributed
by Kluwer Academic Publishers,
P.O. Box 322, 3300 AH Dordrecht, The Netherlands.

Printed on acid-free paper

Printed in the Netherlands.

In memory of my father,
M. Hosein Changizi, M.D. (1936–1997),
who taught me to play devil's advocate.

Contents

Preface

Since brains are not "impossibility engines," they cannot do the logically impossible. Since brains are not infinite in size or speed, they cannot do the computationally impossible. Since brains are found *in* the universe, rather than in some fantasy world, they cannot do the physically impossible. Brains have constraints. And not simply the garden variety Earthly constraints like having to work well at the human body temperature, or having some upper limit in working memory. Brains have "high level" constraints, by which I mean constraints to which nearly any other possible kind of brain will also be subject. Such constraints are typically more at a physics or mathematics level, rather than depending on the particular contingencies of the ecology encountered by any specific kind of brain.

To understand the brain one must, I believe, understand the limits and principles governing all possible brain-like things—objects that somehow instantiate minds. For example, if I tell you I want to know how this computer on which I am typing works, and you tell me about the details of just this kind of computer, you really would have missed the point. That is, unless I already knew all about how computers worked generally, and just wanted to know the specifics about this kind of computer. But if I did not understand how *any* kind of computer works, then to really explain to me how this computer works will require telling me how computers work in general. That is what is interesting about computers and computation: the fundamental questions in computer science are about how computer-like things work generally, not about how this or that computer actually happens to work. Similarly, what is most fascinating about the brain is not that it is our brain (although that helps), but that it happens to be one of presumably an infinite class of brain-like things, and I want to know how brain-like things work.

To do *this*, one must back sufficiently far away from the brains we find here on Earth so as to lose sight of these brains' distracting peculiarities, and consequently to gain a focus on what is important about these Earthly brains. That is, we must view the brain from 25,000 feet up—or, from very high up. At this height, the details of the brain are lost, whether they be ion channels, intricate neural connectivity patterns, or pharmacological effects. The background required of a researcher who wishes to study the brain from this high level is therefore not traditional neurobiology, neuroanatomy, psychology or even computational neuroscience (the latter which almost always focuses on modeling specific, relatively lower-level mechanisms thought to occur in the brain). What is needed is a training in mathematics, computer science and physics, and even an appreciation of conceptual limits from philosophy.

This book serves two purposes. One aim is to illustrate a number of high-level approaches to brain science. These range from an explanation for why natural language is vague, to a solution to the riddle of induction and applications to issues of innateness, to the use of probability and decision theory in modeling perception, to the inevitability of visual illusions for any animal with a non-instantaneous brain, and finally to the morphology, complexity and scaling behavior of many aspects of nervous systems. The second aim is to both encourage others to approach brain science from a high level and to provide, along the way, an introduction to some of the mathematical, computational and conceptual principles needed to be able to think about brains at a higher level.

For the remainder of this preface, I wish (i) to provide a preview of the topics covered throughout the book, and (ii) to communicate the overarching philosophy I take toward the brain sciences, a philosophy that is the theme connecting the diverse topics through which we traverse in this book: that philosophy is that understanding the brain will require ignoring most of the brain's peculiar details and giving greater attention to higher level principles governing the brain and brain-like machines. In an attempt to convince the reader that this philosophy is right, I put forth in this preface an extended allegory concerning futuristic cavemen attempting to reverse engineer the artifacts of a now-dead civilization. It will be rather obvious to us what the cavemen need to know in order to understand these artifacts, and I hope to convince you that in our own attempt to understand the brain, we should expect to need to know similar kinds of things. Namely, we must discover the high level principles governing the brain. The chapters of the book provide examples from my own research of such high level approaches.

Let us start by imagining an intelligent, scientifically primitive society of a post-apocalyptic future. They have no computers, no electricity, no modern architecture, and no principles of physics, mathematics and engineering. They live in caves, eating berries and raw squirrel. One day they stumble onto a magnificent thing like nothing they have seen before: a house. It seems the house had been abandoned at the beginning of the 21st century, many centuries before. Equipped with solar cells, the house and the things inside are still in working order (despite the apocalypse!). They quickly find many other houses in the area, and eventually an entire city of houses and buildings, all miraculously in working order (and also running on solar cells). Unfortunately, they find no libraries.

Always striving for a better life, the cavemen promptly leave their caves to live in the houses, where they find warmth, running water, refrigerators, ovens and video games. They thereby achieve the good life, and their children and children's children are happy. They no longer even look like cavemen, having found ample supplies of 21st century clothing and soap.

But although you can take the caveman out of the cave, it is not so easy to take the cave out of the caveman. They begin this new life in a remarkably ignorant state, very unlike the people who had created the city, namely us. The cavemen quickly guess what toilets and light switches are for, but they have no idea about how such things work. We as a 21st century society, however, *do* know all there is to know about electrical wiring and plumbing in our houses. ... and architecture in our houses and buildings, and computer engineering in our computers, and mechanical engineering in our can openers, and so on. We have entire disciplines of engineering devoted to these kinds of technologies, each one laying the groundwork for how such systems work, and how to make it work best given the circumstances. Furthermore, we grasp the more basic principles of mathematics and physics themselves: e.g., Newtonian mechanics, electricity and magnetism, fluid dynamics, computer science, geometry and calculus. We built from scratch those houses the cavemen live in; and built from scratch the things inside the houses that the cavemen are using. We did not originally build them via mimicry, but, rather, through our knowledge of all the relevant principles involved. This is what I mean when I say that we understand houses and the things inside them, but the cavemen do not.

The cavemen have high aspirations, both materially and intellectually, and are not content to merely live in the nice homes without understanding them. They begin asking how to fix them when they break down. Repairmen of their day pop up, whose aim is to fix failures in the houses. And scientists of their day arise, asking what kind of magic the builders knew, and how can they discover it. These cavemen are not content to remain cavemen on the inside, and to become more, they have a task ahead of them more daunting than they can possibly imagine: to discover all the principles the 21st century builders had discovered over the course of a couple millenia! Of course, the cavemen have a tremendous advantage in that they have countless relics of modern technology in which these principles are "embedded." But their task is nevertheless tremendous, and it should not be surprising to learn that it takes well over a century for them to catch up to where the builders (us) were when they mostly killed themselves off somehow.

What is interesting about this story of futuristic cavemen is that . . .

1. From the point of view of the cavemen, they have a scientific task of understanding the technological artifacts they encounter, and this is analogous to our own 21st century scientific task of understanding the brain and biological systems.

2. And, importantly, we 21st century "builders" are in the special position of actually completely understanding the principles behind the technological artifacts; that is, we have a God's eye view of what the cavemen need to figure out.

As we watch the cavemen begin to study the artifacts, we have the privileged position of seeing how close they are to really "getting it." And by fantasizing about what kinds of stages of discovery the cavemen will go through before achieving full understanding, we can more readily get a handle on what kinds of stages *we* will need to go through before fully understanding brains and biological systems. That is, this story, or thought experiment, is useful because we grasp the kinds of theories the cavemen must (re)discover in order to understand the artifacts, and we may use their intellectual development as a model for the kind of intellectual development we must go through before we will fully comprehend the brain and biological systems. We will further explore this analogy, running futuristic-caveman history forward, looking at the kinds of knowledge they have at each stage, asking what more they need to know, and discussing the connections to our own epistemic position in regards to the brain.

Mapping, a lower level approach

Toward understanding these new-found houses, one of the more obvious tasks the cavemen scientists and engineers think to do is to try to figure out what houses are made of, which parts are attached to which others, how they are attached, what each part in isolation seems to do, and so on. That is, they attempt to "map out" the house, in all its details, from the structural beams and girders, to the windows, doors, hinges and dry wall, to the plumbing and the electrical wiring. They even try to figure out what each part on its own seems to do.

This takes an incredible amount of work—thousands of dissertations are written and tens of thousands of papers are published in the early caveman journals. Eventually they succeed in mapping the entire builder's house, and the details are made publicly available to all the cavemen. Caveman television shows and newspapers declare that all the house repair problems are now on the verge of being solved, and that they have entered an exciting new era in caveman understanding of the technology from the past. "Houses completely understood" emanates from the cavemen, layman and many scientists alike.

And through time, maps are provided for all the different kinds of house and building, from split-level and ranch, to varieties of chapels and movie houses, and so on. With these maps in hand, cavemen scientists can begin to see similarities and differences in the maps of different building types, and can even formulate good guesses as to which kind of building is "more evolved"— i.e., rests on principles developed later in the builders' (our) time.

From the caveman point of view, they have made staggering advances in cracking the magic of the 21st century. And although it is certainly true to say that they have made a great step forward, it is not the case that they are *anywhere near* really understanding the technology.

We know this because *we* happen to entirely comprehend, as a culture, this technology—*we* built it first. And the knowledge needed for us to invent and build this technology encompassed much more than just a map of the parts, connectivity, geometry, and an understanding of what each part seems to do locally. To grasp how all these parts, when correctly assembled, work together, mechanical, electrical and civil engineering are required, each of which is really composed of many disciplines. And for the computer systems found in the homes, computer engineering is required. Principles of architecture, masonry and carpentry must also be known. These disciplines of engineering rest, in turn, upon many areas of physics and mathematics, which required hundreds of years to develop the first time.

These cavemen have just barely begun their scientific adventure—they have only made it to square one.

This is akin to the recent sequencing of entire genomes of various organisms. The analogy is by no means perfect, though, since we have been studying organisms at many different levels for years—e.g., physical chemistry, organic chemistry, genetics, protein biophysics, pharmacology, cell biology, physiology, anatomy, developmental biology and biomechanics. It is not the case that our first step in studying organisms was to sequence the genome! But the analogy can help to put in perspective the importance of the sequencing of the genome. It is analogous to the cavemen for years studying houses at many different levels—e.g., electricity and magnetism, ceramics, electrical engineering, microelectronics, carpentry, and so on—and one group of researchers finally mapping the house. Just as house mapping does not imply any quantum leap in understanding houses, mapping the genome does not entail a quantum leap in understanding organisms.

The same discussion applies to the brain as well. Technologies such as functional magnetic resonance imaging are allowing large-scale mapping of the brain, and decades of single-electrode recording studies (i.e., poking wires here and there in the brain and seeing what makes them light up) have led to mapping of the areas of brains. Anatomical techniques, both morphological and connectivity-based, have also added to our knowledge of the mapping. We know roughly which areas are connected to which others in cat and macaque monkey, and we know the typical connectivity pattern—at a statistical level—in the cortex. We also comprehend aspects of ion channels, synapse modification, and much much more.

All these techniques are helping us to map out the brain, and let us suppose that we do come to know all there is to know about the brain's map—what there is, where they are, what they connect to, and what each neuron or microcolumn or area does. While some might think we would have reached a climax, it seems to me clear that we would have advanced, but we would nevertheless be unenlightened.

The problem for us in understanding the brain is much more severe than the cavemen and buildings. The brain is "more than the sum of its parts" in a much stronger sense than is a house. A computer, also found by the cavemen in the houses, makes a more apt example. The cavemen at some point map out the entire structure and connectivity of the computer, at the level of individual resistors, capacitors and inductors, and at the higher level of transistors and diodes, and at the higher level of logic gates, and at the still higher level of microprocessors, hard drives and random access memory. They are even able to watch what happens among these parts as any software is run.

Will they understand what software is running, much less the principles of computation making the software possible, from their knowledge of the computer map? They will be utterly confused, having no idea how to characterize, parse and interpret the sequential states of the computer. What these cavemen need—what *we* know and they do not—is an understanding of the principles of computation, from general issues of hardware design, to issues of machine language, assembly code, programming languages, software engineering, algorithms, theory of computation, recursion theory and mathematics of functions (as in $f(x) = x^2$).

Accordingly, we need an understanding of analogous levels of theory for brains in order to understand them. Building a brain map will not be a solution—it will be just one of countless important steps along the way, and not a particularly illuminating step at that.

Neural networks, a lower level approach

Let us imagine now that the cavemen, in an attempt to grapple with computers, manage to simulate them with, say, complex arrangements of ropes and pulleys. The ropes and pulleys are arranged to mimic certain key aspects of computer components, and are further set up to connect to one another in a manner like real computers. By eliminating some thought-to-be-irrelevant details, their ropes-and-pulleys simulations aim to capture the key mechanisms underlying computers. And, since the cavemen can easily see which ropes go up and which go down, they are able to predict what would happen with novel computer hardware arrangements, and gain insights into computers by studying a system that they find easier to manipulate and think about.

Have the cavemen thereby figured out computers? While they may have a useful tool for predicting new and useful computer architectures, and for predicting what a given computer will do with a certain input, or for helping them to think about the causal sequences in computers, do they really grasp the point? They have, to first order, just replaced one complex system they do not understand with another one that, although they can think about it a little more clearly, they still do not understand. The cavemen's ropes-and-pulleys constructs are *doing* computations, and the cavemen have no idea about this. While they understand ropes and pulleys, and fathom the workings of several working together, they are at a loss—as would we be—to grasp what fifty or a thousand or a million of them are implementing when working together.

Ropes-and-pulleys machines may be a useful tool for the science of computation, and may even lead to applications in their own right—e.g., maybe the cavemen find a way to predict the weather with them—but these machines do not lead to much understanding of computers. *We* know this because *we* know that what they need to learn is true computer science, from recursion theory and logic, to algorithms, to software engineering, programming languages and more.

A similar story applies for the brain and neural networks or any other lower level simulation-oriented technique. Neural networks—both artificial and biologically minded ones—have proven useful ways of "playing with brains," and have provided applications like predicting the stock market. They do not, however, lead to much understanding of the principles governing the brain and brain-like things.

Associations, a lower level approach

Many of the houses the cavemen live in need repairs. Fixing houses becomes big business. Given that the cavemen do not yet know much about these houses, it may seem that they have little hope of being able to fix them. It turns out, however, that it is not so difficult to become quite good at fixing lots of house breakdowns despite one's ignorance about how houses really work. Let us see what the cavemen do.

Sometimes lights stop working. By experimenting with the houses they learn how to diagnose the source of the failure, and how to fix it. On some occasions the light bulb is the problem. They figure this out by trying the bulb in a socket where a bulb *is* working, and noting that the questionable bulb still does not work. On other occasions, though, the cavemen find that the bulb is undamaged—working fine in other sockets in the house. Something else is thus the problem. They find that in some of these cases, a wire has been cut, and they find that reattaching the wire fixes the problem. Other times they find that a fuse has been blown and must be replaced.

In this experimental fashion, symptoms become associated with diagnoses, and many problems become fixable. Repairmen train hard to make proper diagnoses, and houses are able to be maintained and fixed by them. ...and all this despite their marked ignorance concerning the principles involved.

The fact that people have figured out how to fix many problems in a system is not, itself, a sure-fire test of whether they understand the system. Now, greater understanding of houses *will*, of course, tend to increase the percentage of cases that are repairable, and increase the quality of repairs.

Let us now look at ourselves, and our study of biological systems, rather than cavemen and their study of 21st century technology. First consider cardiovascular problems in humans. In this field, we have made considerable progress in understanding the physiological principles governing the system. Many problems can be fixed, some simply because we have noticed that swallowing x causes a pressure change in y, or what have you, but many problems are fixable because we really have discovered many of the relevant principles and are thereby able to predict from theory how to fix the problem.

In many cases, though, scientists are only able to say that medical problem x is due to property y, and that fixing y tends to fix x. For example, gene y causes or is associated with phenotype x. Or, brain damage of y causes loss of x. Aim: Fix y. Now, even these associations can require the greatest ingenuity to discover, and the most cutting edge technology. Ten million dollars later and one may perhaps have discovered an association capable of saving or bettering thousands of lives. But progress in this fashion—as measured by how many lives or houses benefitting—is not a good measure of the degree to which we understand the system. Just as the cavemen need not understand architecture, engineering or any science to do a great deal of useful house repair, we need not—and do not—understand the genotype-phenotype relation, or the brain-mind relation in order to usefully diagnose and treat lots of medical problems.

These associations eventually contribute to caveman understanding of 21st century technology, and to our understanding the brain, but they do not *comprise* understanding. We and the cavemen will need more than just associations to comprehend our topic.

Causal chains and mechanisms, a lower level approach

The cavemen come to realize that houses and the artifacts inside are mighty complex. When a light is turned on, there is a long sequence of events, each causing the next: light switch gets flicked, which connects the wire-to-light to the wire-to-solar-generator, causing electricity to flow across the wire in the wall to the light socket, and sends current pulsing through the bulb's filament, which excites the gas in the bulb, which in turn emits light. By careful study, the cavemen isolate each point of this chain, thereby coming to know the mechanisms of light bulbs and their switches. Similarly, they determine the mechanisms for toilet-flushing, for hot water baths, for carriage-returns on the computer, and for grandfather clocks.

Do the cavemen thereby understand these technological artifacts? Certainly what they have achieved is an important step, but it is not sufficient for understanding since it gives them only a Rube-Goldberg-level grasp: a hits b, which makes c fall and bump d, which flies over and lands on e, which.... The cavemen can know all about the causal chain for the light bulb even without knowing about the laws of electricity and magnetism, and atomic physics. Or, for a computer, the important thing to know about some computer action is not the particular instantiation of the program, but the more general point that certain functions are being computed. The specific causal chains found in any given computer can even work to obfuscate, rather than to illuminate. Multi-electrode recording from computer chips while a word processor is running will never help us understand word processing principles.

The same is true for the brain. One common goal is to determine the sequence of neural activity patterns that ensue when a person engages in a certain behavior. What areas light up when one maintains something in working memory, and in what order do they light up? Or, what are the effects on neural connectivity when an infant is raised with or without some pharmacological agent? Or, how does behavior change when this or that region of the brain is disabled? And, in each case, what causal chains are involved? Just as was the case for cavemen and 21st century artifacts, these causal chains do not represent final scientific victory over the brain. At best, knowing the causal chains can (a) help understand associations that are useful in repairing systems, and (b) serve as opaque clues toward the general principles underlying the system.

The moral I have been trying to convey in the preceding discussion on "lower levels" is that many or most of the kinds of research on brain and complex biological systems we, as a scientific community, engage in are of the low level sort that leads to little genuine understanding of the essential principles underlying the system. The caveman allegory is useful because the scientific endeavor ahead of them already is completely understood by us, and so we have a good feel for what the cavemen must accomplish before we will be willing to attribute them with success.

We must then only apply this same standard back on ourselves and our own intellectual quest to conquer biological systems. By doing so, we see that much, much more is ahead of us, and much of what is ahead that is key for genuine understanding is not the low level stuff we have discussed, but high level stuff to which we turn next. The discussion that follows serves to both introduce a number of distinct kinds of high level approach, and to briefly outline the topics of the chapters of this book, the book's topic being high level approaches to the brain, carried out via consideration of research on which I have worked over the last decade.

Purpose and function, higher level approaches

The 21st century artifacts were built the way they were for a reason. Some aspects of them were for looks, like wall paper pattern, but many key features were designed so that the device could satisfy some purpose. We have thus far been imagining these cavemen to be human, and if so, they may more quickly gather what toilets are for, that windows are for the view, and that word processing programs allow one to compose messages (supposing they have written language). If the cavemen are, instead, entirely unlike us—intelligent evolutionary offspring of snails—then we can imagine their task of determining the purpose of the 21st century human artifacts to be extremely more vexing. That a chair's purpose is to sit on might be some snail-caveman's dissertation! Discovery of the principles governing a device will require understanding what it was for.

Consider word processors. The snail-cavemen observe the program's behavior: when carriage return is hit, the little flashing thing moves down and to the very left; when 'd' button is pressed, a 'd' appears on the screen and the flashing thing moves to the right; and so on. They find the program code itself, all one millions lines of it. And they persist in examining all the causal chains involved at all the lower levels of hardware. We know that the cavemen will never understand the word processing system unless they figure out its purpose, which is to compose messages. Similarly, to understand fins we need to recognize that they are for swimming, and for eyes that they are for seeing. The brain presumably has many purposes, and to understand it we need to figure them out. Only then can we begin to grasp how that purpose is fulfilled.

Sometimes the explanation for something requires seeing how the purpose is thwarted. If a caveman sets a drinking glass down and it makes a sound that sets off the "clapper"-rigged light, the clapper's purpose has been thwarted since the clapper is meant to turn on or off the light only when someone *claps* twice, not when a glass is set down.

A related idea the cavemen must learn in order to understand a computing machine is the function—the mathematical function—it computes. For example, let us suppose they find object recognition software. They eventually figure out that the purpose of the software is, roughly, to allow robots to navigate through the world. Now they wish to know what function is actually being computed by the machine. Through time, they find that the function computed is this: it takes images as input, and outputs a good guess as to the nature of the world around the machine. This function serves the purpose of navigation, but other mathematical functions could, in principle, also. Understanding the computed function is a necessity if one is interested in understanding the system. Now we may reasonably ask questions concerning what algorithm the machine is employing to compute this function, questions which would hardly have been sensible without knowing what the algorithm was computing. And we may ask what hardware is being used to implement the algorithm.

Since the brain is a computing machine, to understand it it is essential to determine what function it is computing. It is only in light of the computed function that we can hope to comprehend the algorithm used by the brain, and it is only through understanding the algorithm that we can hope to fathom what all the lower level mechanisms are up to, and so on.

Purpose and function will arise at many times throughout this book, but as a specific example, in Chapter 2 we will discuss visual perception, and the classical illusions in particular. Little progress has been made for one hundred years on why illusions occur, and one reason is that too little attention has been paid to determining the function the visual system is trying to compute. I will propose a simple function—namely, accurate percepts at time t of the world at time t—and show that the algorithm it uses to compute this function—a probabilistic inference algorithm—leads to predictable screw-ups, which lead to illusions.

Scaling, a higher level approach

With only one airplane for cavemen to examine, it will be next to impossible to figure out what properties are characteristic of airplanes. What makes an airplane an airplane? What things must it satisfy in order to work? To answer this, it is necessary to figure out the properties true of every airplane; one wants to find the airplane invariants. A good idea is for the cavemen to examine many airplanes and try to determine what is the same among them. For example, door size and steering wheel diameter do not change as a function of the size of the plane—jumbo jets and small propeller planes have roughly the same size door and steering wheel. But wing surface area *does* tend to increase with plane size, suggesting that there is some invariant relationship between the two that must be held in order for planes to function properly. Unbeknownst to the cavemen, the reason the doors and steering wheels are invariant in size is because it is the same sized people who used to fly and ride them. And wings have more surface area in larger planes because the lift must be proportional to the plane's mass; the ratio of lift to plane mass is an invariant across planes.

How plane properties scale with plane size—or more generally, how plane properties scale as other properties change—provides a view into the underlying principles governing a system, principles it may be difficult to impossible to see by looking at any one system.

Scaling studies are, accordingly, useful for us in our quest to understand brain and other biological systems. Chapter 1 discusses studies of mine—each concerning scaling—one concerning the large-scale morphology of limbed animals, another concerning the underlying neuroanatomical wiring principles characterizing the cortex, and a final one dealing with brain and behavioral complexity.

Optimal shape, a higher level approach

The houses the cavemen find are, to first order, all cubes. No house they find is, for example, one hundred feet long and five feet wide. They also note the tendency for bathrooms in houses to be clustered in space—e.g., the master bedroom bathroom is more-often-than-chance adjacent to the upstairs kids bathroom. The cavemen notice certain common patterns on road design, and patterns on the design of microelectronic chips.

If they wish to understand why these things are shaped or arranged like they are, they are going to have to develop and understand principles of optimization, much of it emanating from graph theory, a field of computer science. These 21st century artifacts were built by companies with limited resources, trying to maximize profit. It paid to use the least amount of material in the construction of something.

Houses are typically boxy, rather than extremely long and thin, because houses are built to maximize floor surface area while minimizing the amount of wall material needed. Bathrooms clustered near each other lowers the total amount of plumbing required in the house, making it cheaper to build and easier to maintain. Roads are designed to be tree-like, say within a neighborhood, rather than n different long driveways leading from the neighborhood entrance directly to the n different houses, or rather than a single unbranched meandering road successively passing each house in turn. Branching tree-like design for neighborhood roads tends to minimize the amount of ground taken up by road, thereby decreasing the cost of the road and increasing the amount of land on which houses can be built and sold. Similar observations hold for the design of electronic circuits, where roads are now wires, and houses now electrical components. The 21st century people developed very large scale integrated (VLSI) design techniques to minimize the amount of wire and space needed to embed a circuit.

In order to see that an artifact is optimal in some sense, the cavemen are going to have to understand the space of alternative ways the artifact *could* have been. They need to grasp not just the actual, but the non-actual but still possible. It is only in relation to these non-actual possibilities that "being-optimal" makes sense. To say that a chip is optimally wired means that, over some specified range of possible chip wirings, this chip requires the least wire.

Similar to the 21st century artifacts, biological structures, too, have been under pressure, selective pressure, to be economically constructed. An animal that could use less material for some body part would save on the cost of that material, and could be packaged into a smaller, more economical, body; or, alternatively, the energy could instead be used for other advantageous body parts. In Chapter 1 I discuss some examples from the literature, and focus on a few cases I have researched, from the shapes of neurons, arteries and leafy trees, to the number of limbs and digits, to its relevance in understanding the large-scale organization and scaling properties of the mammalian cortex.

Probability and decision theory, a higher level approach

The cavemen become hooked on computer chess. They are stunned that the computer can behave so intelligently, beating any caveman soundly, so long as the computer chess setting is high enough. How, the cavemen ask, does the computer play chess so well? I, myself, know precious little about chess-playing algorithms, but I am quite sure that, at one level or another, the algorithms implement decision theoretic principles. What does this mean? It means that when the computer is deciding what move to make, it incorporates the probabilities of all the possible moves you will make, along with the severity—the costs and benefits, or disutilities and utilities—of each possibility. The computer weighs all these things and, looking many moves ahead, determines the move that maximizes its expected utility. Thus, if the cavemen are to understand the chess playing algorithm, they must reinvent probability and decision theory (along with other principles employed in chess algorithms).

The brain is in a similar position as the chess program. The brain must attempt to maximize the animal's expected utility—i.e., lead the animal to food, or avoid danger, or obtain sex, etc.—and since the world is unpredictable, it must work with probabilities. The visual system is another case where a decision theoretic approach seems appropriate: the visual system must infer from the retinal "data" an expected utility-maximizing perceptual "hypothesis" concerning what is in the world. In Chapter 2 I will take up the modern rise of decision theoretic models of visual perception.

Conceptual limits on learning, a higher level approach

The cavemen find abandoned artificial intelligence laboratories, and thereby encounter many learning algorithms for machines. What they notice is that the 21st century artificial intelligence community seemed to have a strong desire to design powerful learning algorithms, whether the algorithms were to learn to recognize handwriting, or to learn to perceive objects, or to learn to perceive ripe versus unripe fruits, or to learn to solve novel problems, or to learn to navigate a maze, and so on. Yet, they could not seem to settle on a learning algorithm. To the cavemen it seems intuitively clear, at first glance, that there must be a single learning algorithm which is perfect; a perfect way of learning, so perfect that it learns optimally in any domain. What was wrong with the builders that they could not find a best-possible learning algorithm?

The reason the builders could not find a best-possible learning algorithm is because there is no such thing. The cavemen do not know this yet, but we know this, via our understanding of the intractability of the problem of induction. The problem, or riddle, of induction is that there does not appear to be any particular right way to carry out one's learning. That is, when we have a set of hypotheses and some evidence, we would like to be told exactly what our degrees of belief *should* be in light of the evidence. The problem is that there appears to be no answer; no answer appears to be even possible.

Eventually, the cavemen come to realize the problem of induction, and give up their quest for a perfect learning algorithm. But they do begin to wonder whether all these different learning algorithms may have certain underlying things in common. Are each of these learning algorithms just arbitrary ways of learning in the world? Or are many or most of them similar in that they follow similar rules, but merely differ in some small regard relevant for their specific learning purposes. That is, while there may be no perfect learning algorithm, might there still be fixed principles of learning to which any learning algorithm should still cling?

It turns out that there are similar principles governing different learning algorithms. We understand this because we know that many different kinds of learning algorithm have been built within a probabilistic framework. In these cases learning involves having probabilities on each hypothesis in some set, and updating the probabilities as evidence accumulates. Very many learning algorithms fall within this framework, even if they are not programmed with explicit probabilities. What these learning algorithms share is a commitment to the axioms of probability theory. Probability theory, and a theorem called Bayes' Theorem in particular, is what they share in common. The principles that they share determine how the probabilities of hypotheses should be modified in the light of evidence; thus, although many of the learning algorithms the cavemen find are different, they share the same principle concerning how evidence is to be used. Where the algorithms differ is in a respect called "prior probabilities," which captures the degrees of confidence in the hypotheses before having seen evidence. To understand all this, the cavemen must appreciate the riddle of induction, and probability theory.

Brains learn too. Are brains perfect learning machines, or do brains of different types have ways of learning that are brain-specific? The riddle of induction immediately informs us that brains cannot be perfect learners, since there is no such thing. Brains, then, must come innately equipped with their own learning method; and brains of different types might come innately equipped with different learning methods. We are interested in whether brains, and more generally intelligent agents, may be modeled as following fixed principles of learning, even if they differ in some innate regard. And we are interested to know how much must be innate in a brain in order for a learning method to be specified. Do we have to hypothesize that an agent comes into the world with a full-fledged particular learning algorithm hardwired, and that agents with different learning algorithms have entirely different learning rules they follow? Or are the innate differences between different kinds of learners much more subtle? Perhaps learners differ in some very tiny way, and they follow all the same learning principles. And, if there are any common learning principles, is there some sense in which brains have evolutionarily converged onto the optimal principles? We take up these issues, and a kind of best-we-can-possibly-hope-for solution to the riddle of induction in Chapter 3.

Computational consequences of being finite, a higher level approach

The cavemen eventually figure out how to use computers, beginning with video games and chat rooms, and moving to more productive uses. They begin to notice that software always has bugs. The programs occasionally suddenly crash, or hang, seemingly thinking and thinking without end. What was wrong with the 21st century computer builders of these programs? Why couldn't they do it right?

Because we as a culture know theoretical computer science, we know that it is not due to some problem of ours. *Any* builders of programs, so long as the builder is finite in size and speed, will have this same Achilles' heel. Programs built by finite agents will have bugs, because it is logically impossible for it to be otherwise. If you do not already know about the undecidability theorems to which I am alluding, wait until Chapter 4 where we will talk about them more.

Now switch to the 21st century and our task of understanding the brain. The brain concocts its own programs to run as it learns about the world. Since the brain is finite, it cannot generally build programs without bugs. Our programs in the head generally *must* have bugs, and we can even predict what they should look like. Chapter 4 shows how these bugs are the source of the vagueness of natural language. Also, Chapter 2 shows that perceptual illusions—the kind we have all seen in Psychology 101—are likely to be common to any agent with a non-instantaneous brain.

I hope to have both previewed what the chapters will touch upon, and more importantly to have given the reader reasons, via the post-apocalyptic cavemen allegory, to value higher-level approaches toward the brain. The cavemen will not understand what we built without backing up to 25,000 feet, and we will not understand what evolution built without also backing up to 25,000 feet. Please note, however, that while high level principles are what the cavemen need to understand the artifacts they find, no single caveman is probably going to be able to, alone, make much of a dent. Rather, it will require decades or centuries of research, and hundreds of thousands or more cavemen involved. Similarly, I do not claim that the chapters of this book will make much progress in regard to understanding the brain. What we will need is hundreds of thousands of researchers working at such higher levels, and working probably for multiple generations, before we can expect to see light.

I am indebted to the Department of Computer Science at University College Cork, Ireland, and to W. G. Hall of the Department of Psychological and Brain Sciences at Duke University, for the freedom and time to write this book. I thank my wife Barbara Kelly Sarantakis Changizi for her reflections and support. I have benefitted from the comments of Timothy Barber, Amit Basole, Nick Bentley, Jeremy Bowman, Derek Bridge, Jeffrey Bub, James Cargile, Christopher Cherniak (from whom the term "impossibility engine" emanates), Justin Crowley, James L. Fidelholtz, William Gasarch, Chip Gerfen, Nirupa Goel, William Gunnels IV, WG "Ted" Hall, Brian Hayes, John Herbert, Terence Horgan, Jeffrey Horty, Paul Humphreys, Erich Jarvis, Michael Laskowski, Gregory Lockhead, Michael McDannald, Tom Manuccia, Reiko Mazuka, Robert McGehee, Daniel McShea, Craig Melchert, Michael Morreau, Romi Nijhawan, Surajit Nundy, James Owings, Donald Perlis, John Prothero, Dale Purves, David Rubin, Dan Ryder, Geoffrey Sampson, Carl Smith, Roy Sorensen, Christopher Sturdy, Frederick Suppe, Christopher Tyler, Steven Vogel, Thomas Wasow, Len White, David Widders, Timothy Williamson, and Zhi-Yong Yang.

Mark A. Changizi,   May, 2002
www.changizi.com
changizi@changizi.com

Department of Computer Science	Dept. of Psychological and Brain Sciences
University College Cork—	Duke University
National University of Ireland, Cork	Durham, NC 27708
Ireland	USA

Chapter 1

Scaling in Nervous Networks

Why are biological structures shaped or organized like they are? For example, why is the brain in the head, why is the cortex folded, why are there cortical areas, why are neurons and arteries shaped like they are, and why do animals have as many limbs as they do? Many aspects of morphology can be usefully treated as networks, including all the examples just mentioned. In this chapter I introduce concepts from network theory, or graph theory, and discuss how we can use these ideas to frame questions and discover principles governing brain and body networks.

The first topic concerns certain scaling properties of the large-scale connectivity and neuroanatomy of the entire mammalian neocortical network. The mammalian neocortex changes in many ways from mouse to whale, and these changes appear to be due to certain principles of well-connectedness, along with principles of efficiency (Changizi, 2001b). The neocortical network must scale up in a specific fashion in order to jointly satisfy these principles, leading to the kinds of morphological differences between small and large brains.

As the second topic I consider the manner in which complexity is accommodated in brain and behavior. Do brains use a "universal language" of basic component types from which any function may be built? Or do more complex brains have new kinds of component types from which to build their new functions?

The final topic concerns the nervous system at an even larger scale, dealing with the structure of the nervous system over the entirety of the animal's body. I show that the large-scale shape of animal bodies conforms to a quantitative scaling law relating the animal's number of limbs and the body-to-limb proportion. I explain this law via a selective pressure to minimize the amount of limb material, including nervous tissue (Changizi, 2001a). That is, because we expect nervous systems to be "optimally wired," and because nervous systems are part and parcel of animal bodies, reaching to the animal's extremities, we accordingly expect—and find—the animal's body itself to be optimally shaped.

One feature connecting the kinds of network on which we concentrate in this chapter is that each appears to economize the material used to build the network: they appear to be volume optimal. It is not a new idea that organism morphology might be arranged so as to require the least amount of tissue volume [see, for example, Murray (1927)], but in recent years this simple idea has been applied in a number of novel ways. There are at least three reasons why optimizing volume may be evolutionarily advantageous for an organism. The first is that tissue is costly to build and maintain, and if an organism can do the same functions with less of it, it will be better off. The second reason, related to the first, is that minimizing tissue volume gives the organism room with which to pack in more functions. The third reason is that minimizing tissue volume will tend to reduce the transmission times between regions of the tissue. These three reasons for volume optimization in organisms are three main reasons for minimizing wire in very large-scale integrated (VLSI) circuit design (e.g., Sherwani, 1995); we might therefore expect organisms to conform to principles of "optimal circuit design" as made rigorous in the computer science fields of graph theory and combinatorial optimization theory (e.g., Cormen et al., 1990). ... and we might have this expectation regardless of the low level mechanisms involved in the system.

Y junctions

The first quantitative application of a volume optimization principle appears to be in Murray (1926b, 1927), who applied it to predict the branching angles of bifurcations in arteries and trees (e.g., aspen, oak, etc.). He derived the optimal branch junction angle (i.e., the angle between the two children) to be

$$\cos \theta = \frac{w_0^2 - w_1^2 - w_2^2}{2w_1 w_2},$$

where w_0, w_1 and w_2 are the cross-sectional areas of the junction's parent and two children. One of the main consequences of this equation is that, for symmetrical bifurcations (i.e., $w_1 = w_2$), the junction angle is at its maximum of 120° when the children have the same cross-sectional area as the parent segment, and is 0° when the children's cross-sectional area is very small. [Actually, in this latter case, the branch angle falls to whatever is the angle between the source node of the parent and the termination nodes for the two children.] That is, when trunks are the same thickness as branches the branch angle that minimizes the volume of the entire arbor is 120°. This is very unnatural, however, since real world natural arbors tend to have trunks thicker than branches. And, if you recall your experience with real world natural arbors, you will notice that they rarely have junction angles nearly as high as 120°; instead, they are smaller, clustering around 60° (Cherniak, 1992; Changizi and Cherniak, 2000). *Prima facie*, then, it seems that natural arbors are consistent with volume optimality. Murray also derived the equation for the volume-optimal angle for *each* child segment relative to the parent, and one of the main consequences of this is that the greater the assymmetry between the two children's cross-sectional areas, the more the thinner child will branch at 90° from the parent. We find this in natural arbors as well; if there is a segment out of which pokes a branch at nearly a right angle, that branch will be very thin compared to the main arbor segment from which it came. Qualitatively, then, this volume optimality prediction for branch junctions fits the behavior of natural junctions. And it appears to quantitatively fit natural junctions very well too: These ideas have been applied to arterial branchings in Zamir et al. (1983), Zamir et al. (1984), Zamir and Chee (1986), Roy and Woldenberg (1982), Woldenberg and Horsfield (1983, 1986), and Cherniak (1992). Cherniak (1992) applied these concepts to neuron junctions, showing a variety of neuron types to be near volume-optimal; he also provided evidence that neuroglia, Eucalyptus branches and elm tree roots have volume optimal branch junctions. [Zamir (1976, 1978) generalized Murray's results to trifurcations, applying them to arterial junctions.]

Although it is generally difficult to satisfy volume optimality in systems, one of the neat things about this volume optimization for natural branch junctions is that there is a simple physical mechanism that leads to volume optimization. Namely, the equation above is the vector-mechanical equation governing three strings tied together and pulling with weights w_0, w_1 and w_2 [see Varignon (1725) for early such vector mechanical treatments]. If each of the three junction segments pulls on the junction with a force, or tension, proportional to its cross-sectional area, then the angle at vector-mechanical equilibrium is the volume-optimizing angle (Cherniak, 1992, Cherniak et al., 1999). Natural arbors conforming to volume optimality need not, then, be implementing any kind of genetic solution. Rather, volume optimality comes for free from the physics; natural arbors like neurons and arteries self-organize into shapes that are volume optimal (see, e.g., Thompson, 1992). In support of this, many non-living trees appear to optimize volume just as well as living trees, from electric discharges (Cherniak, 1992) to rivers and deltas (Cherniak, 1992; Cherniak et al., 1999).

In addition to this physics mechanism being advantageous for a network to have minimal volume, natural selection may favor networks whose average path between root and leaves in the network is small—shortest-path trees—and one may wonder the degree to which this mechanism simultaneously leads to shortest-path trees. Such shortest-path trees are not necessarily consistent with volume optimization (Alpert et al., 1995; Khuller et al., 1995), but the near-volume-optimal natural trees tend to come close to minimizing the average path from the root. Considering just a y-junction, the shortest path tree is the one which sends two branches straight from the root; i.e., the junction occurs at the root. An upper bound on how poorly a volume-optimal junction performs with respect to the shortest-path tree can be obtained by considering the case where (i) the root and two branch terminations are at the three vertices of an equilateral triangle with side of unit length, and (ii) the volume per unit length (i.e., cross-sectional area) is the same in all three segments. The volume-optimal branch junction angle is 120° and occurs at the center of mass of the triangle. The distance from the root to one of the branches along this volume-optimal path can be determined by simple geometry to be 1.1547, or about 15% greater than the distance in the shortest-path tree (which is along one of the unit-length edges of the triangle). This percentage is lower if the relative locations of the root and branch terminations are not at the three vertices of an equilateral triangle, or if the volume per unit length of the trunk is greater than that of the branches (in which case the junction point is closer to the root). In sum, natural selection has stumbled upon a simple vector-mechanical mechanism with which it can simultaneously obtain near volume-optimal and near-shortest-path trees.

Multi-junction trees

The applications of a "save volume" rule mentioned thus far were for single branch junctions. Kamiya and Togawa (1972) were the first to extend such applications to trees with multiple junctions, finding the locally optimal multi-junction tree for a dog mesenteric artery to be qualitatively similar to the actual tree. Schreiner and Buxbaum (1993), Schreiner et al. (1994) and Schreiner et al. (1996) constructed computer models of large vascular networks with realistic morphology by iteratively adding locally volume-optimal y-junctions.

Traverso et al. (1992) were the first to consider modeling natural arbors, neural arbors in particular, with multiple junctions using the concept of a Steiner tree (Gilbert and Pollak, 1968) from computer science, which is the problem of finding the length-minimal tree connecting n points in space, where internodal junctions are allowed (i.e., wires may split at places besides nodes). Branch junctions in Steiner trees have angles of 120°, and Traverso et al. (1992) found that some sensitive and sympathetic neurons in culture have approximately this angle.

Most neurons and other natural arbors, however, have branch junction angles nearer to 60° or 70° (Cherniak, 1992; Changizi and Cherniak, 2000); the Steiner tree model is inadequate because it assumes that trunks have the same volume cost per unit length (i.e., same cross-sectional area) as branches, when it is, on the contrary, almost always the case that trunks are thicker than branches. To determine if natural trees have volume-optimal geometries a generalized notion of Steiner tree is needed, one that allows trunks to be thicker than branches. Professor Christopher Cherniak and myself invented such a notion and showed that axon and dendrite trees (Cherniak et al., 1999), coronary arterial trees (Changizi and Cherniak, 2000) and Beech trees (Cherniak and Changizi, unpublished data) optimize volume within around 5%, whereas they are around 15% from surface area optimality and around 30% from wire-length optimality. [The average values for the unpublished tree data are from eight 4-leaf Beech tree arbors, and are 4.63% ($\pm$3.21%) from volume optimality, 10.94% ($\pm$6.56%) from surface area optimality, and 26.05% ($\pm$12.92%) from wire length optimality (see Cherniak et al., 1999, for methods).]

The studies mentioned above concentrated on the morphology of individual natural trees, e.g., individual neurons. We may move upward from individual neurons and ask, How economically wired are whole nervous systems? This has been asked and answered in a variety of ways.

Larger scales

At the largest scale in the nervous system, Cherniak (1994, 1995) showed that animals with brains making more anterior connections than posterior connections should have, in order to minimize volume, the brain placed as far forward as possible; this explains why the brain is in the head for vertebrates and many invertebrates. Radially symmetric animals, on the other hand, are expected to have a more distributed neural network, as is the case (e.g., Brusca and Brusca, 1990, p. 87). In what is to date the most stunning conformance to volume optimality, Cherniak (1994, 1995) showed that, of approximately forty million possible positions of the ganglia in *Caenorhabditis elegans*, the actual placement of the ganglia is the wire-optimal one. He also provides statistical evidence that the placement of cortical areas in cat (visual cortex) and rat (olfactory cortex) are consistent with the hypothesis that the total length of area-to-area connections is minimized.

Ruppin et al. (1993) show that each of the following facts about the brain decrease the overall required volume of the brain: (i) that gray matter is separated from white matter, (ii) that gray matter is a shell on the surface of the brain with white matter in the center (rather than vice versa), and (iii) that the gray matter has convolutions. Van Essen (1997) also argues that the convolutions of the cortex are a fingerprint of low wiring. Wire-minimization has also been used in a number of ways to explain local connectivity patterns in the visual cortex (e.g., stripes, blobs or patches) (Durbin and Mitchison, 1990; Mitchison, 1991, 1992; Goodhill et al., 1997; Chklovskii, 2000; Chklovskii and Koulakov, 2000; Koulakov and Chklovskii, 2001).

Well-connectedness

As we will see later, the neocortical network not only reveals principles of volume optimization, it also reveals principles of well-connectedness, where by that I refer, intuitively, to properties of the network which bear on how "close," in some sense, vertices are to one another. One way to measure neuron-interconnectedness is via the average percent neuron-interconnectedness of neurons, where a given percent neuron-interconnectedness of a neuron is the percentage of all neurons to which it connects. It has been recognized, however, that it is prohibitive to maintain an invariant average percent neuron-interconnectedness as the network size is scaled up (Deacon, 1990; Stevens, 1989; Ringo, 1991), because this requires that the average degree of a vertex [the *degree* of a vertex is the number of edges at a vertex] scales up proportionally with network size, and thus the total number of edges in the network scales as the square of network size. Average percent neuron-interconnectedness is an overly strong notion of neural interconnectedness, however. A weaker measure might be the characteristic path length, which I will call the *network diameter*, which is defined as, over all pairs of neurons, the average number of "edges" (i.e., axons) along the shortest path connecting the pair. Intuitively, network diameter measures how close—in terms of connectivity—neurons are to one another, on average.

For most random networks the diameter is "low." [In a *random network* each pair of nodes has the same probability of having an edge between them.] In particular, the network diameter is approximately $(\log N)/(\log \delta)$ (Bollobás, 1985, p. 233), where N is the network size and δ the average degree. [This also requires assuming $N \gg \delta \gg \log N \gg 1$.] While keeping the average percent neuron-interconnectedness invariant requires that degree (the number of neurons to which a neuron connects) scale proportionally with network size, the network diameter can remain invariant even if the degree grows disproportionately slowly compared to the network size. Suppose, for example, that $N \sim \delta^b$, where $b > 1$; that is, the degree grows disproportionately slowly as a function of network size, but as a power law. [A *power law* is an equation of the form $y = ax^b$, where a and b are constants.] Then in a random network the diameter is approximately

$$\frac{\log(C\delta^b)}{\log \delta} = b + \frac{\log C}{\log \delta},$$

where C is a proportionality constant. In the limit of large brains, the network diameter is invariant, approaching b, even though the degree is scaling disproportionately slowly.

This observation is not particularly useful for biological systems, however, because few biological networks are random. However, in what has now become a seminal paper, Watts and Strogatz (1998) noticed that highly ordered networks can have random-network-like diameter by merely adding a small number of "shortcuts" connecting disparate regions of the network; they named such networks *small world networks*. A firestorm of papers have since been published showing that many natural networks appear to be small world, including nervous systems and social networks. Nature has, indeed, hit upon one useful kind of network, allowing highly ordered clustering, and yet low network diameter. With these ideas in mind, we will be able to see that the mammalian neocortex has low and invariant network diameter, and, in fact, it is approximately 2.

1.1 The mammalian neocortex

Mice, men and whales have brains, and as they are all mammals, their cortex mostly consists of something called the *neocortex*, which is found only in mammals. When researchers talk of the cortex, they are almost always referring to the neocortex. This is the part of the mammalian brain that excites most researchers, as it appears to be the principal anatomical feature separating mammalian brains from other vertebrate brains; it is where the key to mammalian intelligence will be found. The neocortex consists of gray matter which lies on the surface of the brain, and white matter which is in the interior. The gray matter consists of neurons of many types, synapses, and glial cells. The white matter consists of axons which reach from neurons in one part of the gray matter to make synapses with neurons in another part of the gray matter. The neocortical gray matter is characterized by what appears to be layers as one moves from the surface of the gray matter inward toward the center, and six layers are usually distinguished. There are also distinct cortical areas at different locations on the cortical sheet, inter-area connections being made primarily via white matter connections, and intra-area connections being made primarily by local connections not traveling into the white matter.

Although all mammals have a neocortex, many of the basic properties of the neocortex change radically from small brains to large brains. The changes are so dramatic that one might justifiably wonder whether the neocortex in mouse and whale are really the same kind of organ. And if they are the same kind of organ, then why the radical changes? Are there some underlying properties that *are* being kept constant—these properties being the key ones, the ones that really define the neocortex—and the properties that change are changing for the purpose of keeping these key properties invariant? It is these questions we examine in this section. I will describe the ways in which the neocortex changes as it increases in size, and describe a theory (Changizi, 2001b) that aims to explain what the central features of the neocortex are, such that all these other properties must change as they do.

To understand how the neocortical network scales up in larger brains, we need to understand the notion of a *power law*. Power laws are of the form $y = ax^b$, where a and b are constants. a is a proportionality constant, and it often depends on the specific nature of the studied systems. For example, the volume of a cube is $V = D^3$, where D is the diameter; i.e., $a = 1$- and $b = 3$. The volume of a sphere, however, is $V = (4/3)\pi(D/2)^3$, which is $V = (\pi/6)D^3$; i.e., $a = \pi/6 \approx 0.5$ and $b = 3$. The volume of both cubes and spheres scale as the diameter cubed, but the proportionality constants are different. So long as the geometrical shape is similar, the proportionality constant will not change, and it is often appropriate to ignore it, writing $y \sim x^b$ to mean that y is proportional to x^b. Power laws are particularly appropriate for neocortical scaling [and in biology more generally (see Calder, 1996; Schmidt-Nielson, 1984)]. It turns out that many of the properties of the neocortex scale

Table 1.1: *Measured scaling exponents for neocortical variables against gray matter volume V_{gray}. The measured exponents are in most cases acquired from scaling data against brain volume. To obtain exponents against V_{gray}, I have assumed that V_{gray} is proportional to brain volume. This proportionality is empirically justified, as measured exponents for V_{gray} to brain volume are near one:* 0.983 *(Prothero, 1997a),* 0.982 *(Hofman, 1991),* 1.054 *(Hofman, 1989),* 1.04 *(Prothero and Sundsten, 1984),* 1.06 *(Frahm et al., 1982) and* 1.08 *(Jerison, 1982).*

Variable description	Variable	Measured exponent	References
# areas to which an area connects	D	0.30	Here
Neuron number	N	0.62	Jerison, 1973
		0.67	Passingham, 1973
Neuron density	ρ_{neuron}	-0.312	Prothero, 1997b
		-0.28	Prothero, 1997b
		-0.28	Tower, 1954
		-0.32	Tower, 1954
Number of areas	A	0.40	Changizi, 2001b
Thickness	T	0.092	Prothero, 1997a
		0.115	Prothero, 1997a
		0.129	Hofman, 1991
		0.197	Hofman, 1989
		0.08	Prothero and Sundsten, 1984
		0.17	Jerison, 1982
Total surface area	S	0.905	Prothero, 1997a
		0.893	Prothero, 1997a
		0.922	Prothero, 1997a
		0.901	Hofman, 1991
		0.899	Hofman, 1989
		0.89	Hofman, 1985
		0.91	Prothero and Sundsten, 1984
		0.91	Jerison, 1982
Module diameter	m	0.135	Manger et al, 1998
Soma radius	R_0	0.10	Changizi, 2001b
Axon radius	R_i	0.105	Shultz and Wang, 2001
Volume of white matter	V_{white}	1.318	Allman, 1999
		0.985	Prothero, 1997b
		1.28	Hofman, 1991
		1.37	Hofman, 1989
		1.31	Frahm et al., 1982

against gray matter volume as a power law. That is, if Y is the property of interest and V_{gray} is the gray matter volume, then it has been empirically found that $Y = aV_{gray}^b$ for some constants a and b. Ignoring the proportionality constant, we say that $Y \sim V_{gray}^b$. When we say how a neocortical quantity scales up, we can, then, just report the scaling exponent for it against gray matter volume. Table 1.1 shows the scaling exponents measured thus far for neocortex, and which I will explain below. I only show plots here if they have not yet appeared elsewhere in the literature (and this is only for the number of areas to which an area connects, and for module diameter).

Before presenting a theory of neocortical scaling, I begin by making some simplifying assumptions about the neocortical network. Because about 85% of neocortical neurons are pyramidal cells (Schüz, 1998) and only pyramidal cells appear to significantly change in degree of arborization from mouse to whale (Deacon, 1990), it is changes to pyramidal cells that must account for the decreasing neuron density. Accordingly, I will idealize the neocortical network to consist only of pyramidal neurons. Also, because most (over 90%) of the neocortical connections are from one part of neocortex to another (Braitenberg, 1978), the other neocortical connections are probably not the principal drivers of neocortical scaling; I will therefore concentrate on the cortico-cortical connections only, and I will assume that a single pyramidal neuron's axon can innervate only one area.

There are multiple principles shaping the neocortex, and we will see that the exponents are not all due to the same underlying explanation. There are, in fact, three central principles, and they are

- Economical well-connectedness.

- Invariant computational units.

- Efficient neural branching diameters.

The exponents each of these principles explains are shown in Table 1.2. I will take up each principle in turn, and derive the exponents which follow from it.

1.1.1   Economical well-connectedness

The principles

Consider that an area in the gray matter connects to other areas (i.e., it has neurons connecting to neurons in other areas). The fraction of the total number of areas to which it connects is called the *percent area-interconnectedness*. It

Table 1.2: *The exponent predicted by my theory, along with the approximate value for the measured exponent. The exponents are partitioned into three groups, each which is explained by the principle stated above it.*

Variable description	Variable name	Measured exponent	Predicted exponent
Economical well-connectedness $\Rightarrow$			
- Number of areas to which an area connects	D	0.30	$1/3 \approx 0.33$
- Neuron number	N	0.65	$2/3 \approx 0.66$
- Neuron density	ρ_{neuron}	-0.3	$-1/3 \approx -0.33$
- Number of areas	A	0.40	$1/3 \approx 0.33$
Invariant computational units $\Rightarrow$			
- Thickness	T	0.13	$1/9 \approx 0.11$
- Total surface area	S	0.90	$8/9 \approx 0.89$
- Module diameter	m	0.135	$1/9 \approx 0.11$
Efficient neural branching diameters $\Rightarrow$			
- Soma radius	R_0	0.10	$1/9 \approx 0.11$
- Axon radius	R_1	0.105	$1/9 \approx 0.11$
- Volume of white matter	V_{white}	1.3	$4/3 \approx 1.33$

seems a plausible and natural hypothesis that, for an area's efforts to be useful, it must make its results known to an invariant percentage of the areas in the neocortex. That is, suppose that for a mouse brain to work, each area must talk to about one tenth of the areas. Then, so the idea goes, in a whale brain each area must also connect to one tenth of the areas. Whether the percentage is one tenth or one half I do not know; the claim is only that what is good for the mouse in this regard is also good for the whale. This is recorded as the

> *Principle of Invariant Area-Interconnectedness,* which states that the average percent area-interconnectedness remains roughly constant no matter the total number of areas. (See Figure 1.1.)

There is some direct evidence for this hypothesis. From it we expect that the total number of area-to-area connections E should scale as the number of areas A squared; or $A \sim E^{1/2}$. Data exist for only two species—cat and macaque—but we may use disjoint proper subsets of each animal's neocortex as distinct data points. Figure 1.2 shows these data, where the relationship fits $A \sim E^{0.45}$, or closely fitting Hypothesis 1.

Moving to the second invariance principle associated with well-connectedness, areas are composed of many neurons, and thus a connection from one

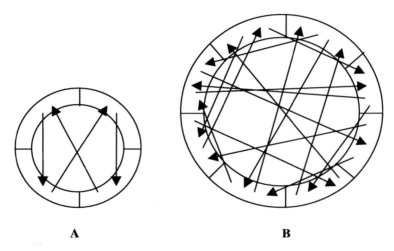

A   **B**

Figure 1.1: *Illustration of invariant percent area-interconnectedness. The average percent area-interconnectedness in a small and large neocortex. The outer part of each ring depicts the gray matter, the inner part the white matter. Each neocortex has multiple areas. (a) Each of the four areas in this small neocortex connects to one other area. The average percent area-interconnectedness is thus 1/4. (b) Each of the eight areas in this large neocortex connects to two other areas. The average percent area-interconnectedness is thus 2/8 = 1/4, the same as for the small brain.*

area to another is always from a neuron in the first area to a certain percentage of the neurons in the second area. We might call this percentage the *percent area-infiltration*. It is, again, natural and plausible to hypothesize that when an area tells another area about its efforts, it must tell a certain invariant percentage of the neurons in the area in order for the area to understand and appropriately respond to the information. That is, if white matter axons in mouse connect to roughly, say, one tenth of the number of neurons in an area, then, so the idea goes, in a whale brain each such neuron connects to one tenth of the neurons in an area. We record this as the

> *Principle of Invariant Area-Infiltration*, which states that, no matter the neocortical gray matter volume, the average percent area-infiltration stays roughly the same. (See Figure 1.3.)

I know of no data directly confirming this principle. It is here a hypothesis only, and it will stand or fall to the extent that it is economically able to account for the observed scaling exponents.

The two above invariance principles (invariant percent area-interconnectedness and invariant percent area-infiltration) concern the degree of well-connectedness of the neocortex, and we might summarize the pair of above principles by a single principle labeled the *Principle of Invariant Well-Connectedness.*

We have not yet made use of the idea of economical wiring, but I mentioned much earlier that the neocortical network appears to be driven, in part, by economical wiring... which leads us to the next principle. All things equal, it is advantageous for a nervous system to use less neural wiring, and as we saw at the start of the chapter many aspects of neuroanatomy and structural organization have been found to be consistent with such wire-optimization hypotheses. With this in mind we might expect that the neocortex would satisfy the Principle of Invariant Well-Connectedness, but that it would do so in a fashion sensitive to the connection costs. In particular, we would expect that the average number of neurons to which a neuron's axon connects—the *average neuron degree, δ*—will not be much greater than that needed to satisfy invariant well-connectedness. The reason for this is as follows: Connecting to more neurons requires a greater number of synapses per neuron, and this, in turn, requires greater arborization—more wire. In terms of scaling, this save-wire expectation can be weakened to the expectation that average neuron degree scales no faster than needed to satisfy invariant well-connectedness. I record this third principle as the

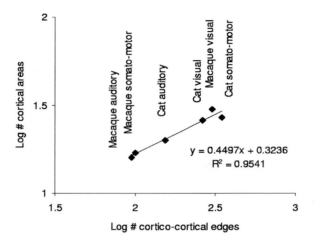

Figure 1.2: Logarithm (base 10) of the number of cortical areas versus logarithm of the number of area-to-area connections, for disjoint proper subnetworks within cat and macaque. Data points are as follows. Cat visual ($A = 26$, $E = 264$), cat auditory ($A = 20$, $E = 153$), and cat somato-motor ($A = 27$, $E = 348$) are from Scannell and Young (1993). Macaque visual ($A = 30$, $E = 300$), macaque auditory ($A = 16$, $E = 95$), and macaque somato-motor ($A = 17$, $E = 100$) are from Young (1993).

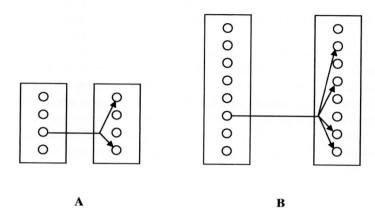

Figure 1.3: Illustration of the invariance of percent area-infiltration. The average percent area-infiltration for small and large areas. Each rectangle depicts an area, and each small circle a pyramidal neuron. (a) Each of these two areas has four neurons, and the left area connects via a pyramidal axon to two neurons in the right area. The percent area-infiltration is $2/4 = 1/2$. The other neurons' connections are not shown. (b) Each of the two areas has eight neurons, and the left area connects to four neurons in the right area. The percent area-infiltration is $4/8 = 1/2$, the same as for the small area.

> Principle of Economical Wiring, which states that the average neuron degree
> scales as slowly as possible consistent with an invariant well-connectedness.

Informally, the conjunction of these three above principles says that, no matter the neocortex size, an area talks to a fixed fraction of all the areas, and when an area talks to another area it informs a fixed fraction of the neurons in the area; furthermore, this is done in a volume-optimal manner. I will call the conjunction of these principles the *Principle of Economical Well-Connectedness.*

Scaling exponents derived from economical well-connectedness

Now let us consider the consequences of this principle. There are a few symbols we will need. First, recall that δ is the average neuron degree, defined as the average number of neurons to which a neuron's axon connects. Let A denote the total number of cortical areas, D the average number of areas to which an area connects, and W be the average number of neurons in an area. The first invariance principle stated that the percent area-interconnectedness is invariant, and this means that D/A is invariant, i.e., $D \sim A$. The second invariance principle stated that the percent area-infiltration is invariant, and this means that δ/W is invariant, i.e., $\delta \sim W$. Since an area connects to D areas and each neuron in an area can connect to neurons in only one area, there must be at least D neurons in an area; i.e., $W \geq D$. The Principle of Economical Wiring stated that δ must scale up as slowly as possible given the other constraints. Since we have already seen that $\delta \sim W$, economical wiring therefore implies that W must scale up as slowly as possible given the other constraints. Since we already saw that $W \geq D$, we can now also say that W scales no faster than D, and thus that $W \sim D$. To sum up for a moment, we have now concluded the following proportionalities:

$$A \sim D, D \sim W, W \sim \delta.$$

By the transitivity of proportionality, it follows that all these are proportional to one another; i.e.,

$$A \sim D \sim W \sim \delta.$$

This will be useful in a moment. Now notice that the total number of neurons N is proportional to the number of areas times the number of neurons per area. That is, $N \sim AW$. But we already have seen that $A \sim W$, and thus $N \sim A^2$, and so $A \sim N^{1/2}$. In fact, it follows that all those four mutually proportional quantities are proportional to $N^{1/2}$. That is,

$$A \sim D \sim W \sim \delta \sim N^{1/2}.$$

In particular, we will especially want to remember that

$$\delta \sim N^{1/2}.$$

I have just related δ to N, and in this paragraph I will relate δ to V_{gray}. With both of these relationships we will then be able to say how N and V_{gray} relate. A combination of empirical and theoretical argument suggests that, to a good approximation, a pyramidal neuron connects to almost as many different neurons as its number of axon synapses (Schüz, 1998). We do not here need to be committed to a claim as strong as this. Instead, all we need is that the neural degree δ is proportional to the number of axon synapses per neuron s. [I am here using the fact that the number of axon synapses scales identically with the total number of synapses per neuron. This must be true since every axon synapse is someone else's dendrite synapse.] Well, since

$$s \sim \frac{\rho_{synapse} V_{gray}}{N},$$

it follows that

$$\delta \sim \frac{\rho_{synapse} V_{gray}}{N}.$$

We may use the fact that $\rho_{synapse}$ is invariant [see Abeles, 1991, and also Changizi, 2001b for an explanation of this that is not presented here] to then say that

$$\delta \sim \frac{V_{gray}}{N}.$$

Thus far, we have seen that $\delta \sim N^{1/2}$, *and* we have seen that $\delta \sim V_{gray}/N$. But then it follows that

$$N^{1/2} \sim V_{gray}/N.$$

Solving for N we can finally conclude that

$$N \sim V_{gray}^{2/3}.$$

It also, of course, immediately follows, that

$$\rho \sim V_{gray}^{-1/3}.$$

And, since $D \sim A \sim N^{1/2}$, we also conclude that

$$D \sim V_{gray}^{1/3}$$

and

$$A \sim V_{gray}^{1/3}.$$

These scaling relationships are very close to the measured ones in Table 1.1. The number of cortical areas increases in bigger brains, then, not because of some kind of pressure to have more specialized areas, but because by not increasing the number of areas the network would become decreasingly well-connected, or would no longer be economically wired. [There are other theories hypothesizing that cortical areas may be due to issues of economical wiring, including Durbin and Mitchison (1990), Mitchison (1991, 1992), Ringo (1991), Jacobs and Jordan (1992) and Ringo et al. (1994).] Note that this theory also predicts that the number of neurons in an area, W, scales with gray matter volume with exponent $1/3$.

Invariant network diameter of 2

I will now show that the Principle of Invariant Well-Connectedness has the re-
markable consequence that the neocortical network has an invariant network
diameter of around 2. (See start of this chapter for the definition of network
diameter.) How may we compute its network diameter? The neocortex is certainly not a random network, so we cannot straightforwardly use the network
diameter approximation for random networks discussed in the introduction of
this chapter. But recall the notion of a small world network introduced there:
because pyramidal neurons usually make long range connections, or "shortcuts," via the white matter, the neocortical network is almost surely a small
world network, and thus would have a network diameter nearly as low as that
for a random network, namely approximately $\log(N)/\log(\delta)$. [This also requires that $N >> \delta >> \log N >> 1$. For the mammalian neocortex this is
true; $N \approx 10^7$ to 10^{11}, $\delta \approx 10^4$ to 10^5 (Schüz, 1998).] The scaling results
described thus far have informed us that $N \sim \delta^2$. The network diameter is,
then,

$$\Gamma \approx \frac{\log(C\delta^2)}{\log \delta} = 2 + \frac{\log C}{\log \delta},$$

where C is a proportionality constant. That is, for sufficiently large V_{gray}, the
neuron degree δ becomes large and thus the network diameter Γ approaches 2;
in the limit there are on average only two edges—one neuron—separating any
pair of neurons. A rough estimate of the constant C can be obtained by com-
paring actual values of the neuron number N and the average neuron degree
δ. For a mouse, $N \approx 2 \cdot 10^7$ and $\delta \approx 8,000$ (Schüz, 1998), so the constant
$C \approx N/\delta^2 = 0.3$. Common estimates for human are around $N \approx 10^{10}$ and
$\delta \approx 50,000$ (Abeles, 1991), making the constant $C \approx 4$. What is important
here is that these estimates of C (i) are on the order of 1, and (ii) are well below the estimates of δ. Thus $(\log C)/(\log \delta) \approx 0$ and the network diameter is
approximately 2. As a point of comparison, note that the network diameter for
C. Elegans—the only nervous system for which the network diameter has been
explicitly measured—is 2.65 (Watts and Strogatz, 1998); its network diameter
computed via the random network approximation is 2.16. This suggests the
conjecture that a network diameter around 2 is a feature common to all central
nervous systems.

1.1.2 Invariant computational units

To derive the scaling exponent for the thickness of the gray matter sheet and the total surface area, it suffices to note another invariance principle to which the neocortex appears to conform. It is the

> *Principle of Invariant Minicolumns*, which states that the number of neurons in a "minicolumn"—which is a neuroanatomical structure lying along a line through the thickness of the gray matter, from pia to white matter—is invariant. (An invariance principle of this form was first put forth by Prothero (1997a).)

The motivation is that, independent of brain size, these minicolumns are fundamental computational units, and that more "computational power" is achieved by increasing the number of such units, not by changing the nature of the fundamental computational units themselves. Evidence exists for this invariance from Rockel et al. (1980). [Rockel et al. (1980) mistakenly concluded that the surface density was invariant, but the latter could only be concluded if the number of neurons under, say, a square millimeter of surface was invariant. This, however, is not the case (Haug, 1987). See Prothero (1997b) for a cogent resolution to this issue.] The line density along a line from pia (the outside boundary of the neocortex) to white matter (the inside boundary), λ, scales as

$$\lambda \sim \rho^{1/3} \sim (V_{gray}^{-1/3})^{1/3} = V_{gray}^{-1/9}.$$

Since the number of neurons along this line is invariant, the sheet must be thickening, namely

$$T \sim V_{gray}^{1/9}.$$

It follows immediately that

$$S \sim V_{gray}^{1-1/9} = V_{gray}^{8/9}.$$

These are very close to the measured exponents, as shown in Table 1.1. The gray matter surface area scales more quickly than $V_{gray}^{2/3}$, then—and thus becomes convoluted—for two principal reasons. First, it is because the neuron density is decreasing—and this, in turn, is because the number of synapses per neuron is increasing in order to economically maintain satisfaction of the Principle of Economical Well-Connectedness. Second, it is because the pia-to-white-matter structure of the cortex remains the same (e.g., same number of neurons in a minicolumn, same number of layers) across mammals. If, instead, the number of neurons along a thin line through the cortical sheet increased in larger brains, the new neurons would not have to spread only along the surface, but could spread into deeper regions of the gray matter; the surface would then not have to scale up so quickly. This, however, would require modifying the basic, uniform structure of the gray matter every time the brain was enlarged; it would demand inventing new basic computational architectures in each brain, whereas by keeping the structure the same, larger brains can work with the same "primitive computational units" as in smaller brains.

A related issue concerns modules found in the neocortex, such as columns, blobs, bands, barrels and clusters.   They are intermediate features, smaller than cortical areas, and larger than minicolumns. The simplest hypothesis is that modules conform to the following invariance principle,

> *Principle of Invariant Modules*, which states that the number of minicolumns in a module is invariant.

The motivation is similar to that for the Principle of Invariant Minicolumns. If this principle holds for neocortex, then from the neuron density decrease it follows that the diameter of a module (when measured along the cortical surface), m, should scale as $V_{gray}^{1/9}$. Manger et al. (1998) measured module size across mammals varying over four orders of magnitude in brain size, and one may compute from their data that the exponent is 0.135 (see Figure 1.4), or very close to the predicted $1/9$. The number of neurons in a module therefore appears to be independent of brain size.

1.1.3   Efficient neural branching diameters

Murray's Law

As far as I know, every kind of tree in nature has thicker trunks when the trunk supports more leaves (Cherniak, 1992; Cherniak et al., 1999; Changizi and Cherniak, 2000). The same is therefore expected of neurons—the soma being the ultimate trunk of a neuron—and certainly appears to be the case (e.g., Cherniak, 1992; Cherniak et al., 1999). But in exactly what quantitative manner do we expect trunk diameter to scale as a function of the number of leaves in the tree? Many natural trees conform to a relationship called Murray's Law (1926a), which says that, for any two depths, i and j, in a tree, the sum of the cubes of the diameters at depth i is identical to the sum of the cubes of the diameters at depth j. So, for example, the cube of a trunk diameter must equal the sum of the cubes of its daughter segment diameters. Murray's Law is expected to apply for any tree where (i) there is laminar fluid flow, and (ii)

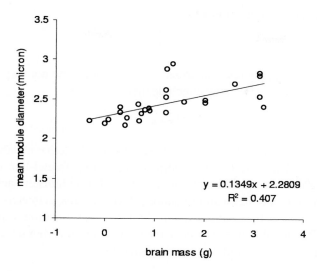

Figure 1.4: Logarithm (base 10) of the mean module diameter versus logarithm of brain size. Data from Manger et al. (1998).

the power required to distribute the fluid is minimized. In fact, it is well known that there is fluid flow in neural arbors (Lasek, 1988), and that the fluid flow is laminar follows from the facts that fluid flow in pipes of diameter less than one millimeter tends to be laminar (Streeter and Wylie, 1985) and that neural arbors have diameters on the micron scale. Murray's Law, in fact, appears to apply to neural trees, as shown in Cherniak et al. (1999). I record this principle as the

Principle of Efficient Neural Branching Diameters, which states that neural segment diameters are set so as to maximize power efficiency.

Soma and axon radius

From this principle—i.e., from Murray's Law—we may derive the expected scaling relationship between trunk diameter, t, and the number of leaves in the tree. Murray's Law states that the trunk diameter, t, cubed should be the same value as the sum of the cubes of all the diameters of the leaf segments in the tree. Let s be the number of leaves in the tree, and d be the diameter of each leaf segment. Then the relationship is,

$$t^3 = sd^3.$$

Given that the leaf segments in neurons—i.e., synapses—do not vary in size as a function of brain size, we may conclude that

$$t^3 \sim s.$$

[West et al., 1997, use elaborations on ideas like this to derive metabolic scaling exponents. Their arguments require space-filling, fractal-like networks, whereas the argument here does not require this. Murray himself back in 1927 might well have noted this scaling feature.] That is, trunk diameter—whether we treat the soma or the source axon as the trunk—of a neuron scales as the 1/3 power of the number of synapses in the neuron. From earlier developments we know that $s \sim V_{gray}^{1/3}$, and thus we may derive that

$$R \sim V_{gray}^{1/9},$$

where I am now using R for trunk radius rather than trunk diameter (since they are proportional). Measured scaling relationships conform well to this, for both soma (or neuron body) radius and for the radius of a white matter axon (see Table 1.1). [Note that Murray's Law states that $t^3 = b_1^3 + b_2^3$, where b_1 and b_2 are the two daughter branch diameters of a trunk with diameter t, and thus, in general, trunk diameter $t \approx 2b$, where b is the average daughter diameter, and thus $t \sim b$. This is why it is justified to treat soma and axon radius as proportional.]

White matter volume

Finally, there is the issue of white matter volume, V_{white}. White matter volume is composed entirely of myelinated (and some unmyelinated) axons from pyramidal neurons sending cortico-cortical connections. Thus, white matter volume is equal to the total number of white matter axons, $N_{whiteaxon}$, times the volume of a white matter axon, $V_{whiteaxon}$. That is,

$$V_{white} = N_{whiteaxon} V_{whiteaxon}.$$

All we need to do is to figure out how these two quantities scale with gray matter volume.

There must be one white matter axon for every neuron, and thus $N_{whiteaxon} \sim N$, and so $N_{whiteaxon} \sim V_{gray}^{2/3}$. The volume of a white matter axon, $V_{whiteaxon}$, is proportional to the length, L, of the axon times the square of its radius, R_1; i.e.,

$$V_{whiteaxon} \sim LR_1^2.$$

White matter axons travel roughly a distance proportional to the diameter of the white matter, and so $L \sim V_{white}^{1/3}$. Also, we saw just above that $R_1 \sim V_{gray}^{1/9}$. Thus,

$$V_{whiteaxon} \sim V_{white}^{1/3}(V_{gray}^{1/9})^2,$$

and so

$$V_{whiteaxon} \sim V_{white}^{1/3} V_{gray}^{2/9}.$$

Recalling that $V_{white} = N_{whiteaxon}V_{whiteaxon}$, we can now combine our conclusions and get the following.

$$V_{white} = N_{whiteaxon}V_{whiteaxon},$$

$$V_{white} \sim [V_{gray}^{2/3}] \cdot [V_{white}^{1/3} V_{gray}^{2/9}].$$

Now we just need to solve for V_{white}, and we can then conclude that

$$V_{white} \sim V_{gray}^{4/3},$$

very close to the measured exponents, as shown in Table 1.1.

White matter volume scales disproportionately quickly as a function of gray matter volume because of the increasing axon radius, and this, in turn, is due to the satisfaction of Murray's law for efficient flow. The exponent would fall from $4/3$ to 1 if axon radius were invariant.

1.1.4 Wrap-up

It is instructive to summarize the principles that appear to govern the neocortex.

1. *Efficiency Principles*

   - *Efficient Neural Diameters*: neural diameters are set for maximum power efficiency for the distribution of materials through the arbor.
   - *Economical Wiring*: invariant well-connectedness is achieved in a volume-optimal manner.

2. *Invariance Principles*

- *Invariant Well-Connectedness*
  - *Invariant Area-Interconnectedness*: the fraction of the total number of areas to which an area connects is invariant.
  - *Invariant Area-Infiltration*: the fraction of the number of neurons in an area to which a white matter axon connects is invariant.
  - (And these lead to an invariant network diameter of 2.)
- *Invariant Computational Units*
  - *Invariant Minicolumns*: the number of neurons in a minicolumn is invariant.
  - *Invariant Modules*: the number of minicolumns in a module is invariant.

Why are these principles advantageous for the neocortex? The answer is obvious for the two efficiency principles. Invariant well-connectedness is useful, lest larger brains have networks that become more and more widely separated, in terms of the average minimal path length between neurons. It is less obvious why a network would maintain invariant computational units. In the next section this will be taken up in more detail, where we will see that in a wide variety of network—including neocortex—there appears to be scale-invariant "functional units," and I will show that this is to be expected if network size is optimized. The basic idea underlying the argument can be seen here in Figure 1.5. Thus, that the neocortex has invariant computational units is derivable from a network optimization principle. This allows us to simplify the above so as to state the least number of principles from which it is possible to explain neocortical scaling.

1. *Efficiency Principles*

   - *Efficient Neural Diameters*: neural diameters are set for maximum power efficiency for the distribution of materials through the arbor.
   - *Economical Wiring*: invariant well-connectedness is achieve in a volume-optimal manner.
   - *Optimal Network Size*: network size scales up no more quickly than "needed" (see next section), from which invariant computational units are derivable.

2. *Invariant Well-connectedness Principles*

   - *Invariant Area-Interconnectedness*: the fraction of the total number of areas to which an area connects is invariant.
   - *Invariant Area-Infiltration*: the fraction of the number of neurons in an area to which a white matter axon connects is invariant.
   - (And these lead to an invariant network diameter of 2.)

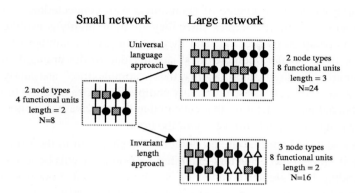

Figure 1.5: *There are broadly two ways to increase complexity in networks, as we will discuss in Section 1.2. Top. In the "universal language approach" there are a fixed number of node types with which all functional units are built. In this case, networks with greater numbers of functional unit types must accommodate the new function types by having longer functional units. In the figure, the small network under the universal language approach begins with 4 functional unit types, each of length 2, and a language of 2 node types; the total number of nodes required for this is 8. In order to accommodate 8 functional unit types in the larger network, the length of functional units must be increased since the expressive power for length- 2 functions has already been exhausted. The larger network, which has 8 functional units, has functional units of length 3, and the total number of nodes required is 24. Bottom. Consider now, in contrast, the "invariant length approach" to complexity increase. Since functional units have an invariant length in this approach, in order to achieve greater numbers of types of functional units, new node types must be invented. In the figure the small network is identical to the small network under the universal language approach. The large network under this approach has, as in the universal language case, 8 functional unit types. However, to achieve it one new node type must be added. The total number of nodes in the larger network is 16, which is smaller than the 24 nodes in the universal language case. The invariant length approach is optimal in the sense that network size grows minimally. Note that it also entails that the number of types must be increased, and we will see in Section 1.2 that this is indeed true for neocortex, and for networks generally.*

1.2   Complexity in brain and behavior

I took up brain scaling in the previous section, and we saw that many of the ways in which larger brains are "more complex" are consequences of brains maintaining an invariant degree of economical well-connectedness. That is, bigger brains *seem* more complex since they are more highly convoluted, they have more synapses per neuron, and they have a greater number of cortical areas; but these greater "complexities" are not due to the brains themselves being "smarter" in any fashion. Rather, these "complexities" are purely due to the brains being bigger. These seemingly complex qualities of larger brains are thus epiphenomenal, where by that I mean that their increase does not signify any increasing functional complexity of the brain.

In this section I concentrate on brain complexity, both in the nervous network itself, and in the behaviors exhibited by brains. I will be interested in understanding how greater complexity is achieved. First, however, we will need to become clear concerning what I mean by complexity.

The central intuitive notion of 'complexity' I rely upon here is that *an entity, or system, is more complex if it can do more kinds of things.* For example, if my radio has one more type of function than does yours—say, scanning—mine is more complex. Note that under this idea of 'complexity,' doing more of the *same* kinds of thing does *not* make something more complex. For example, a book is complex in some sense, but stapling together two copies of the same book does not create a more complex book; in each of these two cases all the same sentences are uttered. Similarly, if two birds have the same song repertoires, they have the same complexity even if one sings twice as often. Complexity, then, concerns the diversity of things done by an entity. Rather than referring to these things that are done by an entity as "things," I will call them *expressions*, and a system or entity of some kind is more complex than another of that kind if it has, or does, more expression types. The number of expression types, E, is thus the complexity of the entity, and I will sometimes refer to this number, E, as the *expressive complexity* of the entity.

The question we are interested in asking in this section is, How does a system, or entity, of a given kind accommodate greater expressive complexity? For example, how is greater song repertoire size handled in birds? And, how is greater brain complexity achieved? The first-pass answer to these questions is that expressions are always built out of lower-level components that come in different types. For example, bird songs are built out of bird syllables of distinct types. And functional expressions of the brain are built out of combinations of neurons of distinct types. Let L be the average number of components in an expression (for some given kind of entity); L is the *expression length*. For example, for bird song L is the number of syllables per song. Also, let C be the total number of component types from which the E expression types of the system are buildable. For bird song, C is the total number of syllable types in the repertoire of a bird (from which that bird's E different songs are built).

If expressions are of length L, and each spot in the expression can be filled by one of C component types, then there are a maximum of $E = C^L$ many expression types buildable. For example, if there are $C = 2$ component types—labeled A and B—and expression length $L = 4$, then there are $E = 2^4 = 16$ expression types, namely $AAAA, AAAB, AABA,..., BBBB$. However, this is insufficiently general for two reasons. First, only some constant fraction α of these expression types will generally be grammatical, or allowable, where this proportionality constant will depend on the particular kind of complex system. The relationship is, then, $E \sim C^L$. Second, the exponent, L, assumes that all L degrees of freedom in the construction of expressions are available, when only some fixed fraction β of the L degrees of freedom may generally be available due to inter-component constraints. Let $d = \beta \cdot L$ (where, again, what β is will depend on the particular kind of system). Call this variable d the *combinatorial degree*. The relationship is, then

$$E \sim C^d,$$

where C and d may each possibly be functions of E. Using the same example as above, let us suppose now that As always occur in pairs, and that Bs also always occur in pairs. The "effective components" in the construction of expressions are now just AA and BB, and the expression types are $AAAA, AABB, BBAA$, and $BBBB$. The number of degrees of freedom for an expression is just 2, not 4, and thus $E = 2^2 = 4$. d is a measure of how combinatorial the system is. The lowest d can be is 1, and in this case there is effectively just one component per expression, and thus the system is not combinatorial at all. Higher values of d mean the system is more combinatorial.

Given this above relationship between expressive complexity E, the number of component types C, and combinatorial degree d, let us consider a few of the ways that a system might increase its expressive complexity.

The first way is the *universal language approach*. The idea here is that there exists a fixed number of component types from which any expression type is constructable. For example, for computable functions there exists such a language: from a small number of basic computable functions it is possible to build, with ever longer programs, any computable function at all. If this universal language approach were taken, then the number of component types, C, would not change as a function of the expressive complexity, E. Something *would* change, however, and that is that the average length, L, of an expression increases as a function of E. In particular, since C is invariant in the equation $E \sim C^d$, it follows that $d \sim \log E$. This approach may be advantageous for systems where new component types are costly, or where there is little cost to increasing the expression length; it is generally difficult to achieve however, since the invention of a universal language is required.

The second way to increase expressive complexity is via the *specialized component types approach*. In this case, for each new expression type, a new specialized set of component types is invented just for that expression type. Here the expression length L may or may not be > 1, but the combinatorial degree $d = 1$. Thus, $E \sim C$. Note that if this possibility holds for a complex system, then a log-log plot of C versus E should have a slope of 1. The advantage of this approach is that expressions are short, and no complex grammatical rules need to be invented. The disadvantage is that the number of new component types must be scaled up very quickly (namely, proportionally with expressive complexity).

The third way to raise E is via the *invariant-length approach*. This is like the previous approach in that the combinatorial degree d (and expression length L) is invariant, except that now it is > 1. Thus, it is combinatorial ($d > 1$), and its "degree of combinatorialness" remains invariant. The expected relationship is the power law $E \sim C^d$, with d invariant and > 1. On a log-log plot of C versus E, we expect a straight line with slope of $1/d$. A log-log straight line with fractional slope means that a small increase in the number of component types gives a disproportionately large number of new expression types; and this is characteristic of combinatorial systems. An advantage to this approach is that the rate at which new component types must be invented is very slow ($C \sim E^{1/d}$, where d is constant and > 1). The disadvantage is that expressions tend to be longer, and that there must exist a relatively sophisticated set of rules, or grammar, allowing the expressions to be built in a combinatorial fashion.

The final way to increase expressive complexity is via the *increasing-C-and-d approach*. This is similar to the previous case, except that now expressive complexity increase is accommodated by increasing C *and* increasing d. If d increases logarithmically with E, then this is the universal language approach from above, where C does not increase. Thus, d must increase sublogarithmically, such as $d \sim [\log E]/[\log \log E]$. In this case, it follows that $C \sim \log E$; C is increasing here less quickly than a power law. As in the previous possibility, a small increase in C gives a disproportionately large increase in E, except that now the size of the combinatorial explosion itself increases as a function of E (since d is increasing). This is a kind of middle ground between the universal language approach and the invariant-length approach.

These are the four key manners in which expressive complexity may be increased in complex systems, and our central question may be stated more rigorously now: In which of these ways do complex systems related to brain and behavior increase expressive complexity? And why? In the next two subsections we discuss behavioral and brain complexity in light of the above ideas. (See Changizi (2001e) for the connections between these above ideas and the notion of 'hierarchy'.)

1.2.1   Complexity of languages and behaviors

Behavior is a complex system, consisting of behavioral expressions, which are built out of multiple behavioral components of some kind. The main question concerning behavior here is, In what manner is behavioral repertoire size—i.e., expressive complexity—increased? That is, which of the earlier approaches to increasing complexity holds for behaviors? Do animals have a "universal language" of component behaviors from which any complex behavior may be built, or do animals with more behaviors have more behavioral component types? We examine this question in three distinct kinds of behavior: human linguistic behavior, bird vocalization behavior, and traditional non-vocal animal behaviors.

Ontogeny of language

Human natural language is a complex system, where components of various kinds are combined into expressions. For example, phonemes are combined into words, and words into sentences. We begin by studying expressive complexity increase during the development of language in children. That is, in light of the earlier approaches to increasing complexity, how do children increase their expressive complexity? We already know a few things about the development of language in children. First, we know that children increase the number of component types (e.g., their phoneme repertoire and word repertoire) as they age. That is, C increases. Second, we know that their ability to string together components increases with age (Pascual-Leone, 1970; Case et

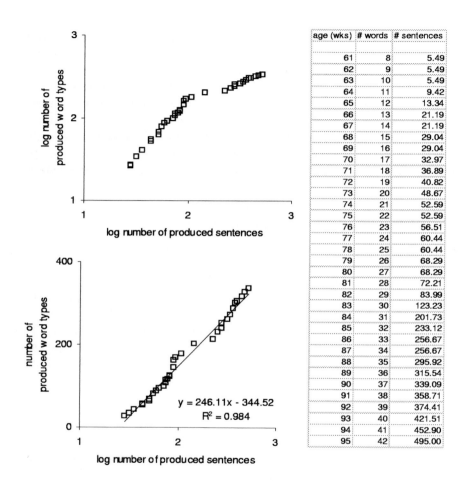

age (wks)	# words	# sentences
61	8	5.49
62	9	5.49
63	10	5.49
64	11	9.42
65	12	13.34
66	13	21.19
67	14	21.19
68	15	29.04
69	16	29.04
70	17	32.97
71	18	36.89
72	19	40.82
73	20	48.67
74	21	52.59
75	22	52.59
76	23	56.51
77	24	60.44
78	25	60.44
79	26	68.29
80	27	68.29
81	28	72.21
82	29	83.99
83	30	123.23
84	31	201.73
85	32	233.12
86	33	256.67
87	34	256.67
88	35	295.92
89	36	315.54
90	37	339.09
91	38	358.71
92	39	374.41
93	40	421.51
94	41	452.90
95	42	495.00

Figure 1.6: Top. *Logarithm (base 10) of the number of word types versus logarithm of the number of sentences, as produced by one child named Damon from 12 to 22 months (Clark, 1993).* Bottom. *Semi-log plot of the same data. Plot is confined to multiword utterance ages, which began at about 14 months. On the right are shown the data; sentence data is fractional because I obtained it via measuring from Clark's plots. [Note that Damon grew up and went on to do some research on aspects of how the brain scales up.]*

al., 1982; Siegel and Ryan, 1989; Adams and Gathercole, 2000, Robinson and Mervis, 1998; Corrigan, 1983). However, from this increasing combinatorial ability we cannot straightforwardly conclude that the child's combinatorial degree, d, will keep pace. Recall that d is measured by the relative rates of growth of the number of component types C and the number of expression types E. A child's combinatorial *ability* could increase, and yet the child could simply choose not to speak much, in which case the combinatorial ability growth would not be reflected in the combinatorial degree measured from the C versus E plot (since E would not increase much). Nevertheless, learning new component types is costly, and efficiency considerations would lead one to expect that new component types are added no more quickly than needed for the given level of expressive complexity. If this were true, we would expect the combinatorial degree to increase as a function of E, and thus the log-log slope of C versus E to fall as E increases. That is, the increasing-C-and-d approach would be used, and $C \sim \log E$. We examine this prediction for the development of words and sentences, and also for the development of phonemes and words.

The number of word types and number of distinct sentences uttered by a single child named Damon for 41 weeks from 12 to 22 months of age (Clark, 1993) are shown in Figure 1.6. One can see that, as expected, the combinatorial degree (i.e., the inverse of the slope in the log-log plot) falls as a function of E. At the start, the combinatorial degree is 1, which means that the child is not yet using words in a combinatorial fashion. Note that the data here are only for the multi-word utterance stage of the child; thus, although the child may *seem* to be using words in a combinatorial manner since his sentences have more than one word, he is not. By the end of the recorded data, the combinatorial degree has increased to about 2.5. This combinatorial degree range is similar to the sentence length range for children of this period (Robinson and Mervis, 1998). Since the combinatorial degree and number of component types are increasing, the increasing-C-and-d length approach is being employed for increasing expressive complexity, and thus we expect $C \sim \log E$. Indeed, a plot of C versus $\log E$ appears linear.

The growth in the number of phonemes and distinct morphemes was compiled from Velten (1943), as produced by a child named Jean from 11 to 30 months of age. [A *morpheme* is the smallest meaningful linguistic unit; or, a word that is not decomposable into meaningful parts.] Figure 1.7 shows the log-log plot of the number of phoneme types versus the number of morpheme types, and one can see that the slope tends to decrease somewhat through

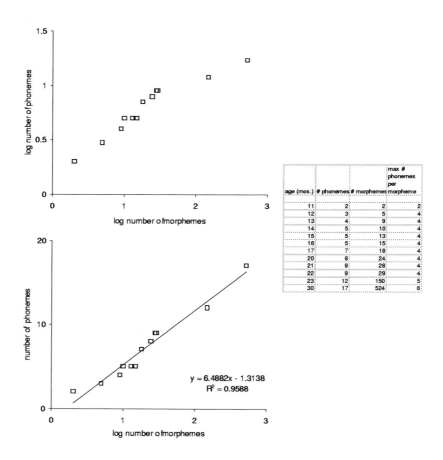

age (mos.)	# phonemes	# morphemes	max # phonemes per morpheme
11	2	2	2
12	3	5	4
13	4	9	4
14	5	10	4
15	5	13	4
16	5	15	4
17	7	18	4
20	8	24	4
21	9	28	4
22	9	29	4
23	12	150	5
30	17	524	6

Figure 1.7: Top: *Logarithm (base 10) of the number of phoneme types versus logarithm of the number of morphemes, as produced by one child named Jean from 11 to 30 months (Velten, 1943).* Bottom: *Semi-log plot of the dame data. Morphemes are the smallest meaningful linguistic unit, and are mostly words in this case. On the right are shown the data.*

development, meaning the combinatorial degree is increasing. The plot of (unlogged) number of phoneme types versus the logarithm of the number of morphemes is comparatively linear, again implicating the increasing-C-and-d approach, as predicted above. The combinatorial degree begins at around 2 ("ma"), and increases to around 4 ("baby"). The number of phonemes per morpheme—i.e., expression length—increases over a similar range during this period (Changizi, 2001d, 2001e).

In each of these language development cases, we see that the expressive complexity is increased via the increasing-C-and-d invariant approach, and that the combinatorial degree appears to increase in proportion to the child's ability to combine components. This accords with the efficiency hypothesis mentioned above, that children will learn component types no more quickly than needed to express themselves.

English throughout history

Here I consider complexity in the English language. Not the complexity of the language of one English speaker as above, but, instead, the complexity of the entire English language. Our "individual" here is the entire English-speaking community. This individual has, over history, said more and more new expression types. Namely, new distinct sentences are being uttered throughout history. How has this increase in expressive complexity been accommodated? Via an increase in the average length of a sentence, or via the addition of new vocabulary words—word types—with which sentences may be built, or via a combination of the two? That is, which of the earlier approaches to complexity increase is employed in the evolution of the English language?

I estimated the growth in the number of English word types by using the Oxford English Dictionary (OED), Second Edition. It is possible to search for years within only the etymological information for all entries in the OED. In this way it was possible to estimate the number of new word types per decade over the last 800 years. To obtain an estimate of the growth rate for the number of sentences the English-speaking entity expresses, I used the number of books published in any given year as an estimate of the number of new sentences in that year. This would be a problematic measure if different books tended to highly overlap in their sentences, but since nearly every written sentence is novel, never having been uttered before, there is essentially no overlap of sentences between books. This would also be a problematic measure if the length of books, in terms of the number of sentences, has been changing through time;

Table 1.3: *The data for the history of English from 1200 to the present.*
Decades covering century or half-century years do not include those years
(since they tend to be overcounted). The new words were measured from the
Oxford English Dictionary, and the number of new books from WorldCat.

decade	# new words	# new books		decade	# new words	# new books
1210	36	3		1610	161	3705
1220	40	5		1620	606	4174
1230	43	4		1630	130	4736
1240	29	3		1640	140	6321
1250	24	0		1650	125	17891
1260	36	5		1660	169	13976
1270	42	1		1670	144	11274
1280	59	3		1680	181	16548
1290	41	8		1690	202	21868
1300	65	10		1700	117	16962
1310	72	12		1710	156	15513
1320	77	4		1720	111	17398
1330	78	5		1730	207	15685
1340	73	5		1740	273	17717
1350	33	5		1750	146	16113
1360	42	9		1760	296	21161
1370	61	3		1770	231	25254
1380	104	8		1780	214	33542
1390	83	7		1790	295	38186
1400	54	7		1800	315	54326
1410	54	7		1810	372	99069
1420	72	12		1820	440	129734
1430	56	6		1830	481	114817
1440	63	5		1840	602	148800
1450	18	4		1850	454	158774
1460	51	10		1860	452	234706
1470	51	13		1870	554	257161
1480	83	38		1880	661	290921
1490	107	79		1890	809	391372
1500	65	103		1900	686	402769
1510	75	139		1910	685	541294
1520	126	134		1920	425	587809
1530	171	265		1930	484	680614
1540	167	468		1940	397	808427
1550	201	626		1950	206	730849
1560	223	814		1960	249	1354217
1570	178	1038		1970	216	2600020
1580	195	1363		1980	120	4594985
1590	163	1688		1990	35	5618350
1600	372	2055				

I have no data in this regard, but it seems plausible to assume that any such trend is not particularly dramatic. The number of new books published per year was obtained by searching for publication dates within the year for literature listed in WorldCat, an online catalog of more than 40 million records found in thousands of OCLC (Online Computer Library Center) member libraries around the world. In this way I was able to estimate the number of new books per decade over the last 800 years, the same time period for which I obtained word type data. These data are shown in Table 1.3.

Figure 1.8 shows the logarithm of the number of new word types and books per decade over the last 800 years, measured as described above. Note that the plot shows estimates for the number of *new* word types per decade, and the number of *new* sentences per decade; i.e., it measures dC/dt and dE/dt versus time. The plot does not, therefore, show the growth in the actual magnitude of the number of word types or the number of sentences. But it is the scaling relationship between the actual magnitudes of C and E we care about, so what can we do with a plot of growth rates over time? Note first that the growth rate for each is exponential (this is because the plots fall along straight lines when the y axis is logarithmic and the x axis not). If a growth rate for some quantity u increases exponentially with time, then this means $du/dt \sim e^{rt}$. And if you recall your calculus, it follows that the quantity itself scales exponentially with time, and, in fact, it scales proportionally with the growth rate: i.e., $u \sim du/dt$. Thus, Figure 1.8 has effectively measured the growth in the number of word types and the number of books. By looking at the growth in the number of word types compared to that for the number of books, we can determine how the first scales against the second.

From the figure we can, then, determine that

$$dC/dt \sim C \sim 10^{0.001725t} \sim e^{0.003972t},$$

and

$$dE/dt \sim E \sim 10^{0.008653t} \sim e^{0.01992t}.$$

We may now solve for C in terms of E, and we obtain

$$C \sim E^{0.003972t/0.01992t} = E^{0.1994}.$$

The number of word types scales as a power law against the number of sentences, and, unsurprisingly, the slope is less than one and thus English is combinatorial. Thus, greater expressive complexity was achieved over the last 800 years not by increasing the combinatorial degree (or average sentence length),

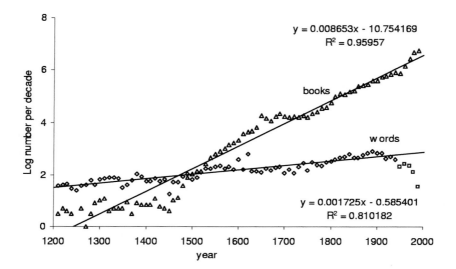

Figure 1.8: Growth rates in the decades from the years 1200 to 1990 for the number of new English word types and the number of new English books. Regression equations and correlation coefficients are shown for each (79 data points each). Unsure etymological dates tend to cluster at century and half century marks and therefore century and half-century marks tend to be overcounted; accordingly, they were not included in the counts. The OED is conservative and undercounts recently coined word types; consequently, the exponential decay region (the last five square data points) was not included when computing linear regression. I do not have any way to similarly measure the number of word type extinctions per year, and so I have not incorporated this; my working assumption is that the extinction rate is small compared to the growth rate, but it should be recognized that the estimated combinatorial degree is therefore an underestimate.

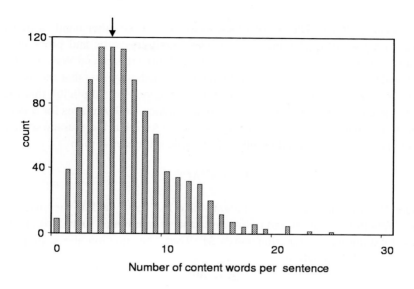

Figure 1.9: Distribution of numbers of content words per sentence in English. Arrow indicates the log-transformed mean. 984 sentences from 155 authors were measured from texts
in philosophy, fiction, science, politics and history. I chose the second sentence on each odd
numbered page. A word was deemed a function word if it was among a list of 437 such words
I generated. A string of words was deemed a sentence if it represented a complete thought
or proposition. So, for example, semicolons were treated as sentence delimiters, multiple sentences combined into one long sentence by ", and" were treated as multiple sentences, and
extended asides within dashes or parentheses were not treated as part of the sentence.

but, instead, by increasing the number of word types with which to build sentences. The scaling exponent of around 0.2 implies an estimated combinatorial
degree of about 5. There appears to be nothing about the English grammar
that implies a fixed combinatorial degree (or sentence length), much less any
particular value of it. What explains this value of 5? [Or, a little more than 5;
see legend of Figure 1.8 concerning word type extinctions.] It cannot simply
be due to the typical number of words in an English sentence, since there are
typically many more words than that, namely around 10 to 30 words (Scudder,
1923; Hunt, 1965).

   To make sense of the combinatorial degree, we must distinguish between
two kinds of word in English: *content* and *function*. The set of content words,
which refer to entities, events, states, relations and properties in the world,
is large (hundreds of thousands) and experiences significant growth (Clark

and Wasow, 1998). The set of function words, on the other hand, which includes prepositions, conjunctions, articles, auxiliary verbs and pronouns, is small (around 500) and relatively stable through time (Clark and Wasow, 1998). The scale-invariant combinatorial degree of English suggests that the average number of words per sentence is invariant. Imagine, for simplicity, that there on average n places for content words in a sentence, and m places for function words, and that these values, too, are invariant. (And thus the average sentence length is $n + m$.) The total number of possible sentences is then

$$E \sim N^n M^m,$$

where N is the total number of content words in English and M the total number of function words. n and m are invariant, as mentioned just above, and so is the total number of function words M. Thus, the equation above simplifies to the power law equation

$$E \sim N^n.$$

Also, note that the number of content words, N, is essentially all the words, since it dwarfs the number of function words; i.e., $C \approx N$. Thus, $E \sim C^n$, and so,

$$C \sim E^{1/n}.$$

That is, the combinatorial degree is expected to be equal to the typical number of *content* words per sentence—not the typical total number of words per sentence—and, up to a constant factor, they may be combined in any order. To test this reasoning, I measured the number of content words in nearly one thousand sentences (see legend of Figure 1.9). The distribution is log-normal (Figure 1.9), and the mean of the logs is 0.7325 ($\pm$0.2987); the log-transformed mean is thus 5.401, and one standard deviation around this corresponds to the interval [2.715, 10.745]. This provides confirmation of the hypothesis that the combinatorial degree is due to there being five content words per sentence.

But *why* are there typically five content words per sentence? One obvious hypothesis is that sentences can convey only so much information before they overload the utterer's or listener's ability to understand or absorb it. In this light, five content words per sentence is probably due to our neurobiological limits on working memory, which is a bit above five (Miller, 1956). The fingerprint of our working memory may, then, be found in the relative rate at which new words are coined compared to the number of sentences uttered by the English-speaking community.

Table 1.4: Number of syllable types and song types for a variety of species of bird.

Species	Number of syllable types	Number of song types	Citation
Turdus nudigenis (Bare-eyed Thrush)	1.20	1.19	Ince and Slater (1985)
Turdus tephronotus (African Bare-eyed Thrush)	3.83	1.26	Ince and Slater (1985)
Turdus iliacus (Redwings)	2.93	1.31	Ince and Slater (1985)
Turdus torquatus (Ring Ouzels)	5.49	3.79	Ince and Slater (1985)
Turdus viscivorus (Song and Mistle Thrush)	43.08	15.74	Ince and Slater (1985)
Turdus pilaris (Fieldfare)	80.65	32.05	Ince and Slater (1985)
Turdus merula (Blackbird)	216.09	38.97	Ince and Slater (1985)
Turdus abyssinicus (Olive Thrush)	43.08	49.99	Ince and Slater (1985)
Turdus migratorius (American Robin)	30.10	71.33	Ince and Slater (1985)
Turdus philomelos (Song Thrush)	309.22	158.77	Ince and Slater (1985)
Catherpes mexicanus (Canyon Wren)	9	3	Kroodsma (1977)
Cistothorus palustris (Long-billed Marsh Wren)	44—118	40—114	Kroodsma (1977)
Cistothorus platensis (Short-billed Marsh Wren)	112	110	Kroodsma (1977)
Salpinctes obsoletus (Rock Wren)	69—119	69—119	Kroodsma (1977)
Thryomanes bewickii (Bewick's wren)	25—65	9—22	Kroodsma (1977)
Thryomanes bewickii (Bewick's wren)	87.5	20	Kroodsma (1977)
Thryomanes bewickii (Bewick's wren)	56	17.5	Kroodsma (1977)
Thryomanes bewickii (Bewick's wren)	50	16	Kroodsma (1977)
Thryomanes bewickii (Bewick's wren)	40.5	10	Kroodsma (1977)
Thryomanes bewickii (Bewick's wren)	35.5	17.5	Kroodsma (1977)
Thryothorus ludovicianus (Carolina Wren)	22	22	Kroodsma (1977)
Troglodytes troglodytes (Winter wren)	89—95	3—10	Kroodsma (1977)
Gymnorhina Tibicen (Australian Magpie)	29	14	Brown et al. (1988)
Paridae bicolor (Tufted titmouse)	3	11.25	Hailman (1989)
Parus wollweberi (Bridled titmouse)	3	3	Hailman (1989)
Serinus canaria (Canary)	101	303	Mundinger (1999)
Empidonax alnorum (Alder Flycatcher)	3	1	Kroodsma (1984)
Empidonax traillii (Willow Flycatcher)	4	3	Kroodsma (1984)

Bird vocalization

Bird songs are built out of bird syllables, and the question we ask is, Do birds with more songs in their repertoire have longer songs, or more syllables, or both? In particular, which of the four earlier approaches to expressive complexity increase is used in bird vocalization?

To answer this I surveyed the bird vocalization literature and compiled all cases where the authors recorded the number of syllable types and the number of song types in the repertoire of the bird. Although song repertoire size counts are common, syllable type counts are much rarer, especially when one is looking for papers recording both. Table 1.4 shows the data and the sources from which I obtained them.

Plotting the number of syllable types, C, versus the number of song types, E, on a log-log plot (Figure 1.10) reveals that (i) they are related by a power law (i.e., the data are much more linear on a log-log plot than on a semi-log plot), and (ii) the exponent is approximately 0.8. That is, $C \sim E^{0.8}$. Since the relationship is a power law, the combinatorial degree is an invariant; i.e., there appears to be no tendency for the combinatorial degree (or expression length) to increase in birds with greater numbers of songs. Instead, greater expressive complexity is achieved entirely through increasing the number of bird syllable types. Since the exponent is about 0.8, the combinatorial degree is its inverse, and is thus about 1.25, which is not much above 1. In fact, it is not significantly different from 1 (Changizi, 2001d). Birds with twice as many songs therefore tend to have roughly twice as many syllable types, and thus bird vocalization may not be combinatorial at all, and, at most, it is not very combinatorial. Birds therefore appear to conform to the specialized component types approach to complexity increase. Using bird vocalization as a model for language is thus inappropriate. Note that the combinatorial degree for bird vocalization is near 1 despite the fact that birds have, on average, around 3 or 4 syllables per song (Read and Weary, 1992; Changizi, 2001d).

Traditional animal behavior

Thus far, the cases of behavior we have discussed have been vocalizations, whether bird or human. Now we wish to consider run-of-the-mill behaviors, and ask how greater behavioral repertoire size is accommodated in animals. Do animals with more distinct behaviors (i.e., more expression types) have more muscles with which the behaviors are implemented? Or do they have the same number of muscles, and behaviorally more complex animals have longer, and

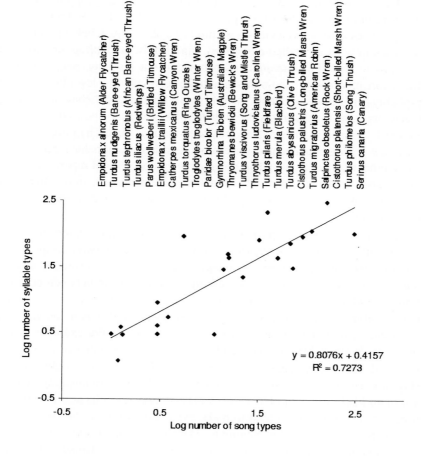

Figure 1.10: *Logarithm (base 10) of the number of bird syllable types versus the logarithm of the number of song types. When a min and a max are given in Table 1.4, 10 to the power of the average of the logged values is used. (The multiple measurements for Bewick's wren are averaged and plotted as one data point.) The slope is not significantly different from 1 (see Changizi, 2001d), suggesting that bird vocalization may not be combinatorial, and thus not language-like.*

more complex, behaviors? (I am assuming that each distinct muscle is its own component type.) What we desire now are data showing how the number of muscles varies as a function of the number of distinct behaviors.

By exhaustively reviewing the animal behavior and ethology literature over the last century, I was able to compile estimates of the behavioral repertoire size in 51 species across seven classes within three phyla. Such behavior counts are recorded in what are called *ethograms*, and I only used ethograms where the aim was to record all the animal's behaviors, not just, say, mating behavior. Behaviors recorded in ethograms tend to be relatively low level behaviors, and probably serve as components themselves in higher level behaviors. I refer to behaviors listed in ethograms as *ethobehaviors*. Table 1.5 shows these data, and Figure 1.11 displays them. There are not enough data in each of these classes to make any strong conclusions concerning the relative ethobehavioral repertoire sizes, other than perhaps (i) that the range of ethobehavioral repertoire sizes for mammals is great, and greater than that for the other classes, and (ii) the number of ethobehavior types for vertebrates tends to be higher than the number for invertebrates.

Recall that our main purpose is to examine how the number of muscles scales with the number of ethobehavior types. There are two reasons to focus on only one class of animals at a time. First, it seems reasonable to expect that if there are universal laws governing the relationship between number of muscles and number of ethobehavior types, the general form of the relationship may be similar across the classes, but the particular constants in the mathematical relationships may depend on the class of animal. For example, perhaps fish with E ethobehavior types tend to have half the number of muscles as a mammal with E ethobehavior types, but within each class the scaling relationships are identical. Second, the community standards for delineating behaviors are more likely to be similar within a class than across classes. For example, it may be that ethologists tend to make twice the number of behavioral delineations for insects than for mammals. Here I examine behavioral complexity within mammals only. One reason to choose this class is because there exists more ethobehavior data here (from 23 species), and it covers a wider range, than the data for the other classes (see Figure 1.11). The other reason is that we also require estimates of the number of muscle types, and I have been able to acquire muscle counts for only a few non-mammals.

Table 1.6 shows the behavioral repertoire sizes for just the mammals, along with estimates of the number of muscles and of index of encephalization (which is a measure of brain mass that corrects for how big it is due merely to the mass

Table 1.5: Number of ethobehavior types (i.e., the number of behaviors listed in the authors' ethogram) for 51 species.

Phylum	Class	Latin name	Name	# etho-behaviors	citation
Arthropoda	Insecta	Apis mellifera	Worker honey bee	30	Kolmes, 1985
		Ropalidia marginata	Social wasps	37	Gadagkar & Joshi, 1983
		Camponotus colobopsis	Mangrove ants	36	Cole, 1980
		Automeris aurantinca Weym	Butterfly	15	Bastock & Blest, 1958
		Pelocoris femoratus	Water bug	22	Brewer & Sites, 1994
		Sceptobiini	Ant-guest beetle	42	Danoff-Burg, 1996
		Stenus	Stenus beetle	73	Betz, 1999
Mollusca	Gastropoda	Aplysia californica Cooper	Sea slug	45	Leonard & Lukowiak, 1986
		Navanax inermis	Sea slug	28	Leonard & Lukowiak, 1984
		Strombidae	Sea snail	7	Berg, 1974
	Cephalopc	Callianassa subterranea	Burrowing shrimp	12	Stamhui et al., 1996
		Eledone moschata	Cuttlefish	14	Mather, 1985
Chordata	Osteichthy	Haplochromis buroni	Mouth-brooding african cichlid fish	19	Fernald & Hirata, 1977
		Lepomis gibbosus, Linneaus	Pumpkinseed sunfish	26	Miller, 1963
		Parablennius sanguinolentus parvicornis	Blennies	40	Santos & Barreiros, 1993
		Pleuronectes platessa L.	Juvenile plaice fish	8	Gibson, 1980
		Colisa	Colisa fish	23	Miller & Jearld, 1983
		Gasterosteus aculeatus	Three-spined stickleback	19	Wooton, 1972
Chordata	Reptilia	Gopherus agassizii	Desert tortoise	80	Ruby & Niblick, 1994
		Caiman sclerops	Caimen	188	Lewis, 1985
		Lampropholis guichenoti	Scincid lizard	45	Torr & Shine, 1994
Chordata	Aves	Ara ararauna and A. macao	Parrot	23	Uribe, 1982
		Melopsittacus undulatus	Budgerigar parakeet	60	Brockway, 1964a, 1964b
		Hydrophasianus chirurgus	Pheasant-tailed and bronzewinged jacana, duck	19	Ramachandran, 1998
		Phalacrocorax atriceps bransfieldensis	Blue-eyed shag (a cormorant)	21	Bernstein and Maxson, 1982
		Coturnix chinensis	Bluebreasted quail	60	Schleidt et al., 1984
		Gallus bankvia	White leghorn-type hen	13	Webster & Hurnik, 1990
		Poephila guttata	Zebra finch	52	Figueredo et al., 1992
Chordata	Mammalia	Alces alces andersoni	North Am. Moose	22	Geist, 1963
		Meriones unguiculatus	Mongolian gerbil	24	Roper & Polioudakis, 1977
		Peromyscus maniculatus gambelii	Deer mouse	29	Eisenberg, 1962
		Dolichotis patagonum	Mara	30	Ganglosser & Wehnelt, 1997
		Rattus rattus	Albino lab rat	43	Bolles and Woods, 1964
		Marmota monax	Woodchuck	43	Ferron & Ouellet, 1990
		Castor canadensis	Beaver	51	Patenaude, 1984
		Sciuridae (four species)	Squirrel	55	Ferron, 1981
		Rattus norvegicus	White rat	33	Draper, 1967
		Spermophilus beecheyi	California Ground squirrel	34	Owings et al., 1977
		Leporidae (family)	White rabbit	30	Gunn & Morton, 1995
		Pteropus livingstonii	Fruit bat	93	Courts, 1996
		Blarina brevicaudo	Short-tailed shrew	54	Martin, 1980
		Mustela nigripes	Black-footed ferret	74	Miller, 1988
		Felis catus	Cat	69	Fagen & Goldman, 1977
		Tursiops truncatus	Bottlenose dolphin	123	Muller et al., 1998
		Calithrix jacchus jacchus	Common marmoset	101	Stevenson & Poole, 1976
		Nycticebus coucang	Malaysian slow loris	80	Ehrlich & Musicant, 1977
		Galago crassicaudatus	Great Galagos	97	Ehrlich, 1977
		Cercopithecus neglectus	De Brazza monkey	44	Oswald & Lockard, 1980
		Macaca nemestrina	Macaque monkey	184	Kaufman & Rosenblum, 1966
		Papio cynocephalus	Baboon	129	Coehlo & Bramblett, 1981
		Homo sapiens	Human child	111	Hutt & Hutt, 1971

of the animal's body). Muscle counts were estimated from atlases of anatomy, and I used the maximum estimate I could find, since lower estimates in an atlas are due to a lack of detail. Here I have listed all the muscle estimates for each mammalian order, only the maximum which was used in the analysis.

- Artiodactyla: 89 (Walker, 1988), 116 (Sisson and Grossman, 1953, ox), 138 (Sisson and Grossman, 1953, pig), 186 (Singh and Roy, 1997), 191 (Ashdown and Done, 1984), 203 (Raghavan, 1964).

- Carnivora: 160 (Sisson and Grossman, 1953), 169 (Bradley and Grahame, 1959), 197 (Reighard and Jennings, 1929), 204 (Boyd et al., 1991, cat), 204 (Boyd et al., 1991, dog), 208 (McClure et al., 1973), 212 (Hudson and Hamilton, 1993), 229 (Adams, 1986), 322 (Evans, 1993).

- Didelphimorphia: 159 (Ellsworth, 1976).

- Lagomorpha: 67 (Busam, 1937), 85 (Chin, 1957), 112 (Wingerd, 1985), 126 (McLaughlin and Chiasson, 1990), 128 (Craigie, 1966), 214 (Popesko et al., 1990).

- Perissodactyla: 146 (Sisson and Grossman, 1953), 172 (Way and Lee, 1965), 194 (Budras and Sack, 1994), 245 (Pasquini et al., 1983).

- Primates: 160 (Schlossberg and Zuidema, 1997), 190 (Stone and Stone, 1997), 228 (Rohen and Yokochi, 1993), 230 (Bast et al., 1933), 255 (Anson, 1966), 267 (Agur and Lee, 1991), 278 Netter, 1997), 316 (Williams et al., 1989).

- Proboscidea: 184 (Mariappa, 1986).

- Rodentia: 104 (Popesko et al., 1990, mouse), 113 (Popesko et al., 1990, hamster), 134 (Popesko et al., 1990, rat), 143 (Popesko et al., 1990, guinea pig), 183 (Howell, 1926), 190 (Hebel and Stromberg, 1976), 206 (Cooper and Schiller, 1975), 218 (Greene, 1935).

Index of encephalization, P, was computed as body mass, M, divided by brain mass, B, to the power of $3/4$; i.e., $P = M/B^{3/4}$. This is appropriate since brain mass scales as body mass to the $3/4$ power (Allman, 1999; Changizi, 2001b). Body and brain masses were acquired from Hrdlicka (1907), Bonin (1937), Crile and Quiring (1940), and Hofman (1982a, 1982b). [These data were first presented in Changizi, 2002.]

Figure 1.12 is reminiscent of Figure 1.8 in that the number of component types (respectively, muscles and words) increases disproportionately slowly compared to the number of expression types (respectively, ethobehaviors and sentences) as a function of some third parameter (respectively, encephalization and time). From the relative rate at which the number of muscles and number of ethobehavior types scale as a function of encephalization, we can compute

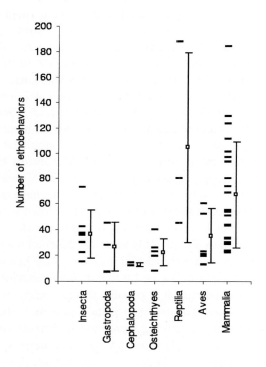

Figure 1.11: *Number of ethobehavior types for each species, and average over all species (and standard deviation) within the seven classes shown.*

the combinatorial degree. In particular, Figure 1.12 shows that $E \sim P^{0.8}$ and that $C \sim P^{0.27}$. From this we may conclude that $E \sim C^3$, and thus the combinatorial degree is roughly 3. However, the data are insufficient to statistically distinguish between whether the combinatorial degree is invariant (as a function of E), or whether the combinatorial degree may be slowly increasing. Greater behavioral complexity is achieved, at least in part, by increasing the number of behavioral component types, or the number of muscles. Muscles therefore do not serve as a universal behavioral language from which any behavior may be built.

The combinatorial degree for mammalian behavior is roughly 3 (possibly not invariant), and there are several interesting implications. (1) Since it is greater than one, it means that behavior is, indeed, language-like. There are many who already believe that behavior is language-like in this sense (Fentress and Stilwell, 1973; Slater, 1973; Dawkins and Dawkins, 1976; Douglas and Tweed, 1979; Rodger and Rosebrugh, 1979; Gallistel, 1980; Lefebvre, 1981; Fentress, 1983; Schleidt et al., 1984; Berkinblit et al., 1986; Greenfield, 1991; Allott, 1992; Bizzi and Mussa-Ivaldi, 1998), but the mere fact that multiple muscles are involved in each behavior is not an argument that behavior is language-like, as we saw in bird vocalization. The results here provide a rigorous test of the language-likeness of behavior. (2) A combinatorial degree of around 3 is surprisingly low, given that behaviors have dozens or more muscles involved. The actual number of degrees of freedom is well below the actual number of muscles involved, and this is due to the stereotyped mutual dependencies between muscles. (3) This value for the combinatorial degree is not too much lower than the combinatorial degree of 5 for human natural language. Since the combinatorial degree for mammalian behavior is effectively an average over many mammals, it is possible that the behavioral combinatorial degree for humans is actually nearer to 5, and that perhaps there are similar neurobiological constraints underlying these values. Preliminary data in my own experiments (Changizi, 2002) show that the combinatorial degree is also around three for the ontogeny of behavior in rats (Figure 1.13), where low-level components were the total number of degrees of freedom exhibited by the joints of the pups (i.e., the behavior of the pup parts), and the high-level behaviors were ethobehaviors. (I expect that my estimates scale in proportion to the true counts, but I do not expect that my counts reflect the actual magnitudes of the repertoire sizes, especially for the low-level components where I suspect severe undercounting.)

Another interesting conclusion we may draw from Figure 1.12 is that ethobe-

Table 1.6: *Number of ethobehavior types, number of muscles, and index of encephalization (i.e., brain size corrected for body size) for mammals.*

Order and *species latin name*	Species common name	# behavior types	behavior citation	index of enceph.	# muscle types	muscle citation
Artiodactyla		27.0		0.0297	203	Raghavan
Alces alces	North Am. Moose	22	Geist	0.0342		
Cephalophus monticola	Duikers	32	Dubost	0.0252		
Carnivora		71.5		0.0862	322	Evans
Felis catus	Cat	69	Fagen & Goldman	0.0888		
Mustela nigripes	Black-footed ferret	74	Miller	0.0837		
Cetacea		123.0		0.1721		
Tursiops truncatus	Bottlenose dolphin	123	Muller et al.	0.1721		
Chiroptera		93.0		0.0679		
Pteropus livingstonii	Fruit bat	93	Courts	0.0679		
Didelphimorphia				0.0185	159	Ellsworth
Insectivora		54.0		0.0490		
Blarina brevicaudo	Short-tailed shrew	54	Martin	0.0490		
Lagomorpha		30.0		0.0345	214	Popesko et al.
Leporidae (family)	White rabbit	30	Gunn & Morton	0.0345		
Perissodactyla				0.0388	245	Pasquini et al.
Primates		106.6		0.1789	316	Williams et al.
Cercopithecus neglectus	De Brazza monkey	44	Oswald & Lockard	0.1454		
Nycticebus coucang	Malaysian slow loris	80	Ehrlich & Musicant	0.1231		
Galago crassicaudatus	Great Galagos	97	Ehrlich	0.0977		
Calithrix jacchus	Common marmoset	101	Stevenson & Poole	0.1445		
Homo sapiens	Human child	111	Hutt & Hutt	0.3502		
Papio cynocephalus	Baboon	129	Coehlo & Bramblett	0.1793		
Macaca nemestrina	Macaque monkey	184	Kaufman & Rosenblum	0.2116		
Proboscidea				0.0731	184	Mariappa
Rodentia		38.0		0.0555	218	Greene
Meriones unguiculatus	Mongolian gerbil	24	Roper & Polioudakis	0.0569		
Peromyscus maniculatus	Deer mouse	29	Eisenberg	0.0569		
Dolichotis patagonum	Mara	30	Ganglosser & Wehnelt	0.0394		
Rattus norvegicus	White rat	33	Draper	0.0337		
Spermophilus beecheyi	Ground squirrel	34	Owings et al.	0.0803		
Rattus rattus	Albino lab rat	43	Bolles and Woods	0.0337		
Marmota monax	Woodchuck	43	Ferron & Ouellet	0.0803		
Castor canadensis	Beaver	51	Patenaude	0.0383		
Sciuridae (four species)	Squirrel	55	Ferron	0.0803		

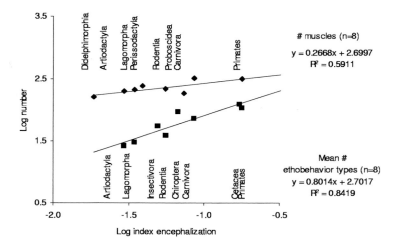

Figure 1.12: A log-log plot of the number of ethobehavior types and number of muscles in mammals, each as a function of index of encephalization.

havioral repertoire size is strongly correlated with index of encephalization. In fact, they are roughly proportional to one another. This provides a kind of justification for the use of encephalization as a measure of brain complexity.

Summing up scaling in languages

Table 1.7 summarizes the results for the behavioral systems we have covered above. One of the first generalizations we may make is that in no case do we find the universal language approach employed. For behavioral complexity across adults—i.e., not the cases of the ontogeny of behavior—the combinatorial degree is, in every case, consistent with its being invariant, implicating the length-invariant approach to complexity increase. We cannot, however, reject the possibility that the combinatorial degree is increasing in mammalian behavior. For the ontogeny of human language, the combinatorial degree clearly, and expectedly, increases as expressive complexity increases, and the relationship thus conforms to a logarithmic law; the increasing-C-and-d length approach is followed. For the ontogeny of rat behavior we are unable to say whether the relationship is a power law or logarithmic, but can conclude that the combinatorial degree is of the same order of magnitude as that for the phylogeny of mammalian behavior.

1.2.2   Scaling of differentiation in the brain

We have looked at the manner in which behavioral complexity increases, and now we consider how the brain itself increases in complexity. When a brain is built to do more things, does it do these "more things" via using the same basic building blocks—the same set of neuron types—but by stringing them together into longer functional expressions, or does it achieve greater complexity via the invention of new kinds of basic building blocks—i.e., new neuron types? Consider digital circuits as an example kind of network. Digital circuits consist of logic gates like AND and OR and NOT. For example, AND gates have two inputs and one output, and output a '1' if and only if both inputs are '1'. OR gates output a '1' if and only if at least one of the inputs is a '1'. And a NOT gate has just one input, and outputs the opposite number as the input. The set of all possible digital circuits is an infinite set of circuits, carrying out infinitely many different digital circuit functions. It turns out that, no matter how complex a digital circuit function is, it can be implemented using just these three logic gate types. (In fact, there exists a single logic gate that suffices to build any digital circuit.) No increase in the number of gate types—i.e., no

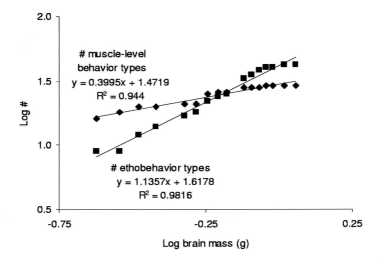

Figure 1.13: Top: *Logarithm (base 10) of the number of muscle-level behavior types versus the logarithm of brain mass (g) for the first 20 days of rat development.* Bottom: *Logarithm (base 10) of the number of ethobehavior types versus the logarithm of brain mass (g) for the first 20 days of rat development. Brain masses taken from Markus and Petit (1987). Ethobehavior types recorded from rat pups during the first 20 days are here recorded, followed by the day of first appearance in at least one pup:* back up, 8; bite cage, 14; bite sib, 15; break self from falling forward, 14; burrow into pile of pups, 1; clean face, 3; clean head, 12; climb wall, 8; dig chips with hands, 13; dig with hind feet, 18; eat chow or poop, 9; fight, 13; free self from pile or mother, 1; grasp bar, 18; grasp feet, 12; head search for nipple, 1; head shake, 4; jump, 15; lick body, 12; lick feet, 8; lick hands, 6; lick sib, 6; lie on back (to lick self), 12; manipulate object, 12; mouth floor, 3; push off pup, 8; righting, 1; run, 12; scratch body with hind leg, 4; scratch ears with front leg, 6; scratch ears with hind legs, 8; seeking nipple, 19; shoveling chips with head, 12; sit on haunches, 12; sleep, 1; sniff air, 10; stand, 14; suckle, 1; turn, 1; walk, 3; walk away from pile, 7; yawn, 1. *Muscle-level behavior types recorded from rat pups during the first 20 days are here recorded, followed by the day of first appearance in at least one pup:* arm lateral push, 2; arm push at elbow, 1; arm push at shoulder, 1; arm push body back, 8; arm stretch, 1; body bend left-right, 1; body bend sit-up, 1; body stretch, 1; body twist, 1; chew, 12; eye open-close, 12; hand grasp, 9; hand to face, 3; head left-right, 1; head twist, 1; head up-down, 1; head rotate, 3; leg burst, 15; leg lateral push, 8; leg push at ankle, 1; leg push at knee, 1; leg stretch, 2; leg to body, 9; leg to face, 8; lick, 6; mouth open-close, 1; suck, 1; tail left-right, 1; tail up-down, 1.

increase in differentiation—needs to occur. For digital circuits there exists a universal language. Perhaps nervous systems are like digital circuits, then: a handful of neuron types are sufficient to carry out any function, and thus brain differentiation remains invariant in more complex brains. The alternative is that there is no universal language employed, and more complex brains have new neuron types.

We address this question first by examining networks generally, rather than just nervous networks. That is, I will present a theory that applies to any kind of network under economic or selective pressure, and then show that many networks, including nervous systems, appear to conform to the theory.

Hypothesis

Nodes in networks combine together to carry out functional expressions, and let L be the average number of nodes involved in an expression. For example, employees in businesses group together to carry out tasks for the business, and neurons in a brain work together to implement brain functions. Let C be, as earlier, the number of component, or node, types; C is a measure of the degree of *differentiation* of the network. If the network can accommodate E distinct expression types, then there must be nodes in the network doing the work. Supposing that, on average, each node can participate in s expressions (where s is a constant depending on the kind of network), the number of nodes in the network, N, must satisfy the inequality

$$N \geq EL/s.$$

For example, if there are $E = 3$ expression types in the network, each of length $L = 4$, and each node can participate in $s = 2$ expressions, then there must be at least $N = 3 \cdot 4/2 = 6$ nodes in the network to realize these expressions.

I am interested here only in networks that are under selective or economic pressure of some kind, and for such networks the following optimality hypothesis plausibly applies (Changizi, 2001d, 2001e; Changizi et al, 2002a): *Network size scales up no more quickly than "needed" to obtain the E expression types.* The motivation for this is that nodes in a network are costly to build and maintain, and network size should accordingly be optimized subject to the functional requirements of the network. Note that networks *not* under selective pressure would not be expected to conform to this hypothesis. For example, a salt crystal is a network with nodes of different types, and the nodes interact with other nodes to carry out functional connective, lattice-related expressions.

Table 1.7: Summary of the kinds of behavior studied. When it was not possible to distinguish between a power law ($C \sim E^a$) and a logarithmic law ($C \sim \log E$), "$\sim$" is written before the rough value of the combinatorial degree to indicate that it might be increasing.

Kind of behavior	Combinatorial degree
Across adults	
Human language over history	Invariant and 5.02
Bird vocalization across phylogeny	Invariant and 1.23
Mammalian behavior across phylogeny	~3.00
Ontogeny	
Ontogeny of language	
- phoneme-morpheme	Increasing from 2 to 4
- word-sentence	Increasing from 1 to 2.5
Ontogeny of behavior	~3

However, salt crystals are not under selective pressure, and the optimality hypothesis does not apply, for a salt crystal that is twice as large will tend to have no more expression types (i.e., no new kinds of interactions among the nodes).

We derived just above that $N \geq EL/s$, and from the optimality hypothesis we may thus conclude that $N \sim EL$. Furthermore, if L were to increase as a function of E, then network size would scale up more quickly than needed, and thus L must be invariant. It follows that

$$N \sim E.$$

For networks under selective pressure, then, we expect network complexity, E, to be directly proportional to network size, N.

How does the network's differentiation, C, relate to network size? Recall from earlier in this chapter that $E \sim C^d$, where d is the combinatorial degree. We may immediately conclude that

$$N \sim C^d.$$

Since we just saw that L must be invariant, d will also be invariant. Therefore, for networks under selective or economic pressure, we predict that network differentiation and size are related by a power law. Do networks under selective pressure—*selected networks*—conform to this prediction? And, in particular, do nervous networks conform to it? We will see that a wide variety of selected network conform to the predicted relationship, and by measuring the inverse of the log-log slope of C versus N we can, as in the earlier cases of behaviors, compute the combinatorial degree, d.

Example networks, and nervous networks

Changizi et al. (2002a) presented data on the scaling of differentiation in a wide variety of networks, and some of the key plots are shown in Figure 1.14; a summary of the studied networks are shown in Table 1.8. The plots on the left in Figure 1.14 are for human-invented networks, and those on the right are for biological networks. Pairs on the same row are analogous to one another. In particular, (i) Legos are akin to organisms in that in each case geographically nearby nodes interact to carry out functions, (ii) universities are akin to ant colonies in that in each case there are individual animals interacting with one another, and (iii) electronic circuits are akin to nervous systems in that each are electrical, with interconnecting "wires."

The data sources are discussed in Changizi et al. (2002a), and I will only mention the neocortex plot here in detail. The data are obtained from Hof and colleagues, who have used immunoreactive staining and morphological criteria to compare the neuron types in mammals from 9 orders (Hof et al., 1999), and in great ape (Nimchinsky et al., 1999). For each mammalian order, indices of encephalization P (i.e., the brain mass after normalizing for body size) were computed from brain and body weights (grams) for all species in that order found in the following references: Hrdlicka (1907), Bonin (1937), Crile and Quiring (1940), Hofman (1982a, 1982b). Since brain mass scales as body mass to the 3/4 power (Allman, 1999; Changizi, 2001a), P is defined as brain mass divided by body mass to the 3/4 power. Averages were then taken within families, and the family averages, in turn, averaged to obtain the average for an order. Index of neuron encephalization Q (i.e., the number of neurons after normalizing for body size) was computed as $Q = P^{2/3}$, since the number of neurons in neocortex scales as brain volume to the 2/3 power (see previous section). Number of neuron types and index of neuron encephalizations are as follows: Monotremata (7, 0.0699), Artiodactyla (8, 0.0860), Dasyuromorphia (7, 0.1291), Insectivora (8, 0.1339), Rodentia (8, 0.1522), Chiroptera (6, 0.1664), Carnivora (9, 0.1830), Cetacea (9, 0.3094), Primate (not great apes)

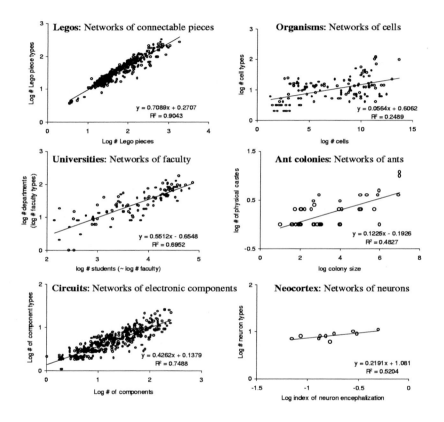

Figure 1.14: Log-log plots of the number of node types versus network size for six kinds of network.

(10, 0.2826), Great Ape (11, 0.4968).

Each of the networks shown in Figure 1.14 and mentioned in Table 1.8 have differentiation increasing as a function of network size. They therefore do not take the universal language approach. Also, the data are consistent with a power law in every case studied thus far, and the logarithmic relationship can be ruled out in the majority of the kinds of network (Changizi et al., 2002a). For neocortex in particular, a logarithmic relationship cannot be excluded (Changizi et al., 2002a) due to the insufficient range. Because of the tendency for selected networks to follow power laws, it seems reasonable to expect that the neocortex does as well, and that with more data a logarithmic relationship could be excluded. In fact, recalling our discussion in the previous section concerning invariant-length minicolumns in neocortex, we have reason to believe that expressions are length-invariant in neocortex, and thus we expect differentiation to scale as a power law of neocortical network size. In sum, then, it appears that, as predicted earlier for network optimality reasons, networks increase in complexity by scaling differentiation as a power law with network size. It also means that in all these networks there are invariant-length expressions; neocortex is hardly, then, unique in this regard.

The combinatorial degree for neocortex is approximately 5—i.e., $N \sim C^5$—and what might this signify? It means that whatever expressions are, there are around five degrees of freedom in their construction. Presumably, most functional expressions in neocortex are carried out by many more neurons than five. That is, it seems plausible that whatever expressions might be, their length L is significantly larger than five. The number of degrees of freedom in an expression may nevertheless be lower than the expression length, as we have seen for human language over history, bird vocalization, and mammalian behavior. What, then, might expressions be given that they have on the order of five degrees of freedom? Consider electronic circuits as an example, where the combinatorial degree is around 2.5. The basic functional expressions here are simple circuits, such as voltage dividers, Zener regulators, and diode limiters (Changizi et al., 2002a), where there are around 2 to 3 electronic components, and this gives the roughly 2 to 3 degrees of freedom, which, in turn, determines the rate at which differentiation scales as a function of network size. For neurons, we must ask what are the functional groupings of neurons in the neocortex? There is no known answer for neocortex here, but one plausible conjecture is the minicolumn, which is a functional grouping of neurons extending along a line through the thickness of the neocortex (Mountcastle, 1957; Tommerdahl et al., 1993; Peters, 1994; Mountcastle, 1997). Minicolumns are invariant in size

Table 1.8: The seven general categories of network for which I have compiled data for scaling of differentiation. The second column says what the nodes in the network are, and the third column gives the estimated combinatorial degree (the inverse of the log-log best-fit slope for differentiation C versus network size N).

Network	Node	Comb. degree
Electronic circuits	component	2.29
Legos™	piece	1.41
Businesses		
- military vessels	employee	1.60
- military offices	employee	1.13
- universities	employee	1.37
- insurance co.	employee	3.04
Universities		
- across schools	faculty	1.81
- history of Duke	faculty	2.07
Ant colonies		
- caste = type	ant	8.16
- size range = type	ant	8.00
Organisms	cell	17.73
Neocortex	neuron	4.56

(see previous section), which is what we expect since the combinatorial degree is invariant. Minicolumns also typically have roughly five layers to them, corresponding to the five cell-rich layers of the neocortex. Perhaps each layer contributes a degree of freedom?

1.3   The shape of limbed animals

Why are limbed animals shaped like they are? Why do animals have as many limbs (digits, parapodia, etc.) as they do? Questions like this are sometimes never asked, it being considered silly, or unscientific, or impossible to answer, or so likely to depend on the intricate ecological details of each individual species that there will be a different answer for each species. Or, if the question is asked using those words, the question will really concern the mechanisms underlying why animals have as many limbs as they do (e.g., certain gene complexes shared by all limbed animals). But the question I asked concerns whether there may be universal principles governing limb number, principles that cut across all the diverse niches and that apply independently of the kinds of developmental mechanisms animals employ.

I began this research (Changizi, 2001a) with the hypothesis that the large-scale shapes of limbed animals would be economically organized. Three reasons motivating this hypothesis were, as mentioned more generally earlier, (1) that animal tissue is expensive and so, all things equal, it is better to use less, (2) that any tissue savings can be used to buy other functional structures, and (3) that economical animal shape can tend to lower information delays between parts of the animal. It is this last motivation that makes this limb problem also a nervous system problem: even if tissue is inexpensive for some species, as long as (i) the animal has a nervous system, and (ii) the animal is under selective pressure to respond to the world relatively quickly, there will be pressure to have a large-scale morphology with low transmission delays.

To make any sense of a hypothesis about optimality, one needs to be precise about what is being optimized. Also, when one says that some shape is optimal, it is implicitly meant that that shape is more economical than all the other shapes in some large class of shapes; so, we must also be clear about what this class of shapes is.

1.3.1 Body-limb networks

To characterize the class of possible (but not necessarily actual) limbed animal shapes, I have developed the notion of a *body-limb network*. The basic idea of a body-limb network is to treat the body and limb tips of an animal as nodes in a network, and the limbs as edges connecting the body to the limb tips. The limbs are required to emanate from the body at points that lie along a single plane—this is the *limb plane*, and the cross-section of the body lying in this plane is what we will represent with our node for the body. More precisely, a body-limb network is any planar network with a central *body node*, and any number of *limb tip nodes* uniformly distributed at some distance X from the body node. Edges are all the same cost per unit length, and may connect any pair of nodes. When an edge connects the body node to a limb tip node, the edge is called a *limb edge*, or a *limb*. Figure 1.15 shows some example body-limb networks. In every network I have ever studied, nodes are points. For the purpose of characterizing animal bodies, this will not do: animal bodies are often not point sized compared to limb length. To accommodate this, body nodes are allowed to have a size and shape. For example, they are circles in Figure 1.15, and a *stretched circle* is shown in Figure 1.16. [Stretched circles are circles that have been cut in two equal halves and pulled apart a stretched-circle length L.] Body-limb networks are general enough to cover both many animal-like networks—e.g., a starfish—and many non-animal-like networks. Body-limb networks with body nodes having stretched-circle shapes have the following important parameters (see Figure 1.16):

- The body radius, R. I.e., the distance from the body's center to the body's edge. This parameter accommodates all those animals for which the body is not negligible compared to limb length.

- The stretched-circle length, L. I.e., the "length" of the body, but where $L = 0$ implies that the body node is a circle. This parameter accommodates long animals, like millipedes.

- The distance from the body node's edge to a limb tip, X. When there are edges from the body to a limb tip, these edges are limb edges, and X is then the limb length. More generally, though, X is the separation between the body-node and the limb-tip nodes. Since a connected body-limb network will always have at least one limb, this distance is always the length of this one limb, at least; accordingly, I will typically refer to it as the limb length.

- The number of limbs, N. I.e., the number of edges actually connecting the body node to a limb tip. To emphasize, N is not the number of limb tip nodes, but the number of limb edges; thus, the number of limb tip nodes must be $\geq N$.

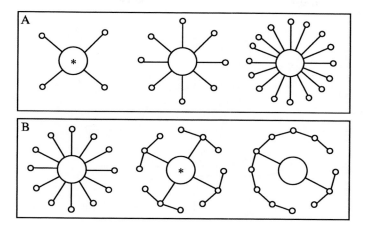

Figure 1.15: *Some example body-limb networks with the same body radius and limb length [as well as the same stretched-circle length (namely 0), see Figure 1.16], and limb length. Body nodes are not required to be points; here they are circles. (A) Three body-limb networks where all the edges are limbs, but where some networks have more limb tip nodes than others. (B) Three body-limb networks where there are just as many nodes as one another, but where some have more limb edges than others. The networks with asterisks on the body-limb node have different numbers of limb tip nodes, but have the same number of limbs.*

To the extent that real limbed animals can be treated as body-limb networks, the treatment is extremely crude. The limb tips in a body-limb network are all equidistant from the body node, whereas real limbed animals often have limb length variation. They are also required to be uniformly distributed around the body, but real animals often violate this. The edges in body-limb networks must have equal costs per unit length, but real animals sometimes have limbs with different cross-sectional areas (and thus different costs per unit length). The positions of the nodes in a body-limb network are all fixed in place, whereas limbed animals move their limbs. Furthermore, although the limbs of an animal might emanate from the animal along a single plane—the limb plane—and although limbs of many animals can, if the animal so wishes, lie roughly flat in that plane, animals rarely keep their limbs within this limb plane. For example, the limbs of an octopus emanate from along the same planar cross-section of the animal, and the limbs *can* lie flat in the plane; but they rarely if ever do. With regard to reaching out into the world, there *is* something special about the plane, special enough that it justifies modeling the shape of

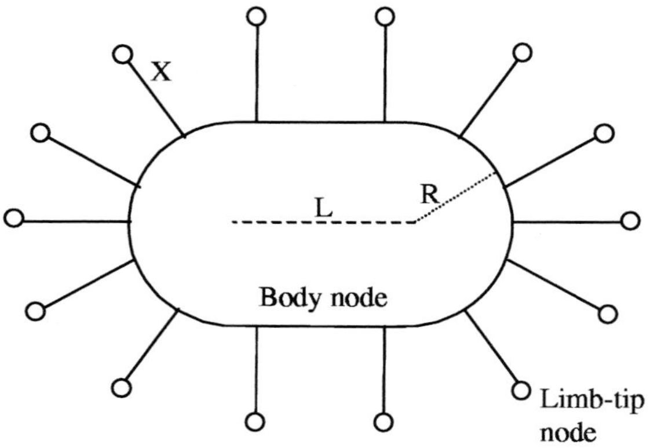

Figure 1.16: An example body-limb network with a stretched circle body node. The limb ratio is $k = X/(R + X)$; the stretched-circle ratio is $s = L/X$.

animals like an octopus as if the limbs are always lying in the plane. Imagine that all the limbs have the same angle relative to the plane; e.g., they are all pointing down and out of the plane, with an angle of 30° with the plane, as is depicted in Figure 1.17. For each such "way of pointing the limbs," let us calculate the total perimeter made by drawing lines connecting the limb tip nodes. Now ask ourselves, At which angle relative to the plane is this perimeter the greatest? Well, it is least when all the limbs are pointing either straight down or straight up; it is greatest when the limbs are lying in the limb plane. Animals have limbs in order to reach out, and since there is more reaching out to do when the limbs are in the limb plane, we might expect that it is the geometry when in the limb plane that is the principal driving force in determining the nature of the network. If an animal's limbs cannot lie in the limb plane— as is, for example, the case for most mammals, who have ventrally projected limbs—then they cannot be treated via body-limb networks as I have defined them. Despite all these idealizations, body-limb networks allow us to capture the central features of the highest level descriptions of limbed animals, and these networks are simple enough that we can easily think about them, as well as answer questions about optimality.

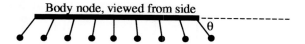

Figure 1.17: Real limbed animals often project their limbs out of the limb plane. [The limb plane is the plane defined by the points where the limbs intersect the body. It is also the plane in which the body node lies.] This figure shows an example "animal" viewed from the side, where all the limbs are pointing below the body at an angle θ relative to the limb plane. The perimeter made by the limb tips is greatest when θ is zero, i.e., when the limbs lie in the plane. There is accordingly the greatest need for limbs in the limb plane, and this is my justification for treating limbed animals as if their limbs lie in the limb plane.

1.3.2 The optimization hypothesis

We now know what body-limb networks are, and how they may be used, to a first approximation at least, to characterize the large-scale morphology of many kinds of limbed animals. They are also sufficiently general that there are many body-limb networks that do not describe real limbed animals. The question now is, If limbed animals *are* economically arranged, then what body-limb networks would we expect to describe them? Or, said another way, which body-limb networks are optimal? To make this question more precise, suppose that an animal has body radius R, stretched-circle length L, and limb length X. Now we consider the class of all body-limb networks having these three values—the class of "R-L-X body-limb networks"—and ask, Which ones are optimal? For example, all the example networks in Figure 1.15 have the same body radius, same stretched-circle length (namely zero), and same limb length; they are therefore all in the same class of body-limb networks from which we would like to find the optimal one. However, rather than asking which such body-limb network is optimal, I will ask a weaker question: How many limbs does an optimal R-L-X body-limb network have? The reason I want to ask this question is that, ultimately, it is the number of limbs that I am interested in. From our point of view, two body-limb networks that differ in their number of limb tip nodes but have the same number of limb edges are the same. For example, the networks with asterisks in Figure 1.15 have the same number of limbs, and so we do not wish to distinguish them.

The answer to the question "How many limbs does an optimal R-L-X body-limb network have?" is roughly that these networks cannot have too

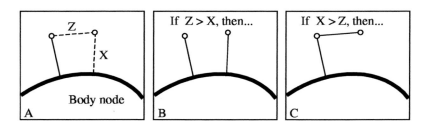

Figure 1.18: *The basic idea behind the argument for why there cannot be too many limbs in an optimal body-limb network. (A) Part of the body node is shown at the bottom, two limb tip nodes on top. One limb edge is presumed to already exist. To connect the network, the other limb tip node must either have an edge straight to the body node, which is of length X, or have an edge to the tip of the existing limb, which is of length Z. (B) When Z > X it is less costly to have a limb go to the limb tip node straight from the body node. (C) But when X > Z it is cheaper to have an edge go to the tip of the existing limb.*

many limbs, where "too many" depends on the parameter values of R, L and X. Figure 1.18 illustrates the argument. The basic idea is that if two limb tip nodes are close enough to one another, then it is cheaper to send an edge directly from one to the other, and to have only one of the limb tips connect to the body node. This occurs when the distance, Z, between the limb tips is smaller than the limb length; i.e., when $Z < X$. However, when limb length is smaller than the distance between the limb tips—i.e., when $X < Z$—it is cheaper to connect the limb tip nodes directly to the body node. That is, it is then cheaper to have a limb for each limb tip node. With this observation in hand, we can say that *an optimal R-L-X body-limb network must have its limbs sufficiently far apart that no limb tip nodes at the end of a limb are closer than X.*

Because our body node shapes are confined to stretched circles, it is not difficult to calculate what this means in regards to the maximum number of limbs allowed for an optimal, or wire-minimal, R-L-X body-limb network. Let us consider the stretched circle's two qualitatively distinct parts separately. First consider the straight sides of the body node. These sides are of length L, and the limbs are all parallel to one another here. It is only possible to fit L/X many limbs along one of these edges. Actually, L/X can be a fraction, and so we must round it down; however, for simplicity I will ignore the truncation from now on, and compute just the "fractional number of limbs." So, along the two sides of a stretched circle there are a maximum of $2L/X$ limbs; letting $s =$

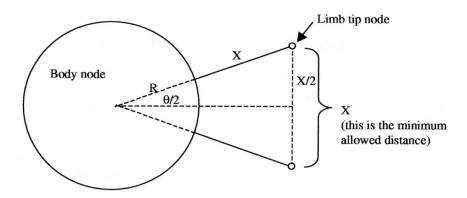

Figure 1.19: *The simple trigonometry involved in computing the minimum allowed angle between two limbs for a circular node. The two nodes cannot be closer than X. We can compute θ/2 as the* $\arcsin[(X/2)/(R+X)]$. *Since the limb ratio—i.e., a measure of how long the limbs are compared to the body—is* $k = X/(R+X)$, *we can rewrite this as* $\theta = 2\arcsin(k/2)$. *(An alternative derivation leads to the equivalent* $\theta = \arccos(1 - k^2/2)$.)*

L/X be the *stretched-circle ratio*, the maximum number of limbs is $2s$. The remaining parts of the body node are two semicircles, which we will imagine pushing together. Limbs on a circular body node poke out radially. Consider the angle between the lines reaching from the body node to two limb tips. What must this angle be in order to make the distance between the two limb tip nodes greater than the limb length X? Figure 1.19 illustrates the simple trigonometry involved. The conclusion is that, for circular body nodes, the angle, θ, between adjacent limbs must satisfy the inequality

$$\theta \geq 2\arcsin(k/2),$$

(or equivalently $\theta \geq \arccos(1 - k^2/2)$), where $k = X/(R + X)$ is the *limb ratio*. The maximum number of limbs that can be placed around a circular body node is therefore

$$\frac{2\pi}{2\arcsin(k/2)} = \frac{\pi}{\arcsin(k/2)}.$$

In total, then, for an R-L-X body-limb network to be optimally wired the number of limbs N must satisfy the inequality,

$$N \leq N_{max} = 2s + \frac{\pi}{\arcsin(k/2)},$$

where $s = L/X$ and $k = X/(R + X)$. Note that this inequality no longer refers to the body radius R, the stretched-circle length L or the limb length X. Instead, it refers only to the stretched-circle ratio s and the limb ratio k. The absolute size of the network therefore does not matter; all that matters are the relative proportions of an animal's body and limbs. It should be noted that this treatment of stretched circle nodes engages in a simplification since I have made the argument for the sides separately from that for the circular ends; a more precise mathematical treatment would determine the maximum number of limbs for the stretched-circle node shape as it is. For our purposes this approximation suffices. The notion of optimality we have employed here is something called a *minimal spanning tree*, or *MST*. *Spanning trees* are networks that connect up all the nodes, but where there are no loops. *Minimal* spanning trees are spanning trees that use the least amount of wire. What we have found thus far is that *if* an *R-L-X* body-limb network is a minimal spanning tree, *then* it must have fewer than N_{max} limbs.

That encompasses the volume-optimality part of the hypothesis. All it concludes, though, is that there must not be more than N_{max} many limbs; it does not predict how many limbs an animal will actually have. This is where I made a second hypothesis, which is that animals are typically selected to maximize their number of limbs subject to the volume-optimality constraint. The simple intuition is that limbed animals have limbs in order to reach out (for many different reasons), and need to "cover" their entire perimeter.

These two hypotheses lead to the prediction that, for those limbed animals describable as *R-L-X* body-limb networks, the number of limbs N satisfies the equation,

$$N = 2s + \frac{\pi}{\arcsin(k/2)}.$$

(I.e., that $N = N_{max}$.) Because the first hypothesis concerns minimal spanning trees and the second concerns maximizing the number of limbs, I have labeled this composite hypothesis the *max-MST hypothesis*. Notice that the max-MST hypothesis says nothing about life as we know it; it is a general hypothesis, so general that one might expect it to apply to any limbed animals anywhere, so long as they are describable by body-limb networks.

Let us ask what this equation means for the relationship between predicted limb number and the body and limb parameters s (the stretched-circle ratio $s = L/X$) and k (the limb ratio $k = X/(R+X)$). First consider what happens as s is manipulated. When $s = 0$ it means that the stretched-circle length is very small compared to the limb length. The consequence is that the stretched-circle term in the equation for the number of limbs drops out, which means that the network can be treated as having a circular body node. As s increases, and keeping R and X constant, the equation is of the form $N = 2s + N_c(k)$, where $N_c(k)$ is a constant referring to the number of limbs for a circle node with limb ratio k. Thus, N increases proportionally with s. For this reason, the stretched-circle length parameter is rather uninteresting; that is, it just leads to the obvious prediction that, for sufficiently large values of s, animals with bodies twice as long have twice the number of limbs. Now consider what happens as the limb ratio is manipulated. When $k = 1$ it means the limbs are very long compared to the body radius, and the number of limbs becomes $N = 2s + 6$. When the body node is circular $s = 0$ and $N = 6$; that is, when the limbs are so long that the body node may be treated as a point, the predicted number of limbs falls to its minimum of 6. As k approaches zero the limbs become very short compared to the body radius. Using the approximation $x \approx \sin x$ for x near 0 radians, it follows that $\sin(k/2) \approx k/2$, and so $\arcsin(k/2) \approx k/2$, and the predicted number of limbs becomes

$$N \approx 2s + \pi/(k/2) = 2s + 2\pi/k.$$

In fact, even when k is at its maximum of 1, $\arcsin(k/2) \approx k/2$; e.g., arcsin $(0.5) = 0.52 \approx 0.5$. The error at this maximum is only about 4%, and the error gets lower and lower as k drops toward zero. Therefore, the approximation above is *always* a reasonable one. When the body node is either a circle or the limb length is very large compared to the stretched-circle length (but still much smaller than the body radius), the equation becomes $N \approx 2\pi/k$. That is, the number of limbs becomes inversely proportional to the limb ratio. In short, when $s = 0$, the number of limbs falls to six for very long limbs compared to the body, but increases toward infinity in a particular quantitative fashion as the limbs become shorter relative to the body. The reader may examine the kinds of body-limb networks that conform to the hypothesis by playing with a little program built by Eric Bolz at www.changizi.com/limb.html.

Before moving to data, it is important to recognize that the hypothesis does not apply to animals without limbs. The hypothesis states that there is a relationship between an animal's number of limbs and its body-to-limb proportion (i.e., limb ratio). Without limbs, the model can say nothing. Alternatively, if having no limbs is treated as having zero limb ratio, then the model predicts infinitely many non-existent limbs. Snakes and other limbless organisms are therefore not counterexamples to the max-MST hypothesis.

1.3.3 Comparing prediction to reality

At this point I have introduced the prediction made by the max-MST hypothesis. With this prediction in hand, I sought to discover the extent to which real limbed animals conform to the prediction. To obtain data for actual body-limb networks, I acquired estimates of the stretched-circle ratio s and the limb ratio k from published sources for 190 limbed animal species over 15 classes in 7 phyla (Agur, 1991; Barnes, 1963; Bishop, 1943; Brusca and Brusca, 1990; Buchsbaum, 1956; Buchsbaum et. al., 1987; Burnie, 1998; Downey, 1973; Hegner, 1933; Netter, 1997; Parker, 1982; Pearse et. al., 1987; Pickwell, 1947; Stebbins, 1954). The studied phyla (classes) were annelids (Polychaeta), arthropods (Myriapoda, Insecta, Pycnogonida, Chelicerata, Malacostraca), cnidarians (Hydrozoa, Scyphozoa), echinoderms (Holothuroidea, Asteroidea), molluscs (Cephalopoda), tardigrades and vertebrates (Mammalia (digits only), Reptilia (digits only), Amphibia). An appendix subsection at the end of this section shows these values. Measurements were made on the photographs and illustrations via a ruler with half millimeter precision. The classes were included in this study if six or more data points from within it had been obtained. Species within each class were selected on the basis of whether usable data could be acquired from the sources above (i.e., whether the limb ratio and stretched-circle ratio were measurable); the number of limbs in the measured animals ranged from 4 to 426. What counts as a limb? I am using 'limb' in a general sense, applying to any "appendage that reaches out." This covers, e.g., legs, digits, tentacles, oral arms, antennae and parapodia. Although for any given organism it is usually obvious what appendages should count as limbs, a general rule for deciding which appendages to count as limbs is not straightforward. Some ad hoc decisions were required. For vertebrate legs only the those of Amphibia were studied, as their legs are the least ventrally projected of the vertebrates. For amphibians, the head and tail were included in the limb count because there is an informal sense in which the head and tail also "reach out". (Thus, amphibians have six "limbs" in this study.) For insects (and other invertebrates with antennae studied), antennae appear to be similar in "limb-likeness" to the legs, and so were counted as limbs unless they were very small (around $< 1/3$) compared to the legs. The head and abdomen of insects were not counted as limbs because, in most cases studied, they are well inside the perimeter of the legs and antennae, and thus do not much contribute to "reaching out" (the head was treated as part of the body). Since I obtained the data for the purposes of learning how body-limb networks scale up when there are more limbs, and since scaling laws are robust to small perturbations in measurement (being plotted on log-log plots), these where-to-draw-the-line issues are not likely to much disturb the overall scaling behavior. Digits are treated in the same manner as other types of limbs, the only difference being that only a fraction of the body (i.e., hand) perimeter has limbs (i.e., digits). Cases of digits were studied only in cases where the "hand" is a stretched circle with digits on roughly one half of the stretched circle. For these cases hands may be treated as if the digits emanate from only one "side" of the node. Digits like those on a human foot are, for example, not a case studied because the foot is not a stretched circle for which the toes are distributed along one half of it. In 65 of the cases presented here the stretched-circle ratio $s \neq 0$, and to observe in a single plot how well the data conform to the max-MST hypothesis, the dependence on the stretched-circle length can be eliminated by "unstretching" the actual number of limbs as follows: (i) given the limb ratio k and the stretched-circle ratio s, the percent error E between the predicted and actual number of limbs is computed, (ii) the predicted number of limbs for a circular body is computed by setting $s = 0$ (and keeping k the same), and (iii) the "unstretched actual number of limbs" is computed as having percent error E from the predicted number of limbs for a circular body. This rids of the dependence on s while retaining the percent error.

After unstretching, each measured limbed animal had two remaining key values of interest: limb ratio k and number of limbs N. The question is now, How do N and k relate in actual organisms, and how does this compare to the predicted relationship? Recall that, for $s = 0$ as in these unstretched animals, the predicted relationship between N and k for limbed animals is

$$N \approx 2\pi/k.$$

If we take the logarithm of both sides, we get

$$\log_{10} N \approx \log_{10}(2\pi/k),$$

$$\log_{10} N \approx -\log_{10} k + \log_{10}(2\pi) = -\log_{10} k + 0.798.$$

Therefore, if we plot $\log_{10} N$ versus $-\log_{10} k$, the predicted equation will have the form of a straight line, namely with equation $y = x + 0.798$. This is shown in the dotted lines in Figure 1.20.

Figure 1.20 shows a plot of the logarithm (base 10) of the number of limbs versus the negative of the logarithm (base 10) of the limb ratio for the data I acquired. If the max-MST hypothesis is true, then the data should closely

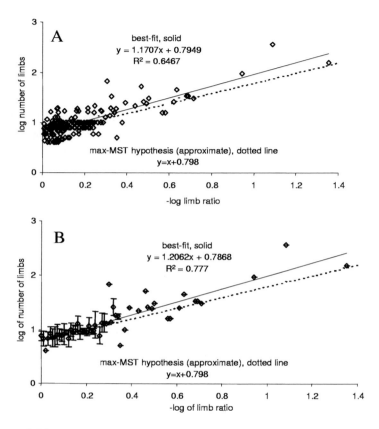

Figure 1.20: (A) *The logarithm (base 10) of the unstretched number of limbs versus the negative of the logarithm of the limb ratio, for all 190 limbed animals. The best fit equation via linear regression is* $y = 1.171x + 0.795$ *(solid line)* ($R^2 = 0.647$, $n = 190$, $p < 0.001$), *and predicted line* $y = x + 0.798$ *(dotted line). The 95% confidence interval for this slope is* [1.047, 1.294]. *The three rightmost data points exert a disproportionate influence on the best-fit line, and removing them leads to the best fit equation* $y = 1.089x + 0.8055$ ($R^2 = 0.487$, $n = 187$, $p < 0.001$), *with a 95% confidence interval for the slope of* [0.900, 1.279]. (B) *The average of* $\log_{10} N$ *values versus* $-\log_{10} k$, *where the* $-\log_{10} k$ *values are binned with width 0.01. Error bars indicate standard deviation (for points obtained from bins with 2 or more cases). The best fit equation is now* $y = 1.206x + 0.787$ *(solid line)* ($R^2 = 0.777$, $n = 52$, $p < 0.001$), *again very close to the predicted line (dotted line). points still exert a disproportionate influence on the best-fit line, and removing them results in the equation* $y = 1.112x + 0.807$ ($R^2 = 0.631$, $n = 49$, $p < 0.001$).

follow the equation $y = x + 0.798$ in the plot, shown as dotted lines. Examination of the plots show that the data closely follow the predicted lines. When $-\log(k) = 0$, $k = 1$, meaning that the body radius R is extremely small compared to the limb length X; and when this is true, the number of limbs falls to a minimum of around six (see legend of Figure 1.20). As $-\log(k)$ increases, the limb ratio decreases toward 0, meaning that the limbs are getting smaller relative to the body radius; and when this is true, the number of limbs increases higher and higher. Not only does limb number clearly increase as limb ratio decreases (and the x axis increases), it appears to be well described by the linear regression equation $\log(N) = 1.171[-\log(k)] + 0.795$ (and, without the three points on the far right, $\log(N) = 1.089[-\log(k)] + 0.8055$). Manipulation of this equation leads to $N = 6.24k^{-1.171}$ (and without the three stray points, $N = 6.39k^{-1.089}$): the number of limbs appears to be roughly inversely proportional to the limb-ratio, with a proportionality constant around 6. This is extraordinarily similar to the predicted relationship which, recall, is $N = 6.28k^{-1}$.

In summary, many limbed animals across at least seven phyla conform well to the max-MST hypothesis, which suggests that their large-scale morphologies are arranged to minimize the amount of tissue needed to reach out in the world; they also appear to have the maximum number of limbs subject to the constraint that they are still optimal trees. And this is despite the complete lack of any details in the hypothesis concerning the ecological niches of the animals, and despite the extreme level of crudeness in the notion of body-limb networks. It is worth emphasizing that, even without the max-MST hypothesis to explain the data, these empirical results are interesting because they reveal that limbed animals follow universal laws relating their body-to-limb ratio to their number of limbs. It happens that this universal law is just what one might *a priori* suspect of limbed animals—as I *a priori* suspected—if they are driven by volume-optimization considerations. It is also worth mentioning that this limb problem is a kind of network scaling problem: the issue is, what changes do body-limb networks undergo as they acquire more limbs? That is, how do animals change as their number of limbs is scaled up? The answer is that limbed animals scale up in such a way as to keep the value $N \cdot k$ invariant; and, in particular, limbed animals satisfy the constraint that $N \cdot k \approx 2\pi$.

Appendix for section: Raw limb data

In this appendix I have included my raw limb data. It appears on the following three consecutive pages, with the phylum, class (with type of "limb" in parentheses), name of animal (species name, or whatever information was available from the source), limb ratio ($X/(R + X)$), stretch ratio (L/X), and the number of limbs. I mention in passing that it may be interesting to look at conformance to this model in two new ways. One, to look at spherical nodes, where the limbs point radially outward in all directions; mathematical research from Coxeter (1962) can be used to determine roughly how many limbs are optimal. [The predicted relationship is $N \approx 4\pi/k^2$, where k is again the limb ratio.] Second, one may look at non-animals, and perhaps even viruses: e.g., the T4 bacteriophage conforms well to the model, having six "lunar-lander-like" limbs attached to a very small "body" (the shaft).

Phylum	Class (limb type)	Name	limb ratio X/(R+X)	stretch ratio (L/X)	# limbs
Annelida	Polychaeta (parapodia)	Glycera americana	0.3658	158.0550	426
		Tomopteris	0.1929	11.3734	52
	(parapodia, long)	Halosydna	0.3214	5.2222	37
	(parapodia, short)	Halosydna	0.2400	7.8333	41
		Syllis cornuta	0.9102	2.6316	30
		Nereis virens	0.4545	23.6000	72
		unnamed	0.3333	69.7500	218
		Nereis diversicolor	0.6087	10.4286	50
Arthropoda	Myriapoda (legs)	Lithobius	0.7865	3.4611	32
		Scolopendra gigantea	0.7037	7.7895	36
		a California centipede	0.4795	11.5429	38
		Scutigera coleoptrata	0.8474	2.5210	34
		Scolopendra cingulata	0.5233	8.5506	42
		Scutigerella	0.6061	9.5333	28
		a millipede	0.6842	8.6923	56
	Insecta (legs and antennae)	Thraulodes salinus	0.7929	0.5732	8
		Pediculus humanus	0.7108	0.0407	6
		Phthirus pubis	0.5725	0.0000	6
		a cockroach	0.9696	0.8784	8
		Microcoema camposi	0.9555	0.5276	6
		Lonchodes brevipes	0.9764	1.1774	6
		Velinus malayus	0.8919	0.1212	6
		an ant	0.8810	0.1622	6
	Pycnogonida (legs)	Nymphopsis spinosossima	0.9189	0.1961	8
		Achelia echinata	0.8882	0.2517	10
		Dodecolopoda mawsoni	0.9556	0.1395	12
		Decolopoda australis	0.9808	0.2157	10
		Tanystylum anthomasti	0.9218	0.3392	8
		Nymphon rubrum	0.9853	0.1660	8
	Chelicerata (legs)	spider larva	0.8123	0.1320	8
		spider nymph	0.8364	0.0372	8
		spider	0.8467	0.0849	8
		Argiope	0.9015	0.0000	8
		Scytodes	0.8551	0.0000	8
		Pardosa amentata	0.8952	0.1064	8
		a generalized spider	0.8571	0.1667	8
		a spider (in amber)	0.8815	0.0640	8
		a crab spider	0.8545	0.0000	8
		Tegenaria gigantea	0.8956	0.0245	8
		Brachypelma emilia	0.7947	0.0397	8
		Buthus martensi	0.8556	0.2412	10
		Ricinoides crassipalpe	0.8464	0.0000	8
		unnamed	0.9256	0.2009	8
		daddy long legs	0.9735	0.0181	8
		Mastigoproctus	0.8773	0.2721	10
		Heterophrynus longicornis	0.9167	0.0121	8
		Stegophrynus dammermani	0.8802	0.0000	8
		Koenenia	0.8477	0.5689	10
		Galeodes arabs	0.8991	0.2801	10
		Chelifer cancroides	0.8985	0.3349	10
		Eurypterus	0.5145	0.0000	10
		Pterygotus buffaloensis	0.7316	0.0000	12
		Limulus	0.9231	0.3750	10
	Malacostraca (legs)	Pachygrapsus crassipes	0.6785	0.0000	10
		Chionoecetes tanneri	0.7745	0.0000	10
		Gecarcoidea natalis	0.6537	0.0000	10
		Carcinus maenas	0.8261	0.0702	10
		Maja squinado	0.7078	0.0183	10
		Callianassa	0.7625	0.4177	10
		Pleuroncodes planipes	0.7874	0.3681	8
		Petrolisthes	0.6676	0.1770	10
		Cryptolithodes	0.8229	0.3472	10
		a crab	0.8012	0.0025	10
		Loxorhynchus	0.7035	0.1533	10
		Pugettia	0.7111	0.1060	10
		Stenorhynchus	0.9554	0.0800	8

Phylum	Class (limb type)	Name	limb ratio X/(R+X)	stretch ratio (L/X)	# limbs
Cnidaria	Hydrozoa (tentacles)	Hydra A	0.9815	0.0000	6
		Hydra B	0.8041	0.0000	8
		Hydra C	0.9620	0.0000	6
		Polyorchis	0.4719	0.0000	33
		Tubularia hydroid adult polyp A	0.6111	0.0000	20
		Tubularia hydroid adult polyp B	0.7536	0.0000	10
		Tubularia hydroid actinula larva	0.8264	0.0000	8
		Tubularia hydroid new polyp	0.9048	0.0000	9
		Tubularia indivisa hydroid	0.8387	0.0000	16
		Niobia medusa	0.5426	0.0000	6
		Sarsia medusa	0.8784	0.0000	4
		Rathkea medusa	0.3137	0.0000	31
		"typical" medusa	0.7957	0.0000	9
		Proboscidactyla	0.2071	0.0000	34
		Obelia medusa	0.3404	0.0000	52
		"typical" medusa	0.4950	0.0000	68
		a hydranth A	0.6747	0.0000	6
		a hydranth B	0.7188	0.0000	11
		a hydranth C	0.6897	0.0000	17
		Linmocnida medusa	0.4750	0.0000	20
		Aglaura medusa	0.2308	0.0000	45
	Scyphozoa (tentacles)	Stomolophus meleagris scyphistoma	0.8511	0.0000	13
		Stomolophus meleagris stobila	0.7391	0.0000	8
		Stomolophus meleagris late strobila	0.8458	0.0000	7
		Stomolophus meleagris ephyra	0.2712	0.0000	16
		Cassiopea andromeda	0.5946	0.0000	8
		Mastigias medusa	0.8485	0.0000	8
		Haliclystis	0.8333	0.0000	8
		Pelagia adult scyphomedusa	0.7748	0.0000	8
	(oral arms)	Pelagia adult scyphomedusa	0.9412	0.0000	4
		Aurelia adult medusa	0.0444	0.0000	154
		Aurelia ephyra	0.2632	0.0000	16
		Aurelia scyphistoma A	0.5652	0.0000	22
		Aurelia scyphistoma B	0.6735	0.0000	17
		Aurelia scyphistoma C	0.8317	0.0000	8
		"typical" medusa A	0.1136	0.0000	96
	(oral arms)	"typical" medusa A	0.8861	0.0000	4
		"typical" medusa B	0.0816	0.0000	368
	(oral arms)	"typical" medusa B	0.9231	0.0000	4
Echinodermata	Holothuroidea (arms)	Cucumaria crocea	0.8864	0.0000	10
		Cucumaria planci	0.8571	0.0000	10
		Enypniastes	0.4643	0.0000	18
		Pelagothuria	0.8958	0.0000	11
		Holothuria grisea	0.5313	0.0000	18
		Stichopus	0.5926	0.0000	10
		Euapta	0.7059	0.0000	8
	Asteroidea (arms)	Luidia phragma	0.7547	0.0000	5
		Luidia ciliaris	0.8590	0.0000	8
		Luidia sengalensis	0.8235	0.0000	9
		Luidia clathrata	0.8281	0.0000	5
		Ctenodiscus	0.7451	0.0000	5
		Astropecten irregularis	0.7800	0.0000	5
		Heliaster microbranchius A	0.2040	0.0000	34
		Heliaster microbranchius B	0.4051	0.0000	25
		Solaster	0.7733	0.0000	10
		Acanthaster planci	0.4500	0.0000	19
		Pteraster tesselatus	0.4455	0.0000	5
		Solaster notophrynus	0.6741	0.0000	7
		Linckia guildingii	0.8752	0.0000	5
		Linckia bouvieri	0.9148	0.0000	5
		Ampheraster alaminos	0.9091	0.0000	6
		Odinia	0.8553	0.0000	19
		a starfish A	0.5476	0.0000	10
		a starfish B	0.8395	0.0000	8
		Freyella	0.8717	0.0000	13
		Crossaster papposus	0.5050	0.0000	13
		Coscinasterias tenuispina	0.8333	0.0000	7
		Coronaster briorcus	0.8182	0.0000	11

Phylum	Class (limb type)	Name	limb ratio X/(R+X)	stretch ratio (L/X)	# limbs
Mollusca	Cephalopoda (arms)	Sepia A	0.7463	0.0000	8
		Sepia B	0.7755	0.0000	8
		Architeuthis	0.8889	0.0000	8
		Octopus	0.8826	0.0000	8
		Octopus dofleini	0.9111	0.0000	8
		Octopus vulgaris	0.9068	0.0000	8
		Loligo	0.8163	0.0000	8
		Loligo pealeii	0.7987	0.0000	8
		Histioteuthis	0.8224	0.0000	8
		a juvenile	0.7000	0.0000	8
Vertebrata	Amphibia (limbs, tail and head)	a salamander A	0.8458	1.0666	6
		a salamander B	0.8128	1.1219	6
		a salamander C	0.8653	1.6132	6
		a salamander D	0.8636	1.2982	6
		a salamander E	0.8854	1.4243	6
		a salamander F	0.8452	1.5385	6
		a salamander G	0.8582	1.3125	6
		a salamander H	0.8582	1.0593	6
		a salamander I	0.8438	1.7284	6
		a salamander J	0.7971	1.0727	6
Tardigrada digit (digits)		Echiniscus	0.6667	0.0000	8
		Halobiotus crispae	0.6667	0.0000	8
		Echiniscoides sigismundi	0.5714	0.0000	16
		Wingstrandarctus corallinus	0.8384	0.0000	8
		Styraconyx qivitoq	0.7857	0.0000	8
		Halechiniscus	0.8530	0.0000	8
		Orzeliscus	0.8000	0.0000	8
		Batillipes	0.8225	0.0000	6
Vertebrata digit	Mammalia (digits)	homo sapien A	0.5879	0.0000	10
		homo sapien B	0.6135	0.0000	10
		homo sapien C	0.5889	0.0000	10
		homo sapien D	0.5782	0.0000	10
		chimpanzee	0.6358	0.0000	10
		Tarsius bancanus	0.7059	0.0000	10
	Reptilia (digits)	Triturus cristatus (rear limb of a newt)	0.6691	0.0000	10
		Triturus cristatus (front limb of a newt)	0.8023	0.0000	8
		Sceloporus occidentalis biseriatus	0.6861	0.0000	10
		Lacerta lepid (a lizard)	0.6067	0.0000	10
		Cnemidophorus tessalatus tessellatus	0.6833	0.0000	10
		Eumeces skiltonianus (a skink)	0.7738	0.0000	10
		Dasia (a skink)	0.8108	0.0000	10
	Amphibia (digits)	a salamander front limb 1	0.7699	0.0000	8
		a salamander rear limb 2	0.4202	0.0000	10
		a salamander front limb 3	0.6329	0.0000	8
		a salamander rear limb 4	0.5868	0.0000	10
		a salamander front limb 5	0.6237	0.0000	8
		a salamander rear limb 6	0.6209	0.0000	10
		Plethodon vandyke (front limb)	0.6053	0.0000	8
		Plethodon vandyke (rear limb)	0.5704	0.0000	10
		Anneides lugubris (front limb)	0.6464	0.0000	8
		Anneides lugubris (rear limb)	0.6578	0.0000	10
		Laeurognathus marmorata (a front limb)	0.6732	0.0000	8
		Laeurognathus marmorata (a rear limb)	0.6552	0.0000	10
		Pseudotriton ruber ruber (a front limb)	0.5769	0.0000	8
		Pseudotriton ruber ruber (a rear limb)	0.5303	0.0000	10
		Plethodon vehiculum (a rear limb)	0.6358	0.0000	10

Chapter 2

Inevitability of Illusions

This chapter primarily concerns a very general constraint on brains: that they take time to compute things. This simple fact has profound consequences for the brain, and vision in particular. I will put forth evidence that it is the visual system's attempting to deal with this computing delay that explains why we experience the classical geometrical illusions. Figure 2.1 shows a sample such illusion; basically, the illusions are those found in any introductory Psychology course. I will also, along the way, briefly discuss a general approach to modeling brain computation: that approach is decision theory, wherein the brain, or some portion of it, is modeled as an ideal rational agent acting to maximize its expected utility on the basis of probabilities concerning the nature of the uncertain world. This is referred to as the Bayesian framework for visual perception, and with it researchers have made some important breakthroughs. We will need to understand it, and its shortcomings, to understand how the visual system copes with the time it takes to compute a percept. I also discuss the difficulties of one of the older and more established inference-based theories of the geometrical illusions. Before proceeding, it is important to understand why there may be computing delays in perception.

Computation is sometimes slow, sometimes fast, but never instantaneous. Computation takes time. Running software on your computer takes time. For example, it takes about one second to start Microsoft Word on my lap top, over two seconds to start Adobe Acrobat, and over half a minute to run LaTex with this book as the input. Despite the orders of magnitude increase in computation speed over the last twenty years since the advent of the personal computer, there seems to always be significant delays for contemporary software. This is presumably because software producers have figured out the time delays

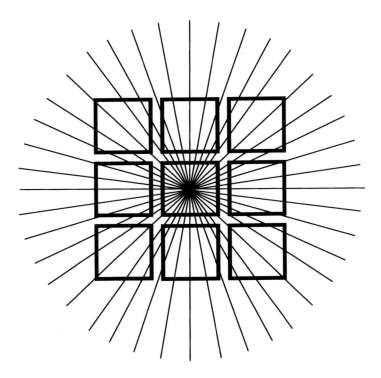

Figure 2.1: Nine perfect, identical squares on a radial display induce an illusion, which is a version of the Orbison illusion.

consumers are willing to put up with and can use this time to carry out more sophisticated computations for the consumer.

Brain computation takes time as well. In addition to the computation delays due to simply traveling through individual dendrite and axon arbors, and to the time it takes signals to traverse synapses, computation delays are also due to the complex time course and pattern of neural firings that actually implement the computation. How much time can the brain afford to take in carrying out its computations? To answer this, consider the brain (and evolution) as the software producer, and the animal (and his genes) as the consumer. The brain will presumably have figured out the time delays the animal is willing to put up with—i.e., delays that the animal is able to deal with without compromising survival too much—so as to be able to use this time to compute more powerful functions of use to the animal. More exactly, the brain and evolution presumably will have discovered how to optimally trade off computation time with computational power. How much time is given to computations in this optimal trade-off will depend on the details of the animal's ecology, but it seems *a priori* unlikely to be exceedingly long—e.g., 10 second delays—or microscopically short—e.g., 0.001 seconds. Because the world changes too much and too unpredictably during a long, say 10 second, interval, long delays will lead to computational solutions that are moot by the time they are computed. Nearly instantaneous computations would avoid this problem, but would leave the brain with too little time to compute much of interest to the animal. Somewhere in between these extremes will be an optimal middle ground, allowing sufficient time for powerful computations, but the time is short enough that the computations are still applicable to the changing world. These considerations are relevant for any brain—Earthly or not—having to deal with an uncertain and dynamic world, so long as they are not literally infinite in computational speed.

One effective possible strategy for a brain to use in its attempt to increase its computation time is to attempt to *correct* for the computation delay (De Valois and De Valois, 1991; Nijhawan, 1994, 1997, 2001; Berry et al., 1999; Sheth et al., 2000; Schlag et al., 2000; Khurana et al., 2000; Changizi, 2001). That is, suppose it would be advantageous to have a time interval Δt to carry out some useful computation, but suppose that Δt is long enough that the world typically has changed to some degree during this time, making the computation moot. What if, to deal with this, the brain took a different tact? Rather than trying to compute something that is useful for dealing with the world the way it *was* when the computation started, the brain might try, instead, to compute a function that *will* be useful for dealing with the world as it probably will be by the time the computation is finished. Such a strategy might be called *latency correction.* To the extent that latency correction is possible, the brain will extend its computation duration to derive more powerful functionality. At some point the computation interval will be so long that latency correction algorithms will no longer reliably work, but such a strategy will buy the brain more time to provide neater software for the animal, thereby increasing the animal's prospects.

Vision is one kind of brain computation that is needed swiftly and is difficult to compute. The visual system computes from the retinal stimulus a perception of the way the world out there is, and since the world is typically in flux either because it is itself changing or because the observer is himself moving, the percept must be computed in a timely manner lest the information from the retinal stimulus be irrelevant. Visual perception is also difficult: it is a classic example of an underdetermined problem, as there is no unique solution to it, there being (infinitely) many possible ways the world could be that would lead to the information on the retina (see also Chapter 3). Our own artificial computer algorithms for vision, despite a few decades of progress, still fall far short of success, where success is defined as the recognition of or navigation within scenes under a wide variety of circumstances. Because vision is difficult, to do a good job at it the visual system would like to have as much time as it reasonably can. In fact, the visual system in mammals *does* take a significant, but not exceedingly long, period of time: there is a latency on the order of magnitude of 100 msec (Lennie, 1981; De Valois and De Valois, 1991; Maunsell and Gibson, 1992; Schmolesky et al., 1998). This is ecologically significant because a lot can happen in 100 msec, or a tenth of a second. Even walking at just one meter per second means that the positions of objects change by 10 cm during that time. If the visual system generated a percept of the way the world probably was when the information was picked up at the retina, the percept would be about the way the world probably was 100 msec in the past. At one m/sec, objects perceived by an observer to be within 10 cm of being passed would, in fact, already have passed the observer... or the observer will have bumped into them. Catching a ball and the other complex activities we engage in obviously worsen this problem.

Latency correction is thus a beneficial strategy, if the visual system can carry it off. That is, the strategy is this: rather than computing a percept of the scene that probably caused the retinal stimulus—a percept that would need to be generated nearly instantaneously to be of much use to the animal—the visual system can, instead, compute a percept of the scene that will probably be out there by the time the computation is finished and the percept is elicited. That is, the visual system attempts to perceive not the past, but, instead, to "perceive the present." In this way the visual system can generate percepts that are typically coincident with reality, but it can also secure itself some elbow room for solving the tough problem of vision.

If a visual system were able to implement latency correction, what kind of algorithm might we expect it to employ? To answer this, let us consider what a latency correction algorithm would have to do. In order to reliably generate a percept at time t of what is out there at time t on the basis of retinal information from $t - 100$msec, the visual system would need to solve the following two conceptually distinct problems.

1. The visual system must figure out what the scene at time $t - 100$msec probably was (e.g., a 10 meter flag pole 5 meters away), and

2. the visual system must determine what scene that scene will probably become by time t (e.g., a 10 meter flag pole 4.5 meters away).

[Note that a *scene* consists of the properties of the objects in the vicinity of the observer, including the observer's viewpoint. Thus, a room viewed from a different position would make for a different scene.]

Each of these problems is an *inference problem*, as it is underdetermined by any information the observer may have. The visual system must infer what might be out there at time $t - 100$msec (the time of the retinal stimulus), even though there are infinitely many scenes that can, in principle, have led to the same information on the retina. And the visual system must also infer how the scene will probably change, even though there are infinitely many ways that the scene might, in fact, change. [I am not claiming that a brain must actually make this distinction between 1 and 2. A brain could solve the latency correction "all at once," but it still would have conceptually dealt with both problems.]

2.1   Visual inferences

Therefore, if the visual system could carry out latency correction, it would have to be good at making inferences. But making inferences is something the visual system actually *is* good at, as has been noticed at least since Helmholtz (1962), and has been taken up by many since (e.g., Gregory, 1997). The visual system appears to act like a scientist, using the evidence present in the retinal stimulus to make a reasoned choice. The visual system also acts like a scientist in that it can learn from past experience. Finally, the visual system is even like a scientist in that it is also simply born with certain biases, or preconceptions, toward some perceptual hypotheses over others. (In fact, it *must* be born with such biases; see Chapter 3.) In this section I discuss two research paradigms within this inference tradition.

Traditional visual inference

My main task for this chapter is to show how the classical geometrical illusions are consequences of the visual system implementing a latency correction strategy. Since, as we discussed earlier, latency correction is something we might expect from any brain with finite computing speed, we also expect any such brain to perceive illusions. But there already exist many theories of the visual illusions; what are wrong with them? First, note that I am only interested here in considering theories of visual perception that concern the purported function computed by the visual system, and also the general kind of algorithm used. I am not interested here in theories about the implementation-level mechanisms found in the visual system (e.g., lateral inhibition, or some neural network). One of the most venerable and most well-entrenched such (functional) theories of the geometrical illusions is what I will call the *traditional inference approach* (Gregory, 1963, 1997; Gillam, 1980, 1998; Rock, 1975, 1983, 1984; Nundy et al., 2000).

Before stating what the general form of this kind of theory is, it is useful to present a sample stimulus with which I will introduce the theory. Consider Figure 2.2, where observers perceive the bold vertical line on the right to have greater angular size than the bold vertical line on the left; this is the illusion. Note that observers *also* perceive the *linear* size of the line on the right to be greater; that is, they perceive that it is a taller object in the depicted scene, when measured by a ruler in, say, meters; and they also perceive that it is farther away. But this latter perception of linear size is not what is illusory about the figure: no one is surprised to learn that observers perceive that the line on the right has greater linear size in the depicted scene. What is illusory is that observers perceive the line on the right to have greater *angular size*—to fill more of the visual field—than the line on the left, despite their angular sizes being identical.

The traditional inference explanation for this illusion states that the line on the right is perceived to be longer because the cues suggest that it probably *is* longer. Describers of the theory will usually also say that such a perception is

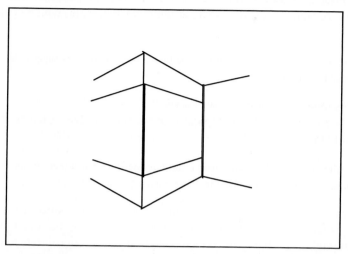

Figure 2.2: *An illusion which is a variant of the Müller-Lyer. The two bold vertical lines are the same angular size, but the right one appears to have greater angular size. One of the most commonly accepted functional explanations for this illusion is an inappropriate inference explanation which says that the line on the right is perceived to be bigger because the cues suggest that it is a bigger line out there. The cues suggest this as follows: the right line is nearer to the vanishing point of the converging lines and thus is probably farther away, and since it has the same angular size as the other line, it follows that it must be bigger. It is "inappropriate" because, in this case, the lines are at the same distance, namely they are both on the page. The deep problem with this explanation is that it equivocates between two notions of perceived size: perception of angular size, and perception of linear (or objective) size. Because the right line probably is bigger in linear size, we should perceive it to be bigger in linear size. Fine. But observers also perceive the right line to be bigger in angular size, and its probably being bigger in linear size does not imply that it is bigger in angular size. They are, in fact, probably the same angular size, since they project identically onto the retina. The traditional inference explanation therefore cannot explain the illusion.*

useful for us in the real-world scene version of Figure 2.2—i.e., when you are standing in front of a real hallway—but when the stimulus is from a piece of paper as it *actually* is in this figure, this perceptual strategy is said to become "inappropriate." There is, however, a deep conceptual problem with this explanation. To start, let us look again at the main statement, which is along the lines of

> The line on the right is perceived to be longer because the cues suggest that it probably *is* longer.

What does the statement mean by 'longer'?

The *first possibility* is that it means 'greater linear size.' That is, the statement would be,

> The line on the right is perceived to have greater linear size (e.g., in meters) because the cues suggest that it probably *is* greater in linear size.

The statement in this case would be fine, as far as it goes, since it is certainly useful to perceive the linear size to be what it probably is. For example, if the line on the right is probably three meters high, then it is appropriate to perceive it to be three meters high. However, this interpretation is no longer relevant to the illusion, since the illusion concerns the misperception of their angular sizes.

The *second possible interpretation* is that 'longer' means 'greater angular size,' in which case the statement becomes,

> The line on the right is perceived to have greater angular size (measured in degrees) because the cues suggest that it probably *is* greater in angular size.

This, however, is no good because the cues do *not* suggest that the line on the right has greater angular size. The lines have, in fact, identical angular size, and the visual system "knows" this since equal angular sizes are unambiguously projected onto the retina. And it is a fallacious argument to say that the angular size of the line on the right is probably greater because it's linear size is probably greater; linearly huge objects very often have tiny angular size (e.g., the moon), and linearly tiny objects often have tremendous angular size (e.g., hold your finger up near your eye).

So far, the traditional inference explanation statement is either irrelevant (the first interpretation) or false because the cues do *not* suggest that the line on the right has greater angular size (the second interpretation).

The *third and final possible interpretation* I will consider is that the *first* occurrence of 'longer' is interpreted as 'greater angular size' and the *second* occurrence of 'longer' is interpreted as 'greater linear size.' That is, in this possibility the statement is equivocating between two meanings of 'longer.' The statement is now,

> The line on the right is perceived to have greater angular size (measured in de-grees) because the cues suggest that it probably is greater in linear size (bigger in meters).

This appears to be the interpretation that people actually have when they utter this view. It is sometimes even phrased as something along the lines of, "the perception of the projective properties of the lines are biased toward the prob-able objective properties of the lines." The statement is not irrelevant as in the first interpretation; this is because the claim concerns the perception of angu-lar size, which is what the illusion is about. The statement also does not err as in the second interpretation by virtue of claiming that the line on the right probably has greater angular size. One preliminary problem concerns what it could possibly mean to bias a projective property toward an objective property; how can something measured in degrees get pushed toward something that is measured in, say, meters? Another issue concerns *how much* the angular size should be increased in the probably-linearly-longer line; the explanation gives us no apparatus by which it is possible to say. I will focus on another problem, which concerns the supposed *usefulness* of such a strategy for vision: of what possible use is it to perceive a greater angular size merely because the linear size is probably greater? The visual system's goal according to these traditional inference approaches is to generate useful percepts, and, in particular, to gener-ate percepts that closely represent reality (because this will tend to be useful). To accurately represent the angular sizes in Figure 2.2 would be to perceive them as being identical in angular size. The visual system would *also* want to perceive them as having different linear sizes, but there is no reason—at least none that this traditional inference explanation gives—for the visual system to misperceive the angular sizes.

It is sometimes said that the illusion is only an illusion because Figure 2.2 is just on a piece of paper. The inferential strategy of increasing the perceived angular size of the line on the right because it is probably linearly longer is inappropriate in this case because, it is said, the figure is just a figure on a page, where the lines in fact have the same linear size. If, the argument continues, the proximal stimulus were, instead, due to a real live scene, then the strategy would be appropriate. Unfortunately, the strategy would be inappropriate in this latter scenario too. To see this, let us imagine that the stimulus is not the one in Figure 2.2, but, instead, you are actually standing in a hallway of the kind depicted, and your eye position is placed in just such a manner that the line on the right has the same angular size as the one on the left. Is there anything "appropriate" about perceiving the line on the right to have greater angular size merely because its linear size is probably greater? It is not clear what would be useful about it, given that its angular size is the same as that of the line on the left, and perceiving their angular sizes to be equal does not preclude perceiving their linear sizes to differ. (E.g., hold your finger out until it fills just as much of your visual field as a tree off in the distance. You now perceive their angular sizes to be identical, but you also perceive the tree to be linearly larger.)

Some may think I have constructed a straw man position for the traditional inference explanation, and that the authors behind such explanations have more sophisticated positions. Perhaps this is so, although I do not think so; I have no interest, however, in whether or not this is *really* the explanation they intended. What is important is that the idea as I stated it is what the "average psychologist and neurobiologist on the street" appear to understand the explanations to be. For example, pulling out the nearest undergraduate perception textbook to me, the cogent author describes the traditional inference explanation for a figure essentially just like Figure 2.2 as follows.

> ...the converging lines are unconsciously interpreted as parallel lines receding into the distance...and the [vertical] lines as lying in the same receding...plane as the converging lines.... The unconscious perception of differential depth leads to the conscious perception of differential size: The [right] line would have to be longer because it...connects the receding parallel lines, whereas the lower one is not even close [Palmer, 1999, p. 324].

Now, as we will see later in this chapter, there *is* a good reason to perceive the angular size of the line on the right to be greater: namely because its angular size probably *will* be greater by the time the percept is actually generated (due to the observer's probable forward movement toward the vanishing point). However, the traditional inference explanation of the geometrical illusions provides no such reason, and is thus, at best, an explanation that provides no real explanation for why the visual system would generate the illusions.

In addition to the above conceptual difficulties, the traditional inference approach has more run-of-the-mill difficulties in explaining the illusions. As one example, consider the Orbison illusion where the square is directly below the vanishing point (see Figure 2.1). The square in this case appears to project as a trapezoid, with its longer edge on top. To explain this in the traditional inference manner, one needs to argue that the top edge of the projection is actually due to a real world line that is bigger in meters than the bottom edge of the square. For this to be the case, the projected square would have to be due to a real world trapezoid with its top edge tilted backward. The difficulty is: Why would such a tilted trapezoid be the probable source of a square projection? This is a highly coincidental, or non-generic (Freeman, 1994), projection for such an object. It seems obviously much more probable that the source of the perfectly square projection is a square in the observer's fronto-parallel plane and near the vanishing point. But in this case, the top and bottom of the object are identical in length, and so the traditional inference approach predicts no illusion. Other explanations by the traditional approach require similarly improbable sources. For example, in the Hering illusion (Figure 2.19), the probable source of two vertical lines on either side of the vertical meridian cannot be that they are two vertical lines, for then the distance in meters between each line would be the same and the traditional inference account would predict no illusion. Instead, for traditional inference to work here, the probable source would have to be that the two lines bend away from the observer, and as they bend away, they also get farther apart in meters; and all this in just such a manner that they happen to project perfectly straight. With this strange source, the lines are farther apart in meters when nearer to the vanishing point, which is why they are perceived to bow out according to the traditional inference approach. However, it seems much more plausible that the probable source of the two lines is that they are two vertical lines.

2.1.1 The standard Bayesian approach

In recent years the visual-system-as-inference-engine approach has been reinvigorated by a *Bayesian approach* to inference. There are many ways of modeling inference, but the Bayesian framework is a particularly good one. I will not discuss it in detail here, but will only try to communicate what is so good about it. [See Chapter 3 for an introduction to the Bayesian framework.]

The basic idea is that an agent has a numerical degree of confidence in each of the perceptual hypotheses, the hypotheses which are mutually exclusive. These degrees of confidences are modeled as probabilities, where each hypothesis has a probability in the interval from 0 to 1, the sum of the probabilities over all the hypotheses equals 1, and the probability of no hypothesis being true is 0. A probability of 1 for a hypothesis means that the agent has complete confidence in the hypothesis. A probability of 0 means the agent has complete confidence that the hypothesis is *not* true. The Bayesian framework tells us how these probabilities should be altered when evidence, or retinal information, is accumulated. This approach is, in a certain sense, optimal, because if you do *not* follow this approach, then others can dupe you out of all your money; I am here intimating an important result called the *Dutch Book Theorem*, or the Ramsey-de Finetti Theorem (Ramsey, 1931; de Finetti, 1974; see also Howson and Urbach, 1989, pp. 75–89 and 99–105, for discussion).

It is not only a nice framework because of this kind of optimality argument, it is also nice because it makes certain conceptual distinctions that allow us, the scientists, to make better sense of the inferential process. In particular, the framework distinguishes between

- *prior probabilities*, which are the probabilities in the hypotheses before seeing the evidence,

- *likelihoods*, which are the probabilities that the evidence would occur given that a hypothesis were true, and

- *posterior probabilities*, which are the probabilities in the hypotheses after seeing the evidence.

The reader is invited to read the introduction to the Bayesian framework in Chapter 3, but it is not necessary to cover it in any detail here.

The main idea to get across is that the inference-engine idea appears to apply well to the human visual system, and has been taken up during the 1990s within the Bayesian framework (Knill and Richards, 1996), where considerable success has been made: e.g., the perception of 3D shape (Freeman, 1994), binocular depth (Nakayama and Shimojo, 1992; Anderson, 1999), motion (Kitazaki and Shimojo, 1996), lightness (Knill and Kersten, 1992) and surface color (Brainard and Freeman, 1997).

In fact, if the visual system truly can be described within a probabilistic framework, then the proper treatment is a decision theoretic one, where the brain is treated as attempting to maximize its expected utility. That is, perception is an act, and an agent cannot decide how to act purely on the basis of the probabilities of hypotheses. For example, suppose there are two main possibilities concerning the scene that caused the retinal stimulus: the first is that there is a bed of flowers, and the second is that there is a tiger. Even if a flower bed is more probable than the tiger, the costs are so high for not recognizing a tiger that the perception that maximizes your expected utility may be the tiger perception. We would therefore expect that visual perception should be modifiable by modifying only the utilities of the observer, and evidence exists that even appetitive states such as thirst can modulate low-level perceptions such as transparency (Changizi and Hall, 2001).

Although the Bayesian approach has allowed significant advances in understanding visual perception, there is a difficulty with the way in which it is typically conceived. It is always assumed, either explicitly or implicitly, that the visual system is attempting to generate a percept of the scene that probably caused the retinal stimulus. That is, the "standard Bayesian perception approach" is to assume that the perceptual hypotheses are about the various possible scenes that are consistent with the retinal stimulus actually received. So, for example, when we say that a stimulus is bistable (such as the Necker cube, which is just a line drawing of a wire cube), we mean that the visual system jumps back and forth between two percepts of scenes that are consistent with the stimulus. The possible percepts are confined to percepts of scenes that could have caused the retinal stimulus. It is not, then, possible within the standard Bayesian perception approach to have percepts of scenes that are not even consistent with the retinal stimulus. The standard Bayesian approach can only accommodate *consistent perception.* Note that misperception can be consistent perception, since the perception could be of something that is not actually there, but is nevertheless consistent with the retinal stimulus. Many of our perceptions are consistent with the retinal stimulus, and the standard Bayesian approach is fine in such cases. For example, the examples of Bayesian successes I mentioned earlier—perception of 3D shape, binocular depth, motion, lightness and surface color—appear to be consistent perceptions. E.g., for the motion aperture phenomenon there are many different possible motions consistent with a line moving behind a circular aperture; it is a case of consistent perception since subjects appear to perceive one of the possibilities consistent with the stimulus.

The difficulty for the standard Bayesian perception approach lies in the fact that there are many perceptual phenomena where the observer perceives a scene that could *not* have caused the retinal stimulus. That is, there are cases of *inconsistent perception.* For example, the geometrical illusion from the earlier Figure 2.1 is an example of inconsistent perception. The angles of the squares in the figure project toward your eye at nearly 90°, supposing you are looking straight at it and are not too close. Yet many or all of the projected angles are perceived to be significantly different from 90°. Why is this a case of inconsistent perception? Because the actual stimulus does *not* project (much) differently than 90°.

[Note that if proximal stimuli possessed significant errors, it *would* be possible for the standard Bayesian perception approach to handle inconsistent perception. For example, suppose that an object projects toward an observer with an angular size of θ, but that the retina records this angular size with error according to some normal distribution. Then an ideal probabilistic engine would realize that projected lines in the world can have angular sizes markedly different from the angular size measured by the retina, and could sometimes generate perceptual hypotheses inconsistent with the proximal stimulus, but hopefully consistent with the true angular size. However, this does *not* appear to be relevant for the retina and visual system; at least, any error for angular sizes (and projected angles) are negligible.]

To help drive home the point, consider Figure 2.3. There is an object X in the lower half of the figure, and whatever it may be, it is projecting toward your eye as a perfect square. No, you certainly are not perceiving it to project as a perfect square, but we'll get to that in a moment. First, let us ask about what the three-dimensional shape and orientation of object X are. Well, there are infinitely many possible three-dimensional shapes and orientations for X that would allow it to project toward you as a square. For example, it could simply be a square in your fronto-parallel plane; or it could, instead, be a trapezoid with its longer edge on top and tilted away from you until it projects as a perfect square. And so on. So long as the three-dimensional shape and orientation you perceive is consistent with its projecting toward you as a square, then you are having a consistent perception. Now, however, let us ask about what the *projected* shape of object X is. Despite the fact that X may be infinitely many different things, all those things would *still* project as a square, so the projected shape of X is, in fact, unambiguously a square. To perceive the projected shape in a manner consistent with the stimulus, you must perceive X to project toward you as a square. The problem is that we *don't* perceive X to project toward us as a square, despite the fact that it does project toward us as a square. Instead, we perceive X to project toward us as a trapezoid. This is inconsistent perception.

In fact, all the classical geometrical illusions are cases of inconsistent perception: in each case, observers perceive a projected angle, an angular size, or an angular separation to be inconsistent with the information in the proximal stimulus. Note that all these are cases of perception of projective properties, and projective properties are more likely to change quickly in time, and thus ripe for latency correction. In Figure 2.3, observers misperceive all these three

Figure 2.3: One perfect square on a radial display induces an illusion. There are many possible three-dimensional orientations and shapes for the square-like object that are consistent with the stimulus; i.e., where the object would project toward the observer as a perfect square. Perceiving any one of these possible objective orientations and shapes would be a case of consistent perception. However, since the square-like object in the figure actually projects as a perfect square, it is not consistent with the stimulus that it projects in any other way. Nevertheless, we perceive it to project not as a perfect square, but as a trapezoid. This, then, is a case of inconsistent perception.

projective properties. (1) It has misperception of projected angle because observers perceive the top two angles to project differently (namely, smaller) than the lower two angles, when they in fact all project identically. (2) It has misperception of angular size because observers perceive the top side of the box to have longer angular size than the bottom, but they have the same angular size. (3) And it has misperception of angular separation because observers perceive the higher parts of the sides to have greater angular separation than the lower parts of the sides, but the angular separations are in fact identical. [(2) and (3) are essentially the same kind of misperception, but I have distinguished them here because in some classical illusions it is more natural to think in terms of one over the other.]

It was only after understanding that certain kinds of illusions are cases of inconsistent perception that I both realized the inadequacy of the standard Bayesian approach to perception, and was propelled toward a nonstandard Bayesian approach to perception: latency correction.

The principal feature making the standard Bayesian approach "standard" is that, as mentioned, it presumes that the visual system is trying to choose among hypotheses concerning the scene out there at the time the retinal stimulus occurred. What if, however, the visual system is not trying to use the evidence to figure out what *was* out there when the stimulus occurred, but, instead, is trying to use the evidence to determine what is *going* to be out there by the time the percept actually occurs? That is, what if the visual system is implementing latency correction? For latency correction, the perceptual hypotheses the visual system is picking from are not hypotheses about what was out there when the retinal stimulus occurred, but hypotheses about what will be out there when the perceptual computations are completed.

With this alternative Bayesian approach for perception, it becomes possible for inconsistent perception to occur. Why? Because now it is quite possible that the scene probably out there at the time t the percept is generated is different than any possible scene that could have caused the retinal stimulus (which occurred at $t - 100$msec). That is, it is entirely possible that the probable scene out there at time t is causing a new retinal stimulus that is different from the one at time $t - 100$msec. For example, in Figure 2.3, imagine that the object X at the bottom actually is a perfect square in your fronto-parallel plane, but a little below the horizon. Furthermore, suppose you are moving toward the center point of the radial display. How would the projection of the square change as you move forward? Well, the top would project larger—i.e., have greater angular size—than the bottom because you are closer to the top than the bottom.

That is, object X would project trapezoidally in the next moment. If, upon being presented with Figure 2.3 as a retinal stimulus, your visual system infers that you are moving toward the center point, then the elicited percept will, if a latency correction strategy is being employed, be of object X projecting trapezoidally. This is, in fact, the central idea behind my latency correction theory of the classical geometrical illusions, which we take up in detail in the next section.

The latency correction (nonstandard Bayesian) approach to perception does not predict *only* inconsistent perceptions; consistent perceptions are still possible. Consistent perceptions *better* be possible, since many of our perceptions *are* (or at least appear to be) consistent. In what circumstances would latency correction lead to consistent perception? That's easy: any time the probable scene properties causing the stimulus are probably unchanging. What kinds of properties do not typically change much in the short term? Although *projective properties*—how objects project toward the observer, either geometrically or spectrally—change very quickly through time since they depend on the observer's position relative to the objects, *objective properties*—the properties of objects independent of their relationship to other things—do not typically change much through time. For example, the angular size of a flag pole is a projective property, as it depends on how far you are from it. Accordingly, it often changes in the next moment as you move, projecting either larger or smaller. The linear size of a flag pole, however, is an objective property, as it is, say, 10 meters high independent of where you stand with respect to it. In the next moment the linear size is very unlikely to change. Accordingly, we expect that perception of the flag pole's linear size will be a consistent perception because latency correction will generate a percept of a 10 meter pole, which is still consistent with the retinal stimulus. In fact, the cases where the standard Bayesian approach has mainly excelled are in applications to the perception of objective properties, like surface color and object recognition.

To sum up some of our discussion, a latency correction approach to vision can explain the existence of inconsistent perceptions; the standard Bayesian approach cannot. This latency correction approach is a nonstandard Bayesian approach, which means (i) it is a Bayesian approach, acquiring all of its powers and benefits, but (ii) it has a slightly different view concerning the kind of perceptual hypotheses the visual system is looking for... namely, it is looking for perceptual hypotheses about the way the world *is*, not about the way the world *was*. Furthermore, in the special case of perception of objective properties this alternative, latency correction, Bayesian approach collapses to the standard Bayesian approach, thereby squaring with the Bayesian approach's many successes.

2.2   A simple latency correction model

In this section I describe my model for how latency correction leads to the classical geometrical illusions (Changizi, 2001; Changizi and Widders, 2002). The following section applies the model to the illusions.

Recall that the latency correction hypothesis is, in my statement of it, as follows:

> On the basis of the retinal information the visual system generates a percept representative of the scene that will probably be present at the time of the percept.

The 'probably' that appears in the statement means that the statement is a probabilistic hypothesis, a Bayesian one in particular (but not a standard Bayesian one where the percept would represent the scene probably causing the proximal stimulus). And as mentioned earlier, we may conceptually distinguish two problems the visual system will have to solve.

1. First, the visual system must figure out what scene probably caused the proximal stimulus.

2. And, second, the visual system must figure out how that scene will change by the time the percept is elicited.

Again, this does not mean that the visual system's algorithm or lower-level mechanisms must distinguish these things, only that whatever function the visual system is computing, it would, in effect, have to solve both of these problems. This conceptual distinction is helpful for us scientists who wish to make predictions from the latency correction hypothesis: to predict what percept a visual system will generate given some proximal stimulus, we can subdivide our task into two smaller tasks. Namely, we must, for the geometrical illusions, try to determine what the probable scene is that would cause the geometrical figure, and then try to determine how the observer will typically move in the next moment (i.e., by the time the percept occurs).

We first need a way of deciding what the probable scene is for simple geometrical figures like those in the classical geometrical stimuli. That is, we need a way of figuring out what a figure probably depicts. Before I describe a

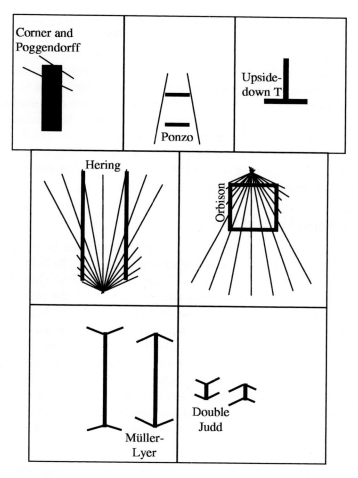

Figure 2.4: *Eight classical geometrical illusions. Corner Poggendorff: the line through the corner of the rectangle appears to be bent. Poggendorff: the line through the rectangle appears to be two, parallel, non-collinear lines. Hering (also a variant of the Zöllner stimulus): the two parallel lines appear to be farther apart as one looks lower. Upside-down 'T': the horizontal bar appears to be shorter than the same-length vertical bar resting on top of it. Orbison: the right angles near the top appear to be acute, and the right angles at the bottom appear to be obtuse. Ponzo: the higher horizontal line appears to be longer than the same-length lower one. Double Judd: the vertical shaft of the left figure appears higher than the same-height one on the right. Müller-Lyer: the vertical shaft on the left appears longer than the same-length one on the right. See Coren and Girgus (1978) for references; see Greene (1988) for the corner Poggendorff.*

model for helping us do this, let us see some of the classical figures, as shown in Figure 2.4.

The first feature to notice is that these most famous classical geometrical illusions consist entirely of straight lines. The second thing to notice is that, in addition to many oblique lines, there are also many horizontal and vertical lines, many more than we would expect if lines were thrown onto the page with random orientations. Finally, we can see that for many of the illusions there is a subset of the obliques that seem to all point toward the same point. All these features suggest that there may be simple rules for determining what the figures depict. That is, that there may be simple rules for determining what kind of real world line is the source of any given projected line.

Three principal kinds of line

The question I ask now is, For each kind of projected line in a figure, what kind of real world line or contour probably projected it? To help us answer this, let us look at a geometrical figure, namely Figure 2.5, where there are so many cues that it is obvious what the source lines of the projected lines are. The figure clearly depicts a room or hallway. It is my hypothesis that the projected lines in the geometrical stimuli are typically caused by lines and contours in "carpentered" environments like rooms and hallways. Furthermore, I hypothesize that observers typically move down hallways and rooms; they do not tend to zigzag wildly, nor do they tend to move vertically. The focus of expansion—the point of the forward-moving observer's visual field from which objects are expanding radially outward—is thus the vanishing point.

There are three kinds of line in the scene depicted in Figure 2.5: x lines, y lines and z lines.

- x lines are the lines that lie parallel with the ground, and perpendicular to the observer's direction of motion.

- y lines are the lines that lie perpendicular with the ground, and are also perpendicular to the observer's direction of motion.

- z lines are the lines that lie parallel with the ground, and are parallel to the observer's direction of motion.

Note that these kinds of line are defined in terms of the observer's probable direction of motion, which, again, is toward the vanishing point. In my simple model, I will assume that these are the *only* kinds of line in the world; I call them the *principal lines*. All we really need to assume, however, is that these

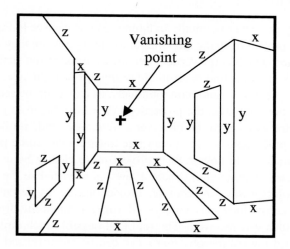

Figure 2.5: A sample geometrical figure showing the probable kind of source line for each line segment in the stimulus. The assumed observer direction of motion in such a stimulus is toward the vanishing point. The classical geometrical figures will be interpreted in this fashion.

three kinds of line are sufficiently more frequent in our experiences than other kinds of line that, in simple geometrical stimuli, one of these kinds of line is always the probable source.

Given that there are just three kinds of line in the world, we can ask of each of them, How do they typically project toward the observer? Once we have learned how each kind of line typically projects, we can work backward and ask, Given the projection, which kind of line probably caused it?

How do principal lines project?

x lines typically project horizontally in figures, as one can see in Figure 2.5. In particular, they project from the left straight to the right when they are near the *vertical meridian*, which is the vertical line drawn through the vanishing point in the figure. When an x line is off to either the left or right side, however, x lines begin to project more and more obliquely, as can again be seen in Figure 2.5. In fact, at the right side, the projections of x lines begin to point toward a vanishing point way off to the observer's right side; and, similarly, on the left side x line projections begin to converge toward a vanishing point on that side. We can understand how x lines project more clearly by considering a *projection sphere*. A projection sphere allows us to visualize the way things in the world project toward an observer. Projections are, by definition, devoid of depth information; they only possess information about the direction from which the stimulus was received. The set of all possible such directions from the outside world toward the observer's eye can be encapsulated as a sphere with the observer's eye at its center; each point on the sphere stands for a different projection direction from the outside world. Figure 2.6 shows how the three kinds of line may project toward the observer within our simple model. The "x line" sphere in Figure 2.6 shows how x lines project. Every x line segment lies along some great circle extending from the left pole to the right pole. The contour on the sphere that goes through the focus of expansion (the cross) is the way the horizon, for example, projects toward the observer, and is called the *horizontal meridian*. If the observer is about to cross over railroad tracks, the tracks project like the contours on the lower half of the sphere; as the observer nears the tracks, they project progressively lower and lower, eventually projecting along the very bottom of the projection sphere. As long as x line segments are near the *vertical meridian* (which is the contour extending from directly overhead, through the focus of expansion, and down to directly below the observer), they project horizontally onto the projection sphere, and parallel to one another. However, in the left and right peripheral parts of the sphere, x lines begin to converge toward the left and right poles; they no longer project horizontally, and they are no longer parallel to one another. We must be careful, however, because x lines *do* project horizontally in the periphery *if* they happen to lie along the horizontal meridian.

How do y lines project? From Figure 2.5 one may see that y lines typically project straight from the bottom to the top (i.e., non-obliquely), and that they are parallel to one another. Although it is not all that common in our experience, if a y line segment is very high above or very low below an observer, they begin to project obliquely, are no longer parallel to one another, and begin to converge toward the top or bottom pole, respectively. We can make this more precise by looking at the y line projection sphere in Figure 2.6. Every y line segment lies along some great circle extending from the top or North pole to the bottom or South pole. Suppose you are floating in front of a pole that goes infinitely far above you and infinitely far below you. If you are moving directly toward it, then it projects as the contour on the sphere that goes through the focus of expansion. Suppose now that you are going to pass the pole on your right. As you near it, the pole will project progressively more and more along the right side of the projection sphere (which is on the left in the figure). As long as the y segments are relatively near the horizontal meridian, they project nearly purely up and down, and are parallel to one another, as they all are in Figure 2.5. When the y line segments are in the upper or lower periphery, however, they begin to converge toward a pole of the sphere, and are no longer parallel to one another. y lines in the periphery can still project non-obliquely if they happen to lie along the vertical meridian.

Finally, how do z lines project? From Figure 2.5 we can see that z lines project obliquely, and that they share a vanishing point, namely at the focus of expansion. The z line projection sphere in Figure 2.6 encapsulates how z lines project. Each z line segment lies along a great circle from the focus of expansion all the way to the focus of contraction (which is directly behind the observer). For example, if you are walking on a sidewalk, the sides of the sidewalk project on the lower left and lower right of the projection sphere. z lines typically project obliquely, but a z line can project horizontally if it happens to lie along the horizontal meridian, and it can project vertically if it happens to lie along the vertical meridian.

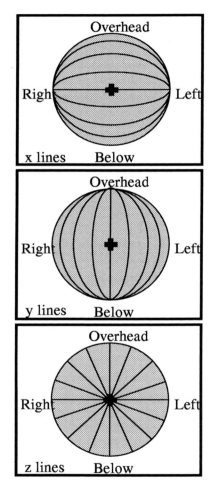

Figure 2.6: *Three projection spheres showing, respectively, how* **x** *lines,* **y** *lines and* **z** *lines project toward an observer. The focus of expansion is shown as the cross. Note that each of these figures depicts a convex sphere (even the "z line" one), and the contours are on the near side.*

An aside on the non-Euclidean visual field

Before using these insights on how principal lines project to determine the probable source of a projected line, there is a neat observation we may make from the discussion thus far. Suppose you are about to cross railroad tracks. The projection of each of the two tracks is a straight line in your visual field (each follows a great circle on the projection sphere). Furthermore, these two projected lines are parallel to one another when at the vertical meridian of your visual field. However, the two projected lines become non-parallel to one another in the periphery of your visual field, and eventually even intersect. How is it possible that two straight lines which are parallel at the vertical meridian can intersect one another? Can this really be?

It *can* really be, and it is possible because of the non-Euclidean nature of the geometry of the visual field. The geometry that is appropriate for the visual field is the surface of a projection sphere, and the surface of a sphere is not Euclidean, but, well, spherical. There are three main kinds of geometry for space: elliptical, Euclidean (or flat), and hyperbolic. Spherical geometries are a special case of the elliptical geometries. In Euclidean geometry, the sum of the angles in a four-sided figure (a quadrilateral) is 360°; in elliptical it is more, and in hyperbolic it is less. Let us ask, then, what the sum of the angles in a four-sided figure in the visual field is. A four-sided such figure is built out of four segments of great circles. Figure 2.7 shows an example four-sided figure on a projection sphere. In particular, it is a square. It is a square because (i) it has four sides, (ii) each side is a straight line (being part of a great circle), (iii) the lines are of (roughly) the same length, and (iv) the angles are (roughly) the same. Notice that each angle of this square is bigger than 90°, and thus the square has a sum of angles greater than 360°. The visual field is therefore elliptical.

One does not need to examine projection spheres to grasp this. If you are inside a rectangular room at this moment, look up at the ceiling. The ceiling projects toward you as a four-sided figure. Namely, you perceive its four edges to project as straight lines. Now, ask yourself what each of its projected angles is. Each of its angles projects toward you greater than 90°; a corner would only project as exactly 90° if you stood directly under it. Thus, you are perceiving a figure with four straight sides, and where the sum of the angles is greater than 360°. The perception I am referring to is your perception of the projection, not your perception of the objective properties. That is, you will *also* perceive the ceiling to objectively be a rectangle, each angle having 90°. Your perception of

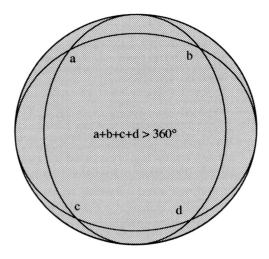

Figure 2.7: *Four great circles on a sphere (or on a visual field). In this case they make a square on the sphere; that is, the figure is four-sided, each side is straight and of equal length, and the angles are equal. Each angle of the square is greater than 90°, however. [To see this, the reader must judge the angle on the sphere, not the angle on this page.] Thus, the sum of the angles of the square is greater than 360°, which means the space must have elliptical geometry. In particular, it is spherical.*

the objective properties of the ceiling is Euclidean, or at least approximately so. Your perception of the way the ceiling projects, however, conforms to elliptical geometry. [There is a literature which attempts to discover the geometry of our perception of three dimensional space, and it is argued to be hyperbolic. This is an entirely different issue than the one we are discussing, as we are focusing just on the perception of projective properties (without depth information).]

It is often said that non-Euclidean geometry, the kind needed to understand general relativity, is beyond our everyday experience, since we think of the world in a Euclidean manner. While we may think in a Euclidean manner for our perception of the objective lines and angles, our perception of projective properties is manifestly non-Euclidean, namely spherical. We therefore *do* have tremendous experience with non-Euclidean geometry, it is just that we have not consciously noticed it. But once one consciously notices it, it is possible to pay more attention to it, and one then sees examples of non-Euclidean geometry at one's every glance.

Given a projection, which principal line is source?

We have seen earlier the way that the three principal kinds of line—x, y and z lines—project toward an observer. Now we wish to utilize this knowledge to ask the "inverse" question: Given some projected line in a proximal stimulus, which of the three kinds of line is the probable source?

Observers typically are looking forward as they move, and it is therefore reasonable to assume that, unless there are cues to the contrary, a projected line is probably not in the extreme peripheral regions of the visual field. This fact is useful because examination of the regions relatively near the focus of expansion (i.e., not in the periphery) of the projection spheres in Figure 2.6 reveals some simple regularities. The only kind of line projecting obliquely in this regime of the projection sphere is the z line, and all z lines converge to the same vanishing point (which is also the focus of expansion since observers are assumed to move parallel to the z axis). As a consequence of this, we may state the following rule.

> Rule 1: *If there is a single set of oblique projected lines sharing a vanishing point, then they are probably due to z lines.*

The only kind of line projecting horizontally in this regime, and *not* lying on the horizontal meridian, is the x line. Therefore...

Rule 2: *A horizontal projected line that does not lie on the horizontal meridian is probably due to an x line.*

Both x and z lines can cause horizontal projected lines lying on the horizontal meridian, and thus the following rule applies.

Rule 3: *A horizontal projected line that* does *lie on the horizontal meridian may be due to either an x line or a z line.*

The only kind of line projecting vertically in the relatively-near-the-focus-of-expansion regime is the y line, and the following rule therefore applies.

Rule 4: *A vertical projected line that does not lie on the vertical meridian is probably due to a y line.*

Analogously with Rule 3, we also have. . .

Rule 5: *A vertical projected line that* does *lie on the vertical meridian may be due to either a y line or a z line.*

One kind of proximal stimulus we will want to decipher is one where there are two sets of converging projected lines with distinct vanishing points. Because observers tend to look where they are going, one of these sets probably consists of projected z lines, and it will probably be the set with lines for which it is most salient that they converge to a vanishing point. The other set of converging lines consists of either projected y lines (which would point toward a vanishing point above or below the focus of expansion) or projected x lines (which would point toward a vanishing point to the left or the right of the focus of expansion). This is recorded as the following rule.

Rule 6: *When there are two sets of projected lines with different vanishing points, the set with the more salient vanishing point probably consists of projections of z lines, and the other of either x or y lines, depending on where they point.*

These rules are all consequences of the simple model of three kinds of principal lines and forward motion parallel to the z axis.

Recall that, for us scientists to make predictions using the latency correction hypothesis, we must determine the probable scene causing the proximal stimulus, and we must determine how that scene will probably change in the next moment. Although we now have machinery enabling us to infer the probable scene, we have not yet addressed the latter. How a scene will change in the next moment depends on where the observer is moving toward, and how fast. Where the observer is moving can be determined by the vanishing point of the z lines; since the observer moves along the z axis, the vanishing point of the projected z lines is also the focus of expansion, or the direction of motion. Therefore, once we have discovered which projected lines in the proximal stimulus are due to z lines, we have also discovered the direction of motion. Because of the importance of this consequence, I record it as a final rule.

> Rule 7: *The probable location of the focus of expansion is the vanishing point of the projected z lines.*

The observer's speed can be set to some reasonable value; I typically set it to 1 m/sec. I also often assume in simulations a latency of 50 msec, which is an underestimate.

One important aspect of the probable scenes that this simple model does not accommodate is distance from the observer. If all the probable sources were as in the model, but were probably a mile away, then we can expect no change in the nature of the projections in the next moment. It is reasonable to assume that latency correction will be primarily tuned to nearby objects, objects that we can actually reach, or that we might actually run into. Accordingly, it is plausible that the visual system interprets these geometrical stimuli as scenes having a distance that is on the order of magnitude of meters away (rather than millimeters or hundreds of meters).

How general is this model?

The model I have proposed above requires that the ecological environment of the observer have an abundance of x, y and z lines, where the observer moves parallel to the z lines. I have called this a "carpentered world assumption" (Changizi, 2001c), but how much does my explanation depend on this assumption?

First consider z lines. One of the principal roles they will play in the explanation of the geometrical illusions is that they provide the cue as to the location in the visual field of the focus of expansion. That is, the visual system figures out where the observer is probably going on the basis of where the vanishing point is for the z lines. However, there need not actually be any z lines in the world for there to be, for the moving observer, projections which are like the projections of z lines. If an observer is moving forward in an unstructured environment, the optic flow itself will cause "optical blur" projected lines, and these will converge to the focus of expansion (Gibson, 1986). Thus, my explanation does not require that the ecological environment actually possess a propensity for z lines. The projected z lines may be due not to z lines at all, but to optic flow; radial lines mimic optic flow, and may trick the visual system into believing it is probably moving forward.

For x and y lines, all that my model really requires is that the probable source of a horizontal projected line (not on the horizontal meridian) is an x line, and that the probable source of a vertical projected line (not on the vertical meridian) is a y line. It could, for example, be the case that x lines and y lines are relatively infrequent, but that they are still the most probable source of horizontal and vertical projected lines, respectively.

It is also worth noting that a propensity for y lines does not require a carpentered world assumption. Forests, for example, have a propensity for y lines; gravitation makes the y axis unique, and any gravitational ecology will probably have a propensity of y lines. x lines, on the other hand, really do seem to require a carpentered world assumption; e.g., although the forest will have a propensity for there to be lines parallel to the ground, which is half the definition of an x line, it will not have a propensity to lie perpendicular to the observer's direction of motion. The model therefore *does* depend, in this regard, on the carpentered world assumption. Humans raised in non-carpentered environments would, then, be expected to have a different repertoire of geometrical illusions, which appears to be the case (Segall et al., 1966).

2.3   Explaining the geometrical illusions

In this section I explain how the latency correction hypothesis explains the classical geometrical illusions. The first subsection answers the question, What is the probable scene underlying the geometrical stimuli? This includes determining what the lines are and where they are with respect to the observer's direction of motion. The next subsection tackles the geometrical illusions that are misperceptions of projected angle, which includes the corner, Poggendorff, Hering and Orbison. The following subsection explains the illusions of angular size or angular distance, which includes the double Judd, Müller-Lyer, Hering, Orbison and the upside-down 'T'. The final subsection tests a psychophysical prediction of the latency correction hypothesis, providing further confirmation.

2.3.1   The probable source and focus of expansion

The rules developed in the previous section can now be applied to the illusions from Figure 2.4, both in determining what are the probable sources of the stimuli, and in determining what is the probable direction of motion for the observer. This is our task in this subsection.

The probable sources

Figure 2.8 shows the same illusions as in Figure 2.4, but each projected line has been labeled with the probable kind of source line via the earlier rules. The explanations for the probable sources are as follows.

- No vertical line in any of the illusory figures has cues suggesting it lies along the vertical meridian, and thus each is probably due to a y line.

- Of all the horizontal lines, only the one in the upside-down 'T' illusion possesses a cue that it might lie along the horizontal meridian. The cue is that there is a 'T' junction, and such junctions are typically due to three-dimensional corners (i.e., an x-y-z corner). The horizontal segment of the 'T' junction is probably, then, due to two distinct segments, one the projection of an x line, and one the projection of a z line. That is, it is probably a corner that is being viewed "from the side." I have arbitrarily chosen the left segment to be the projection of an x line, but the cues in the upside-down 'T' illusion (which consists of just the upside-down 'T') do not distinguish which is which.

- All the remaining horizontal projected lines are parts of stimuli without any cues that they lie along the horizontal meridian, and so are thus due to x lines.

- All that is left are the obliques. In the Hering, Orbison, Ponzo, Corner and Poggendorff there exists just one set of converging obliques, and they are thus probably due to z lines.

- In each of the Müller-Lyer and the Double Judd there are two sets of converging projected lines: one set consists of the four inner obliques (the ones *in between* the two vertical lines), and the other set consists of the four outer obliques (the ones not in between the two vertical lines). The four inner obliques are more salient and clustered, and appear to share a vanishing point more clearly than do the outer ones. The inner obliques are therefore probably due to z lines. Since the outer obliques have a vanishing point horizontally displaced from the vanishing point for the inner obliques, the outer obliques must be due to x lines. [While this serves as an adequate first approximation, greater analysis in fact reveals that the outer obliques probably do not share a vanishing point at all (and thus they cannot all be principal lines). Consider just the Müller-Lyer Figure for specificity. Lines in the world project more obliquely as they near their vanishing point (see Figure 2.6). The two outer obliques on the left are far in the visual field from the two outer obliques on the right; if they were projections of the same kind of line in the world, then they would *not* project parallel to one another, one pair being considerably closer to the vanishing point (for that kind of line) than the other. But the outer obliques on the left *are* parallel to the outer ones on the right, and thus they cannot be projections of the same kind of line, and they do not point to a single vanishing point. Only the four inner obliques are approximately consistent with a single vanishing point.]

The probable focus of expansion

Now that we know what the probable sources are for the eight illusory proximal stimuli, we can use the information about the projected z lines to determine the focus of expansion. That is, the z line vanishing point is the focus of expansion (see Figure 2.8). Figure 2.9 shows the earlier figures, but where the illusions now share the same focus of expansion, and Figure 2.10 shows the key features of each illusion embedded in a display which provides a strong cue as to the focus of expansion.

- For the Hering, Ponzo, Orbison and Müller-Lyer stimuli there is exactly one focus of expansion determined by the projections of the z lines, and Figure 2.9 shows them embedded in a radial display at the appropriate location with respect to the focus of expansion. Notice that for the Müller-Lyer the fins act as cues as to the location of the focus of expansion, and that in Figure 2.10, where the radial display does the cueing work, the fins are no longer necessary for the illusion.

- The projected z lines for the double Judd are so similar in orientation that they may converge either up and to the right of the figure, or down and to the left; that is, the focus of expansion may be in one of these two spots. I have placed the fin-less version of the double Judd in Figure 2.10 into these two positions with respect to the focus of expansion. Note that the illusion is qualitatively identical in each case to the earlier one (since the cues to the focus of expansion are provided by the radial display rather than the fins).

- The corner and Poggendorff illusions could be placed anywhere in the display so long as the projected z line points to the probable focus of expansion; I have chosen one spot arbitrarily. Any conclusions I draw later will not depend on this choice.

- The upside-down 'T' illusion could be placed on either side of the vertical meridian, so long as the horizontal segments lie along the horizontal meridian. I have arbitrarily chosen one spot. Any conclusions I draw later will not depend on this choice.

At this point in this section I have used the model to determine the probable source and focus of expansion given a sufficiently simple geometrical proximal stimulus. The model has concluded that each of the eight classical geometrical illusions I have been discussing are probably caused by a particular kind of source in the world, and are probably located in a certain position in the visual field with respect to the focus of expansion. These conclusions were summarized in Figure 2.8.

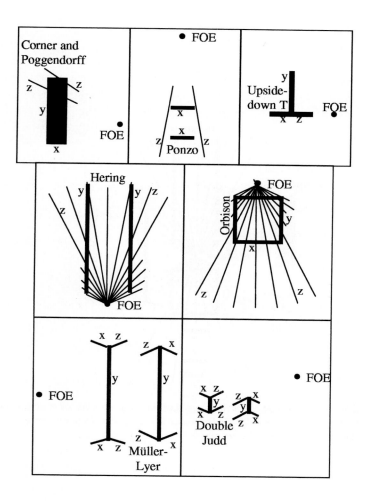

Figure 2.8: Eight classical geometrical illusions, now showing for each projected line the probable kind of source line, *x*, *y* or *z*. The probable focus of expansion is also shown in each case.

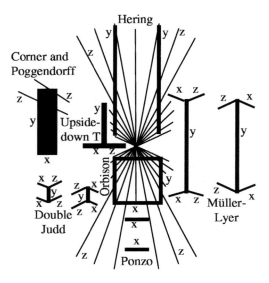

Figure 2.9: *The same eight classical geometrical illusions, showing, as in an earlier figure, for each projected line the probable kind of source line,* ***x****,* ***y*** *or* ***z****. They have been placed such that their focus of expansion is the same.*

We still have not explained the illusions, however. Recall that, under the latency correction hypothesis, in addition to determining the probable scene causing the proximal stimulus—which is what we have done thus far—we must also figure out how that scene will probably change by the time the percept occurs. Well, since we know the probable scene, and we know which direction the observer is probably moving, all we have to do is to determine how the sources will project when the observer is moved forward a small amount.

2.3.2 Projected angle misperception

One kind of illusion concerns misperception of projected angle. First, let me be clear about what I mean by perception of projected angle. If you look up at a corner of the room you are in, you will notice that you perceive there to be three right angles; this perception is the perception of the objective angles. You simultaneously perceive there to be three obtuse angles summing to 360°; this perception is the perception of the projected angles. It is the latter that is relevant for the geometrical illusions.

The corner, Poggendorff, Hering and Orbison can be treated as misperceptions of projected angle. In the corner and the Poggendorff the angles appear to be nearer to 90° than they actually are. The same is true for the angle between the vertical line and the obliques in the Hering illusion. In the Orbison illusion, the right angles appear to be bent away from 90°. How do we make sense of these projected angle illusions? And why are some misperceived towards 90° and some away from it?

First, let us distinguish between two kinds of projected angle. Since there are just three kinds of line in my model, the only kinds of angle are those that result from all the possible ways of intersecting these kinds of line. They are x-y, x-z and y-z angles; these are the *principal angles*. That is, x-y angles are any angles built from an x line and a y line, and so on. The latter two kinds of angle are actually similar in that, having a z arm, the plane of these angles lies parallel to the observer's direction of motion. I call x-z and y-z angles xy-z angles. x-y angles, on the other hand, lie in a plane perpendicular to the observer's direction of motion, and must be treated differently.

xy-z projected angles

Note that the corner, Poggendorff and Hering illusions have angle misperceptions where the angles are xy-z angles, and the misperception is that observers perceive the projected angles to be nearer to 90° than they actually are. Why is

Figure 2.10: *Each illusion from Figure 2.9 is "transferred" into a stimulus with strong cues as to the location of the focus of expansion. The same kinds of illusion occur, suggesting that it is cues to the location of the focus of expansion that is of primary importance in the illusions. In the case of the double Judd and Müller-Lyer figures, the probable location of the focus of expansion is the same as in Figure 2.9, but now its location is due to the radial lines rather than the fins; because the double Judd is also consistent with being in the upper right quadrant, it has been transferred there as well as in the bottom left quadrant. The corner and Poggendorff stimuli could be placed anywhere in the radial display so long as radial lines traverse them in the appropriate fashion.*

this? The latency correction hypothesis says it is because in the next moment the angles will project nearer to 90°, and thus the misperception is typically a more veridical percept. [It is inappropriate in the case of a static stimulus.] But *do xy-z* angles actually project more like 90° in the next moment? Yes, and there are a number of ways to understand why.

The most obvious way to see this is to hold something rectangular, maybe an index card, out in front of you, below, and to the right of your eyes. Hold the card out flat (i.e., parallel with the ground), and orient it so that the near edge is perpendicular to your line of sight. This is depicted in Figure 2.11 (A). Observe how the four right angles of the card project toward you. The projected angles nearest and farthest from the vertical meridian—i.e., angles *a* and *d*—are both acute, and the other two are obtuse. What happens to these projected angles if you move your head and eyes forward as if you are going to pass the card? If you move so far that the card is directly below and to the right of your eyes—i.e., you are just passing it—you will see that the four angles all project as 90°. Thus, as you move forward, each of these *x-z* projected angles changes toward 90°, eventually becoming exactly 90° when you pass it. The same observation applies any time an observer moves forward in the vicinity of *x-z* angles (e.g., a rug, or the ceiling). The same observations also apply for the projection of *y-z* angles, one case which is depicted in Figure 2.11 (B). A real world example is when you walk past a window: all the projected angles begin either very acute or very obtuse, but as you near the window, they progressively project more and more as 90°.

Another way of comprehending how *xy-z* angle projections change is to examine projection spheres upon which *xy-z* angles have been projected. Figure 2.11 (C) and (D) show such projection spheres; (C) consists of the intersections of *x* and *z* line projections, and (D) of *y* and *z* line projections. Recall the nature of optic flow on the projection sphere: flow begins at the focus of expansion and moves radially outward toward the periphery of the sphere. Thus, projected angles nearer to the focus of expansion are the way they project when they are farther away from the observer, and projected angles in the periphery of the projection sphere are the way they project when the observer is nearer to passing the angle. For example, in Figure 2.11 (C) the reader may see asterisks at four projected angles along the same radial line. The asterisk nearest the focus of expansion is roughly 60°, the next one a bit bigger, the next still bigger, and the last one is 90°. [Recall that to judge these angles the reader must judge the angle on the sphere, not on the page.] We could have, instead, put our asterisks on the other side of the projected *z* line, and we would have had

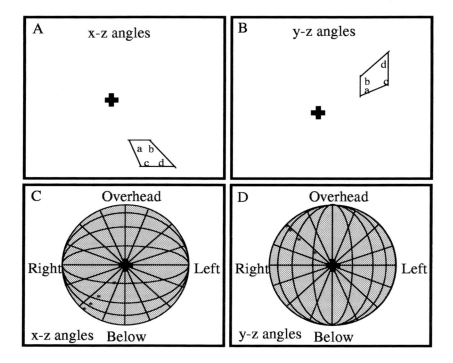

Figure 2.11: (A) Depiction of the view of a rectangular card below and to the right of an observer's direction of motion (represented by the cross). The card is lying parallel to the ground, with one axis parallel to the observer's direction of motion, and the other perpendicular to the direction of motion. The card's angles are thus **x-z** angles. (B) This depicts the analogous card as in (A), but now the angles are **y-z**. (C) A projection sphere upon which **x** and **z** lines are projected; their intersections are **x-z** angle projections. Notice how, along any radial line, the angles of intersection between **x** and **z** lines become more like **90°** in the periphery (see the asterisks); that is how they change in the next moment, since the angles move toward the periphery as the observer moves forward. (D) A projection sphere upon which **y** and **z** lines are projected; their intersections are **y-z** angle projections. Notice how, along any radial line, the angles of intersection between **y** and **z** lines become more like **90°** in the periphery (see the asterisks); this is how they change in the next moment.

the projected angles starting from around 120° and falling to 90°. A similar account applies to the projections of y-z angles as shown in Figure 2.11 (D).

In short, xy-z angles project more like 90° as observers move forward. If a proximal stimulus has cues suggesting that a projected angle is due to an xy-z angle, then latency correction predicts that observers will perceive the angle to be more like 90° than it actually is. That is, people should perceive the projected angle to be "regressed" toward 90° (Thouless, 1931a). The corner, Poggendorff, and Hering illusions each had projected xy-z angles, and each is perceived to be nearer to 90° than it actually is. These illusions are, therefore, consistent with latency correction.

The Poggendorff has another salient illusory feature in addition to the projected angles being perceived nearer to 90° than they are: the two oblique lines are collinear, but do not appear to be. Each oblique line appears to, intuitively, undershoot the other. Latency correction explains this illusory feature as follows. Suppose that a single z line lies above you and to your left along the wall (perhaps the intersection between the wall and the ceiling). Now also suppose that there is a black rectangle on your upper left, but lying in your fronto-parallel plane. That is, the rectangle is made of x and y lines. Suppose finally that the rectangle is lying in front of the z line. The projection of these objects will be roughly as shown by the Poggendorff illusion in Figure 2.9. I say "roughly" because the projection will not, in fact, be as in this figure. Consider first the projected angle the z line will make with the right side of the rectangle. Suppose it is 60°; that is, the (smaller) y-z angle on the right side of the rectangle is 60°. What will be the projected angle between that same z line and the other vertical side of the rectangle? The part of the z line on the other vertical side of the rectangle is farther away from the focus of expansion and more in your periphery. Thus, this more peripheral y-z angle will be nearer to 90°; let us say 63° for specificity. That is, when the *same* z line crosses through or behind a rectangle as constructed, the projected angles will *not* be the same on either side. Now, the two projected angles in the Poggendorff figure *are* the same on either side, and thus the projected lines on either side *cannot* be due to one and the same z line. Instead, the more peripheral y-z projected angle, being farther from 90° than it would were it to be the projected angle made with the z line from the other side, must actually be a line that is physically higher along the wall. The visual system therefore expects that, in the next moment (i.e., by the time the percept is generated), the oblique projected line on the left should appear a little higher in the visual field compared to the extension of the oblique line on the right (since differences in visual field position are accentuated as an observer moves forward).

x-y projected angles

The Orbison illusion primarily concerns the misperception of the four projected angles, each which is 90°, but which observers perceive to be greater or lower than 90°. The squares in the Orbison illusion are composed of x and y lines (Figure 2.8, and see also Figure 2.1), and we must ask how the projections of x-y angles change as observers move toward the focus of expansion (which is the vanishing point of the projected z lines in the Orbison figure).

The most straightforward way to understand how x-y angles change is to hold up a rectangular surface like an index card in your fronto-parallel plane, with one axis vertical and the other horizontal, and move your head forward. When the card is sufficiently far out in front of you, each of its four angles projects nearly as 90°. As you move your head forward as if to pass it, the angles begin to project more and more *away* from 90°. Some angles begin to project more acutely, and some more obtusely. When you are just about to pass the card, some of its angles will project closer and closer to 0°, and the others will project closer and closer to 180°. If the card is in your lower right quadrant, as is depicted in Figure 2.12 (A), two angle projections fall toward 0°—*b* and *c*—and two rise toward 180°—*a* and *d*. If, instead, it is directly below you, the top two angle projections fall to zero and the bottom two rise to 180°. If it is directly to your right, then the near two go to zero and the far two straighten out. If the card is directly in front of you—i.e., each of its four angles is in each of the four quadrants of your visual field—then each angle gets more and more obtuse as you move forward. For example, as you walk through a doorway, each of the corners of the door projects more and more obtuse, so that when you are just inside the doorway each corner projects as 180° (and the doorway now projects as a single line all the way around you).

We may also comprehend how x-y angles change in the next moment by examining a projection sphere on which x and y lines have been projected, as shown in Figure 2.12 (B). If the reader follows the asterisks from the focus of expansion outward, it is clear that these x-y angles begin projecting at approximately 90° and as the observer moves forward and the angle moves peripherally the projections become greater and greater. Following the '#'s shows the same thing, except that the projected angles get smaller and smaller as the observer moves forward.

In sum, x-y angles project further away from 90° in the next moment; they

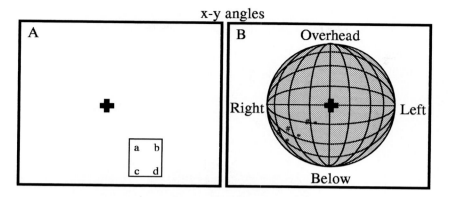

Figure 2.12: (A) Depiction of the view of a rectangular card below and to the right of an observer's direction of motion (represented by the cross). The card is lying upright and in the observer's fronto-parallel plane. The card's angles are thus x-y angles. (B) A projection sphere upon which x and y lines are projected; their intersections are x-y angle projections. Notice how, along any radial line, the angles of intersection between x and y lines become less like 90° in the periphery (see the asterisks and '#' signs); that is how they change in the next moment since the angles move toward the periphery as the observer moves forward.

are "repulsed" away from 90° instead of regressed toward as in xy-z projected angles. Which direction a projected x-y angle will get pushed away from 90° depends on the angle's orientation and its position relative to the focus of expansion. Figure 2.13 shows how one kind of x-y line changes its projection in the next moment, and one can see that it depends on the quadrant. The figure also shows the magnitude of the projected angle change as a function of position along the x and y axes on a plane one meter in front of the observer. For x-y angles of other orientations the plot looks similar, except that it may be rotated by 90°. Figure 2.14 (A) summarizes the directions which x-y projected angles change in the next moment as a function of position in the visual field with respect to the focus of expansion.

The latency correction hypothesis predicts that if cues suggest that a projected angle is due to an x-y angle, then observers will misperceive the angle to be whatever it will probably be in the next moment (by the time the percept is elicited). Figure 2.14 (B) shows the same squares as in (A), but now embedded in a radial display which provides strong cues as to the location of the focus of expansion. In every case, observers misperceive the right angles in (B) in the direction predicted by latency correction in (A). This is just a special case

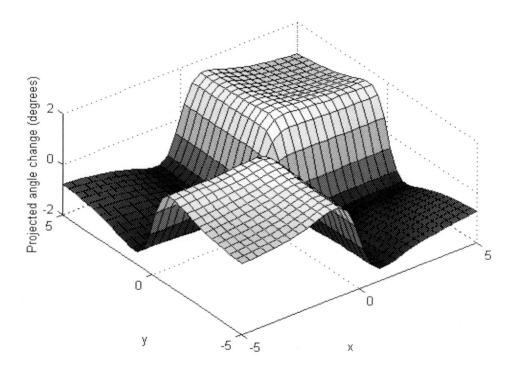

Figure 2.13: *The change in projected angle as a function of position with respect to the focus of expansion on a plane one meter ahead and perpendicular to the direction of motion, for an x-y angle with one arm pointing up and another arm pointing right. This assumes the observer moves forward 5 cm (or that the latency is 50 msec and that the observer is moving at 1 m/sec.)*

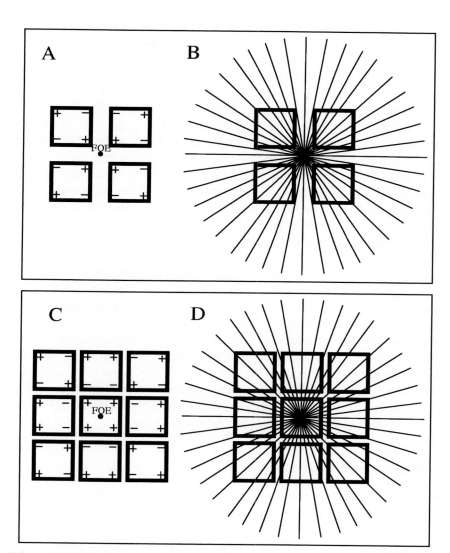

Figure 2.14: (A) *Four perfect squares. If an observer were moving toward the cross at the center, the projected angles for each angle would change away from* **90°***, and the pluses and minuses show the direction of change. (B) The same squares embedded in a stimulus with strong cues that the focus of expansion is at the center. Subjects perceive the angles to be different than* **90°** *as predicted by the directions shown in (A). (C) and (D) are the same, but with more squares.*

of the Orbison illusion, and thus the Orbison illusion is consistent with latency correction.

Before we leave the Orbison illusion, I should note that there is another illusion also called by the same name (and also discovered by Orbison (1939)), and shown on the left in Figure 2.15. The illusion is primarily that the two projected angles nearer to the center of the circles appear to be a little obtuse, and the other two a little acute. Recall that when the square is on the right side of the other Orbison display with the radial lines (on the right in Figure 2.15) the two near-the-center angles are perceived to be a little acute, and the two farther ones a little obtuse. That is, the concentric-circle version of the Orbison leads to qualitatively the opposite illusion of the radial-line version. The square in the concentric-circle Orbison looks more like the square on the opposite side of the radial-line Orbison (see Figure 2.15), although it is less dramatic. My model has no apparatus at this point to accommodate concentric circles, but it is unclear what kind of ecologically valid scenario would lead to such a stimulus. My hunch at the moment is that concentric circles have no strong association with some probable source, and that, instead, the visual system is noticing that in the vicinity of the square there are many oblique lines, and they are all pointing to the right, and pointing toward the horizontal meridian. The oblique lines are not straight, and they do not point to a single vanishing point, but to the extent that the obliques all point to the right and toward the horizontal meridian, the visual system may guess that the focus of expansion is more probably somewhere on the right side of the square. This would explain why observers misperceive this square in a qualitatively similar manner as the square on the left in the radial-line Orbison.

Ambiguous projected angles

We have to this point showed how the latency correction hypothesis can explain the misperception of projected angles with cues as to whether they are due to xy-z angles or x-y angles. When cues suggest a projected angle is due to an xy-z angle, observers perceive the projected angle to be nearer to $90°$, just as predicted by latency correction. And when cues suggest a projected angle is due to an x-y angle, observers perceive the projected angle to be farther from $90°$, also as predicted by latency correction.

But even ambiguous projected angles—i.e., projected angles for which there are no cues as to what kind of angle caused it—are misperceived toward $90°$ (Fisher, 1969; Bouma and Andriessen, 1970; Carpenter and Blakemore,

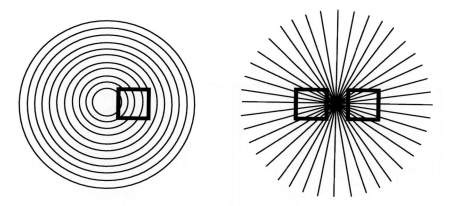

Figure 2.15: *The concentric-circle version of the Orbison illusion, and the radial-line version.*

1973; Nundy et al., 2000). The illusion magnitude can be as great as a couple degrees or so. For example, the projected angle in a 'less than' symbol—'<'— is perceived nearer to 90° by observers. It is as if the visual system decides that ambiguous projected angles are probably xy-z angles, since it is misperceiving them like it misperceives projected angles with cues they are xy-z angles. Why should this be?

Let us ask, what *is* the probable source of an ambiguous projected angle? There are no cues, but it may still be the case that in the absence of cues one kind of angle is much more probable. If you are in a rectangular room, look over at one of the walls and suppose you are about to walk straight toward the wall. Consider one of the corners of the wall you are approaching. It consists of three right angles, an x-y angle, an x-z angle, and a y-z angle. Two of these three angles are therefore xy-z angles, and therefore xy-z angles tend to be around twice as frequent in our experience. If a projected angle is ambiguous, then without knowing any more information about it we should guess that it is an xy-z projected angle. Another difference between x-y angles and xy-z angles is that the former project nearly as 90° when they are in front of an observer, only projecting much different from 90° when the observer is about to pass the angle (Figure 2.16). xy-z angles, on the other hand, project in all ways when out in front of an observer (Figure 2.16). This may be made more precise by placing an observer at random positions within a room with the three

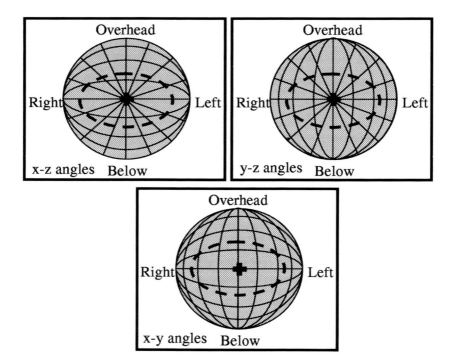

Figure 2.16: Projection spheres for the three kinds of angle. The top two are the two kinds of *xy-z* projected angles, and the bottom is for *x-y* projected angles. The dashed ellipse in each identifies the region of the projection sphere observers typically view (i.e., observers tend to look in the direction they are going, and look left and right more than up and down). The reader may observe that within the ellipse *xy-z* angles project in all sizes, acute to obtuse, but that *x-y* angles project only very near **90°**.

angle types stuck in one spot, and seeing how often each projects a given angle. Figure 2.17 shows this, and we can see that when a projected angle is near 90° it is probably due to an x-y angle, but otherwise it is probably due to an xy-z angle.

In sum, then, ambiguous projected angles are probably xy-z angles if they are acute or obtuse. We therefore expect the perception of ambiguous projected angles to be like the perception of projected angles with cues that the projected angle is due to an xy-z angle, and, as mentioned earlier, this is indeed what observers perceive. In particular, the latency correction hypothesis can predict how much the projected angle should be misperceived depending on its angle. Figure 2.18 shows how, on average, the projected angle changes in the next moment as a function of the starting projected angle. We see that there is no projected angle change when the angle is very close to 0°, 90° or 180°; the projected angle change is maximally positive somewhere in between 0° and 90°, and maximally negative somewhere in between 90° and 180°. The latency correction hypothesis therefore predicts that this will be the shape of the psychophysical function for observers of ambiguous projected angles. The inset of Figure 2.18 shows how observers misperceive projected angles as a function of the angle, and the psychophysical function is indeed very similar to that predicted by the latency correction hypothesis.

2.3.3 Angular size misperception

We have now seen that the illusions of projected angle—the corner, Poggen-dorff, Hering and Orbison illusions—are just what we should expect if the visual system engages in latency correction. We have not, however, touched upon the double Judd, the Müller-Lyer or the upside-down 'T' illusion. Each of these illusions involves the misperception of an angular distance or an angular size. Even the Hering can be treated as a misperception of angular distance, since the angular distance between the two lines appears to be greater nearer the vanishing point (see Figure 2.19 for two versions of the "full" Hering illusions). The Orbison, too, can be classified as a misperception of angular size since the sides of the squares are not all perceived to be the same. In this subsection I describe how latency correction explains these angular size illusions.

Projected x and y lines

How do the angular sizes of projected x and y lines change as an observer moves forward? Let us focus on how x projections change, and what we learn

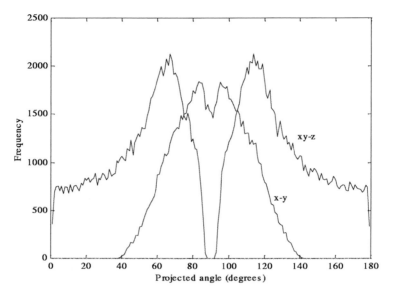

Figure 2.17: *Histogram of counts for the projections of x-y and xy-z angles. One can see that x-y angles rarely project angles much differently than 90°; most acute and obtuse projected angles are due to xy-z angles. The curves were generated by placing a simulated observer at 10^5 positions near an angle of the specified kind (10^5 for each of x-y, x-z and y-z). Each placement of the observer consisted of the following. First, a random orientation of the principal angle was chosen. For example, for an x-z angle there are four orientations: +x and +z, +x and −z, −x and +z, and −x and −z. Second, the angle's vertex was placed at the origin. Third, a position for the simulated observer was determined by randomly choosing values for x uniformly between 0.1 m and 1 m to one side of the angle, values for y uniformly between 1 m above and below the angle, and values for z uniformly between 0.5 m and 1 m in front of the angle. The simulation was confined to these relatively nearby positions since one might expect that veridical perception of nearby objects matters more in survival than veridical perception of objects far away. The nature of my conclusions do not crucially depend on the particular values used in the simulation.*

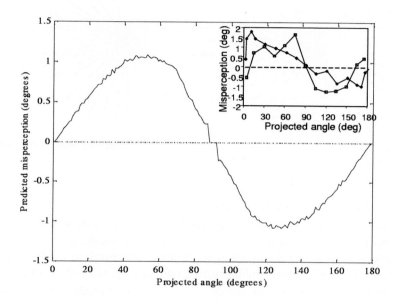

Figure 2.18: Average projected angle change as a function of the pre-move projected angle, for principal right angles lying in a plane parallel to the direction of motion (**xy-z** angles). One can see that the latency correction hypothesis predicts that, for projected angles that are probably due to **xy-z** angles, acute projected angles are overestimated and obtuse projected angles are underestimated. The graph was generated from the same simulation described in Figure 2.17, except that for each placement of the observer, the observer was then moved along the z-axis toward the angle (i.e., **z** got smaller) at a speed of 1 meter/sec for (a latency time of) 0.05 sec. The particular position of the peak is not important, as it depends on the allowed range of pre-move positions in the simulation. Inset shows two plots of actual misperceptions for subjects. Diamonds are averages from one representative non-naive subject (RHSC) from Carpenter and Blakemore (1973, Figure 3), and squares are averages from six naive subjects from Nundy et al. (2000, Figure 5).

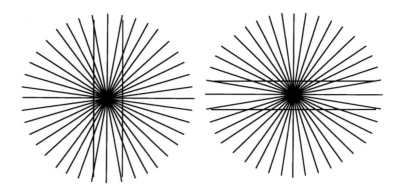

Figure 2.19: *Two versions of the Hering illusion. The perceived angular distance between the two lines is greater near the middle than near the edges.*

will immediately apply to y line projections as well.

Consider the angular distance between the sides of a doorway at eye level. As you approach the doorway, the angular distance between the sides increases. When you are just inside the doorway the angular distance is at its maximum of 180°. Consider how the angular distance between the sides of the doorway a little above eye level changes as you move forward. As before the angular distance increases, but it now does more slowly, and when you are just inside the doorway, the angular distance reaches its maximum at a value below 180°. The farther above or below eye level you look, the slower do the sides of the doorway expand as you approach. The angular distance between the sides of a doorway is really the length of a projected x line, namely an imaginary line extending between the two sides. The same is true for projected y lines: the angular distance between the top and bottom of the doorway increases as you approach, and does so most quickly for the angular distance between the part directly above and below your eye.

This may also be understood via examining projection spheres, as shown in Figure 2.20. In each sphere of the figure there are three pairs of squares and circles, the inner-most, the next-farther-out, and the outer-most pairs. They represent three snapshots of the horizontal (A) or vertical (B) angular distance between the points. Focusing just on (A), one can see that although the inner-most pair of squares and circles have about the same horizontal angular distance

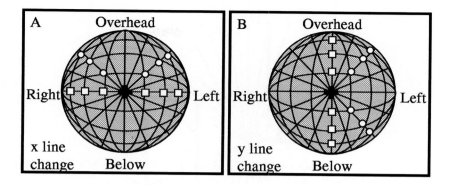

Figure 2.20: Projection spheres with x, y and z line projections. (A) This aids us in under- standing how angular sizes of x line projections change as an observer moves forward. The inner-most pair of squares and circles depict the sides of a doorway that is far in front of an observer, the squares are at eye level (i.e., lying on the horizontal meridian) and the circles above eye level. The angular distance between the two squares is about the same as that be- tween the two circles. But as an observer moves forward, in the next moment the sides of the door expand, the sides at eye level project as the next-farther-out pair of squares, and the sides above eye level project as the next-farther-out pair of circles. The horizontal angular distance between the squares is now greater than that between the circles. Similarly, in the next moment the sides are depicted by the next pair of squares and circles. (B) Identical to (A) but shows how vertical angular distances grow most quickly when they lie along the vertical meridian.

between them—this corresponds to the sides of a doorway far in front of an observer, the squares at eye level and the circles above eye level—by the time the observer approaches, the horizontal angular distance between the squares has grown considerably more than the horizontal angular distance between the circles.

There is one major summary conclusion we can make concerning how pro- jected x lines change as observers move forward:

> *The angular distance between any point and the vertical meridian increases as observers move forward. Furthermore, this angular distance increase is maximal for points lying along the horizontal meridian, and falls off as the point gets farther away from the horizontal meridian.*

This statement is just another way of saying that as you approach a doorway, its sides bow out most quickly at eye level (and less and less quickly the further it is from eye level). The analogous conclusion holds for y lines.

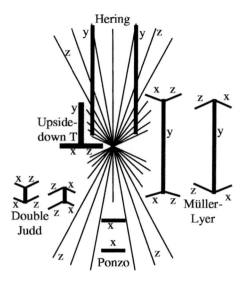

Figure 2.21: *The geometrical illusions which rely on misperception of angular distance are shown again here for convenience.*

> *The angular distance between any point and the horizontal meridian increases as observers move forward. Furthermore, this angular distance increase is maximal for points lying along the vertical meridian, and falls off as the point gets farther away from the vertical meridian.*

These conclusions are sufficient to explain the angular size illusions shown in Figure 2.21, except for the upside-down 'T' illusion (which I take up in the next subsubsection). I will explain each in turn.

- *Double Judd*: The double Judd illusion consists of two projected *y* line segments, projections which do not cross the horizontal meridian (see Figure 2.21). It suffices to treat each segment as if it were a point. We are interested in the angular distance between each segment and the horizontal meridian. They are, in fact, the same in the figure. However, the conclusion above states that in the next moment the segment nearer to the vertical meridian—i.e., the inner segment—will have a greater distance from the horizontal meridian than the other segment. The latency correction hypothesis therefore predicts that observers will perceive the segment that is nearer to the vertical meridian to have greater angular separation from the horizontal meridian. And this just is the illusion for the double Judd illusion: the inner segment of each pair in Figure 2.21 appears to be farther away from the horizontal meridian. [A similar explanation would work if the double Judd stimulus were rotated **90°**.]

- *Müller-Lyer*: The Müller-Lyer illusion consists of two projected y line segments, projections which *do* cross the horizontal meridian. Consider just the tops of each projected y line. The top of the projected y line on the left in Figure 2.21 is nearer to the vertical meridian than the top of the other projected y line, and so it will move upward more quickly in the next moment. Thus, the angular distance between the top of the left projected y line and the horizontal meridian should appear to observers as greater than that for the right projected y line. The same also holds for the lower halves of each projected line, and thus the total angular distance from the top to the bottom of the left projected line will be longer in the next moment than that of the right projected line, and thus should be perceived in that way if latency correction applies. And, of course, this is the illusion in the case of the Müller-Lyer. The same explanation holds for the variants of the Müller-Lyer in Figure 2.22.

- *Ponzo*: The explanation for the Ponzo illusion follows immediately from the argument for the Müller-Lyer illusion, except that it concerns the distance from points to the vertical meridian.

- *Hering*: In the Hering illusion in Figure 2.21, there are two projected y lines on either side of the vertical meridian. The angular distance between the lines is perceived to depend on how high one is looking above or below the horizontal meridian. At the horizontal meridian the perceived angular distance between the two projected y lines is greatest, and it falls as one looks up or down. The conclusion concerning x lines above explains this: points on one of the Hering lines nearer to the horizontal meridian will, in the next moment, move away from the vertical meridian more quickly. [A similar explanation would hold if the Hering had been presented as two projected x lines lying on either side of the horizontal meridian.]

We see, then, that one simple latency correction rule underlies these three, seemingly distinct, classical geometrical illusions.

Projected z lines

The angular size and distance illusions discussed above concerned the angular sizes for x and y lines. What about the angular size of z lines? Consider how projected z line segments change as an observer moves forward. When the segment is very far away, it projects small, and as you near it it projects larger. This is no different from the behavior of x and y lines. Consider, though, how a z line projection changes when you are already relatively nearby. It still projects larger in the next moment. This is partly because it is closer, but also partly because it begins to project more perpendicularly toward the observer. Consider, as a contrast, how an x line segment lying on the horizontal meridian and to one side of the vertical meridian projects as an observer near it moves

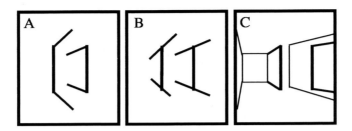

Figure 2.22: *The angular size of the vertical bold lines are the same in each figure, but the left one appears larger because the cues suggest that the focus of expansion is to the left, and thus the left one will grow more quickly in the next moment. Note that in (C) the illusion is the opposite of the standard Müller-Lyer: the fins-in line appears longer than the fins-out line.*

forward. Eventually, the x line begins to project less perpendicularly toward the observer—i.e., less of the line is facing the observer. When the observer passes the x line, its angular size will have fallen to zero. For the z line segment, however, when the observer passes it, its angular size will be at its maximum.

With this under our belts we can ask and answer the question of how the probable source of the upside-down 'T' illusion will change in the next moment. Recall that the source of the 'T' is a corner made of an x, y and z line, whose point lies on the horizontal meridian, and thus so does the x and z line. The probable focus of expansion is somewhere on the same side as the z arm, but past the tip of the projected z arm (e.g., see Figure 2.21). The angular size of the horizontal bar is due to the sum of the angular sizes of the x line and the z line, these lines being at right angles to one another in the world. Suppose each line has a length of L meters. It's angular size could then be mimicked by a single straight real world line (it is not a principal line) going from the tips of each line that is the square root of $L^2 + L^2$, or $1.414L$. The y line must, then, be approximately $1.414L$ meters long as well, since it projects the same angular size and is approximately the same distance away. Consider now what happens when the observer is about to pass the corner. Since the x line is to one side of the vertical meridian, its angular size has fallen to $0°$. The angular size of the z arm is at its maximum, however. The bottom of the y arm rests on the horizontal meridian, and it will therefore not get smaller in the last moments before passing it, but, instead, will increase to its maximum. Since the z line is of length L and the y arm length $1.414L$, and since each is about the same distance from the observer, the angular size of the y arm will be about 1.41 times as large as the angular size of the z arm. This is how the corner will project when the observer is just passing it, but the more general conclusion is, then, that the total angular size of the bottom of the 'T' grows less quickly than does the angular size of the y line. Latency correction therefore predicts that observers will perceive the vertical line to have greater angular size, as is the case.

A new illusion

In the explanation of the upside-down 'T' illusion, we learned that, when relatively nearby, x line segments lying on the horizontal meridian and on one side of the vertical meridian—like the one in the upside-down 'T' illusion—increase their angular size more slowly than do z line segments lying in the same part of the visual field. We can use this observation to build a novel illusion. Figure 2.23 shows two identical horizontal lines lying on the horizontal meridian, one on each side of the vertical meridian. The one on the left has cues suggesting it is due to an x line, and the one on the right has cues that it is due to a z line. Although they are at equal distances from the vertical meridian, the z line appears to have greater angular size, as latency correction predicts. (The bold vertical lines are also identical in angular size to the horizontal lines.)

2.3.4   Psychophysical confirmation

It is possible to summarize the explanation for all those illusions that did not rely on misperception of the angular size of z lines; i.e., all the illusions except for the upside-down 'T' and the new illusion just discussed above. Figure 2.24 shows how much a point in an observer's visual field moves away from the horizontal meridian in the next moment. The figure for movement away from the vertical meridian is identical, but rotated 90°.

I will describe how this one plot explains most of the illusions discussed thus far.

- *xy-z projected angles, including corner, Poggendorff, Hering, and ambiguous angle perception*: The two white dots in Figure 2.24 can be thought of as the endpoints of an x line extending between them. The figure indicates that the dot nearer to the vertical meridian will move up more than the other dot in the next moment. Consider the angle this projected x line segment makes with a projected z line (which goes radially outward from the focus of expansion in the figure). The projected x line will, after the move, make an angle with the projected z line that is more near 90° than it originally was. This

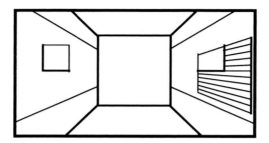

Figure 2.23: Two predicted illusions. First, the left horizontal line appears to have smaller angular size than the right one, but they are identical. The reason is that the right one is probably due to a *z* line (being part of the flag on the wall), whose angular size will increase in the next moment more than that of the *x* line on the left. Second, and for the same reason, the horizontal line on the right appears to have greater angular size than the adjacent vertical line, but the two lines on the left appear roughly identical (and, the predicted perception on the left is that the vertical line should be a little larger than the horizontal line).

captures the regression to $90°$ phenomenon we discussed earlier. The explanation for y-z projected angle illusions is similar, but relies on the plot that is rotated by $90°$.

- *x-y projected angles, and the Orbison illusion*: We just learned that the projected x line extending between the two points in Figure 2.24 will "lift up" on its left side (i.e., its left side will acquire greater angular distance from the horizontal meridian than the right side). Consider the projected angle the x line makes in the next moment with a y line. The projected angle begins at $90°$, but gets pushed away from $90°$ in the next moment. Projected y lines also change, and change so as to accentuate this projected angle change.

- *x and y angular distances, including the double Judd, Müller-Lyer, Ponzo and Hering illusions*: When we just consider the two dots in Figure 2.24, we have the raw material of the double Judd illusion, and the plot states that the one nearer to the vertical meridian moves away from the horizontal meridian more in the next moment, which agrees with perception. Not only does the dot on the left appear higher, the angular distance between it and the horizontal meridian appears greater, which is essentially the Müller-Lyer, Hering and Ponzo illusion.

Since Figure 2.24 encapsulates most of the predictions my model of latency correction has made, it would be nice if we could test observers to see if their perceptions of the angular distance between each dot and the horizontal merid-

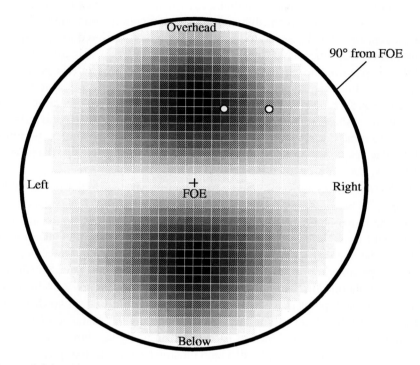

Figure 2.24: *Change in angular distance from the horizontal meridian as a function of position within the visual field. Rim of circle is* **90°** *from the focus of expansion. Plot uses a linear gray scale, with white representing zero degrees angular distance change, and black representing approximately two degrees. The two dots are props referred to in the text. By rotating the plot by* **90°** *, one obtains the plot for the change in angular distance from the vertical meridian as a function of position within the visual field.*

ian fits this composite prediction. This is what an undergraduate student and myself did, with intriguing and encouraging results (Changizi and Widders, 2002).

Using a computer, two dots were placed on a radial display of black lines, the whole display was 20 cm in diameter, and subjects typically sat about one to two feet from the screen (this was not experimentally controlled). The dots were kept horizontally separated by about 2 cm, and were red to be easily distinguished from the radial lines. They were moved as a pair to each of 300 different positions in an 18 by 18 grid in the radial display (six positions at the extremity of each quadrant lie outside the radial display and were not measured). For each position, the subject was asked to move the outer dot (the one farther from the vertical meridian) up or down until its perceived angular distance from the horizontal meridian was the same as that for the less peripheral dot. The resolution was roughly a third of a millimeter. (See Changizi and Widders, 2002, for detailed methods.)

The data from subjects is not of a form directly predicted by the plot in Figure 2.24 because the subjects were judging the *difference* in angular distance from the horizontal meridian, whereas the plot measures how much any given point will move upward in the next moment. Instead, the predictive plot we want is the one that records, for each point in the visual field, how much more the less peripheral dot will move away from the horizontal meridian than the more peripheral dot. This plot can be obtained from Figure 2.24 by simply taking, for each point in the visual field, the next-moment angular distance of the less peripheral dot minus the next-moment angular distance of the more peripheral dot. This is shown in Figure 2.26; this figure shows the predicted strength of the vertical angular distance illusion as a function of position in the visual field. This one plot encapsulates the predicted illusion magnitude for nearly all the illusions discussed in this section. If the visual system follows a latency correction strategy, then we expect it to approximate the predicted plot, at least to first order; this plot is the fingerprint of latency correction. The predicted plot assumes that all points are equidistant from the observer, whereas in reality it may be that points at different positions in the visual field have different probable distances from the observer. However, the basic "bauplan" of the predicted plot is expected to be followed, even if not the particulars.

Figure 2.27 shows averages from the above described experiment for myself, David Widders, and one naive undergraduate (NG), along with the average of our averages. In each case, the psychophysical results have the latency correction fingerprint, which provides further strong confirming evidence for

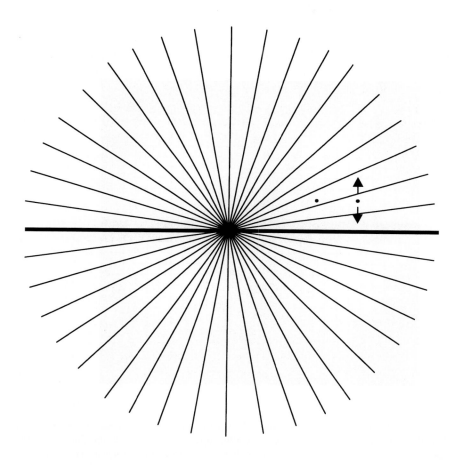

Figure 2.25: *An example of the stimulus used in the psychophysical test of the latency correction hypothesis. The arrows indicate that the more peripheral dot could be moved up and down by the subject.*

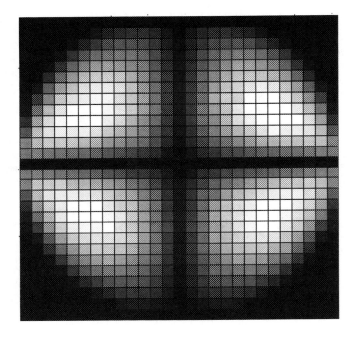

Figure 2.26: *This plot shows the predicted misperception for the vertical angular distances from the horizontal meridian for two horizontally displaced dots, as a function of position of the points in the visual field. This plot is the predicted fingerprint of a latency correction strategy for vision. The plot is generated by assuming that all dots are at the same distance from the observer. Whiter here means greater predicted illusion.*

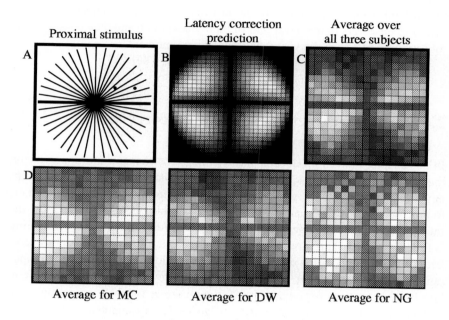

Proximal stimulus   **Latency correction prediction**   **Average over all three subjects**

Average for MC   **Average for DW**   **Average for NG**

Figure 2.27: *(A) The general kind of stimulus used in the experiment is repeated here for convenience, as is (B) the predicted plot for the latency correction hypothesis. (C) The average of the average results over the three subjects. (D) Three experimental plots for three subjects individually, the first two (non-naive) averaged over four experiments, and the last (naive) averaged over two runs. The range of misperceptions for the three subjects are approximately, in centimeters: MC [−0.07, 0.13], DW [−0.04, 0.14], NG [−0.03, 0.11], and average of the averages [−0.04, 0.11]. Experiments for angular-distance-from-vertical-meridian perceptions were similar. Note that the predicted plot ranges over the entire visual field, whereas the experimental results are for some subset of it. Whiter means greater illusion. For each plot, zero misperception is represented by whatever gray level lies along the horizontal and vertical meridians (where subjects experience no illusion). [I thank Nirupa Goel for being the naive subject here.]*

latency correction. Not only do observers experience illusion gradients in the areas predicted, but the illusion magnitude tends to be more clustered, in any given quadrant, nearer to the horizontal meridian, which is also a property found in the predicted plot. Our experimental results are qualitatively identical for the perception of angular distance from the vertical meridian. We have also noticed a tendency for greater illusion magnitudes in the bottom half of the visual field, which may be due to the fact that, on average, objects tend to be nearer to observers in the lower half of their visual field, and they consequently move more quickly in the visual field in the next moment.

2.4   Further directions for latency correction

In this chapter I have introduced a basic strategy for vision, a strategy so useful that we might expect any kind of computational system to utilize it. That strategy is latency correction: rather than carrying out computations whose intent is to provide a solution relevant to the problem that initiated the computation, the intent is, instead, to provide a solution that will be relevant when the computing is finally finished. This strategy is useful because it allows an optimal tradeoff between fast computation and powerful computation. More powerful computations can be carried out if the system has more time, and the system can buy itself more time for computing if it can correct for this computing time, or latency. I have concentrated only on vision in this chapter, and provided evidence that the visual system utilizes a latency correction strategy. The evidence thus far has concerned the perception of classical geometrical illusions; we have seen that observers perceive the projected angles and angular sizes of scenes not as they actually project, but as they probably will project in the next moment, i.e., at the time the percept is actually elicited.

The explanatory value of the latency correction hypothesis is, I believe, much greater than just explaining the classical geometrical illusions or cases such as the flash-lag effect (which I will mention below). I believe that a considerable fraction of all visual illusions may be due to latency correction; in particular, I believe that all inconsistent perceptions are due to latency correction. There is much work ahead of us in understanding the consequences of latency correction, and before leaving this chapter I will discuss preliminary ideas and research in progress.

Motion-induced illusions

Evidence for latency correction in the literature has, except for my own work, concentrated on motion-induced illusions. (The stimuli in the work I have described here are all static.) The most famous effect is called the flash-lag effect, where an unchanging object is flashed in line with a continuously moving object such that, at the time of the flash both objects are identical (MacKay, 1958; Nijhawan, 1994, 1997, 2001; Schlag et al., 2000; Sheth et al., 2000). [There has also been a fireworks-like debate about this interpretation (Baldo and Klein, 1995; Khurana and Nijhawan, 1995; Whitney and Murakami, 1998; Purushothaman et al., 1998; Lappe and Krekelberg, 1998; Krekelberg and Lappe, 1999; Whitney et al., 2000; Eagleman and Sejnowski, 2000; Brenner and Smeets, 2000; Khurana et al., 2000).] The continuously moving object appears to be "past" the flashed object, even though they are identical. In the first flash-lag effect, the continuously moving object is a rotating bar, and the flashed object is a light that flashes in line with the moving bar; observers perceive the flashed light to lag behind the moving bar. [Some of this extrapolation may even be carried out by retinal ganglion cells (Berry et al., 1999).] Sheth et al. (2000) showed that the effect holds for other modalities besides perceived position. The continuously changing stimulus may be in the same position, but changing in luminance from dim to bright, and the flashed stimulus has, at the time of its appearance, the same luminance as the other stimulus; in this case observers perceive the changing stimulus to be brighter than the flashed one. It also works for hue and other modalities. Other evidence for latency correction can be found in Thorson et al. (1969) who have shown that when two very nearby points are consecutively flashed, motion is perceived to extend beyond the second flashed point. Also, Anstis (1989) and DeValois and DeValois (1991) have shown that stationary, boundaryless figures with internal texture moving in a direction induce a perceived figure that is substantially displaced in the same direction (see also Nishida and Johnston, 1999; and Whitney and Cavanagh, 2000).

One difficulty with these illusions is that they cannot be shown in a book; one needs a computer display or real live moving objects to see them. There are, however, two illusions from the literature that are motion-induced and *are* able to be displayed here. Furthermore, although neither illusion was introduced by the authors for the purposes of latency correction, there is a relatively straightforward latency correction explanation for both.

The first is due to Foster and Altschuler (2001) and is called the Bulging

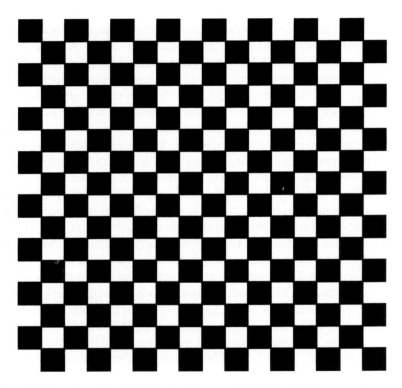

Figure 2.28: *The Bulging Grid illusion (Foster and Altschuler, 2001) occurs when you move your head quickly toward the checkerboard. In addition to the perception of a bulge, the projected angles and angular sizes change just as in the Orbison illusion. The Orbison illusion does not require observer motion because the radial lines provide the cue as to the focus of expansion.*

Grid (Figure 2.28). Before I tell you what the illusory aspects of it are, note that it is essentially a bunch of squares, or projections of x and y lines. The Orbison illusion (see Figure 2.1), recall, was when the cues suggest that the probable source consists of x and y lines, and the radial lines in the Orbison acted as cues to the observer's direction of motion—the vanishing point was the probable focus of expansion. The Bulging Grid figure is, in a sense, then, like the Orbison illusion, except that it does not possess any cues as to the probable focus of expansion. Well, there is one obvious way to create a probable focus of expansion: move your head toward the image. There is arguably no better cue to a focus of expansion than optical flow emanating radially from some point. We should predict that if an observer moves his head toward it, he should experience an Orbison-like illusion. Indeed this is what occurs. Try it. Forget for the moment about the bulge, and focus just on the perception of the projected angles and the angular sizes. The angles change away from 90° as in the Orbison illusion, and in the same ways. [What about the bulge? I have no good answer to this as of yet, although I can make two observations. First, if one overlays the bulging grid (or any grid of squares) with a radial display, one also perceives a bulge (albeit smaller). This suggests that the radial lines are indeed serving to cue direction of motion. Second, a bulge *is* consistent with the misperceived projected angles, although it is inconsistent with the actual projected angles. That is, the actual projected angles are best explained by a flat grid in front of the observer, and so I would expect that they would be perceived as such. Instead, the perception of a bulge suggests that it is as if the visual system determines the perceived projected angles according to latency correction, and *then* uses these angles to compute the probable depths. At any rate, more thinking on this is needed.]

The next motion-induced illusion worth bringing up is one by Pinna and Brelstaff (2000), and is displayed in Figure 2.29. In my opinion is it is *the* most striking illusion ever; plus it requires no complex computer display or stereoscopic glasses, etc. It was invented by them without latency correction in mind, but it has a relatively straightforward latency correction explanation. When you move toward the point at the center that point becomes the probable direction of motion. What kind of scene probably generated this stimulus? One intuitive conjecture is that the observer is walking down a circular tube or tunnel. Consider just the inner ring. The oblique lines pointing roughly toward the center do not actually point at the center. If they did, then the lines would probably be the projections of z lines. Instead, they point inward and a little counter-clockwise. What kind of line would project this way? Answer:

A line that was painted on the inside wall of the tube, but was spiraling around it and going down the tube simultaneously. That is, if there were lines on the inside wall of the tube that wrapped counter-clockwise around the tube once every, say, ten meters, the nearest segments of those lines would project just like the nearly-radial lines of the inner ring. We must also suppose that, for whatever reason, the observer is only able to see the nearby parts of these spiraling lines. Now that we have an idea of what kind of scene might cause such a stimulus as the inner ring, we can ask how that scene will project as the observer moves forward. In the next moment, the spiraling lines nearest the observer will no longer be at the same positions, but will, instead, have rotated or spiraled counter-clockwise a little. That is, as the observer moves forward, the spirals will move counter-clockwise around him. And this is exactly the illusion we experience here. The illusion may, then, be due to the visual system attempting to engage in latency correction; but it is inappropriate here since there is no tube. The explanation is similar for the outer ring, and is similar for when you move your head away from the stimulus rather than toward it. Much work is needed to examine in detail such images and whether latency correction really is the explanation. At this time, it is just highly suggestive and encouraging.

Brightness and color

There is considerable evidence since Helmholtz that, when cues make it probable that a surface has a certain reflectance, brightness and color judgements—to be distinguished from lightness and surface color judgements—are influenced away from the actual luminance and chromaticity in the proximal stimulus and partially towards the probable reflectance, or partially towards the "typical" or "generic" luminance and chromaticity emitted by the probable surface (Arend and Reeves, 1986; Arend and Goldstein, 1990; Arend et al., 1991; Arend and Spehar, 1993; Adelson, 1993; Kingdom et al., 1997). There has been little success, however, in explaining this "regression toward the 'real' object" (Thouless, 1931a, 1931b) phenomenon. The reason it has been difficult to explain these brightness and color illusions is that they are cases of inconsistent perception (see Subsection 2.1.1). For example, in brightness contrast (Figure 2.30) two identical gray patches are surrounded by, respectively, dark and light surrounds. The patch in the dark surround is perceived to be lighter *and* brighter than the other patch.

That the patch is perceived to be lighter—i.e., perceived to have greater

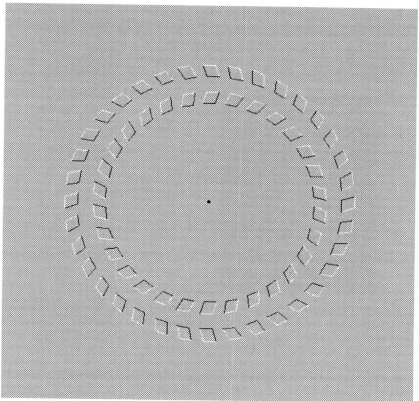

Figure 2.29: In this striking illusion from Pinna and Brelstaff (2000), you should move your head either toward or away from the figure while focusing on the point at the center.

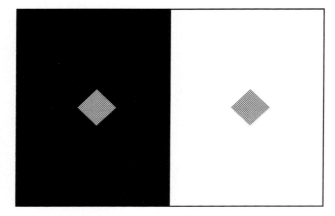

Figure 2.30: Demonstration of lightness and brightness contrast. The gray patches are identical, but the one in the dark surround appears to have greater reflectance than the one in light surround. This is called lightness contrast, and is easily accommodated by an inference approach (namely, a "subtracting the illuminant" account going back to Helmholtz). The display also demonstrates brightness contrast, where the patch in dark surround appears to send more light to the eye (i.e., appears to have greater luminance) than the patch in light surround.

reflectance, or greater ability to reflect more light—is easily explained by an inference or Bayesian framework. The explanation is that the dark surround suggests that its patch is under low illumination, and the light surround suggests that its patch is under high illumination. Since the patches have the same luminance—i.e., they send the same amount of light to the eye—the patch in the dark surround must be a more reflective object. This is sometimes referred to as the "subtracting the illuminant" explanation. The same idea applies for perception of surface color, which refers to the perception of the reflectance of the object, where now we care about the full spectral reflectance properties, not just the amount of light reflected.

However, the explanation for why the patch in the dark surround is perceived to be *brighter*—i.e., perceived to have greater luminance—is *not* explainable by the traditional inference or Bayesian account. The reason is that the luminances of the two patches are probably identical; the retina "knows" this. Yet the brain generates a percept of the patch in dark surround having greater luminance than the patch in light surround. The brain therefore generates a percept that is inconsistent with the proximal stimulus. As we discussed earlier in this chapter, inconsistent perception *can*, in principle, be accommodated within a latency correction approach. This observation led me to look for latency correction explanations for brightness illusions. Similar arguments lead us to the same conclusion for the perception of color—perception of the chromatic quality of the light sent to the eye, or perception of the chromaticity—as opposed to the perception of surface color.

Latency correction is, indeed, suggestive of an explanation for brightness (and color) contrast illusions. As an observer walks through the world, the luminance and chromaticity received from any given surface can change radically as a function of the surface's angle with respect to the observer. It is reasonable to assume that the following is true:

> If a surface currently has a luminance/chromaticity that is atypical for it, then
> the luminance/chromaticity is probably going to become more typical in the next   ‑
> moment, not less.

For example, if a surface with high (low) reflectance currently has low (high) luminance, then it is more probably going to have higher (lower) luminance in the next moment. Similarly, if the chromaticity from a red surface is currently yellowish (because of a yellow illuminant), the chromaticity is more probably going to become less yellowish and more reddish in the next moment. Latency correction accordingly predicts that if cues suggest that the actual lumi-

Figure 2.31: The Kanizsa square. An illusory square is perceived, along with a illusory luminance contours at its top and bottom.

nance/chromaticity is atypical for the probable source, an observer will perceive a brightness/color representative of a luminance/chromaticity more toward the typical luminance/chromaticity of the probable source, as that is more probably what will be present at the time the percept is elicited. This prediction is qualitatively consistent with actual psychophysical trends in the perception of brightness and color, as cited above.

Even the illusory contour phenomenon is a case of inconsistent perception, as one perceives luminance contours despite the proximal stimulus being inconsistent with luminance contours. Illusory contours are perceived along the edges of objects that are probably there, like in the Kanizsa square (Figure 2.31). This may be expected within latency correction, however, since although there is no luminance discontinuity at the time of the stimulus, since there probably *is* a surface discontinuity, it is probable that in the next moment there *will* be a luminance discontinuity. That is, a surface discontinuity without a luminance discontinuity is a rare situation, and it is much more likely to have a luminance discontinuity by the time the percept occurs, so the visual system includes one.

At best, though, this is all just encouraging; it does not provide any strong evidence that latency correction explains brightness and color illusions. These illusions are merely roughly what one might, *prima facie*, expect if latency correction were true. What I need are more detailed models akin to what I have put forth for geometrical stimuli. Such theoretical work is in progress.

I leave this subsection with a very exciting illusion that strongly suggests that brightness illusions *will* fall to latency correction explanations. It is a motion-induced illusion like the Bulging Grid and the spiral discussed earlier—each seemingly explainable by latency correction. This illusion, however, is a brightness illusion, also with a latency correction explanation. The illusion is due to David Widders, an undergraduate student of mine, and is shown in Figure 2.32. Move your head towards the center of the figure, and you should perceive the middle to become brighter and the dark edges to become brighter. The brightness appears to flow radially outward. If the probable scene causing such a stimulus is a tunnel with an illumination gradient along it (due to, say, some light at the end), then latency correction would predict such an outflowing brightness illusion, since that is how the luminances would be in the next moment. If you move your head backward the effect is the opposite. Even more interestingly, the illusion works for hue and other gradients as well, and is especially stunning on a computer or a glossy printout. We are further examining motion-based illusions of this kind within a latency correction framework.

Representational momentum

There exists a literature, possibly of great relevance to latency correction, called "representational momentum" (see, e.g., Freyd, 1983a, 1983b; Freyd and Finke, 1984, 1985; Hubbard and Ruppel, 1999, and references therein). The phenomenon is as follows. Freyd (1983b) showed subjects images taken from a scene possessing motion. The images possessed ample cues as to the motion in the scene, and two images were chosen from successive moments in the scene, so that one image, A, obviously just preceded the next, B. Subjects were presented two images in succession, and asked to say whether the two were the same or different. When the images were presented in the order they actually

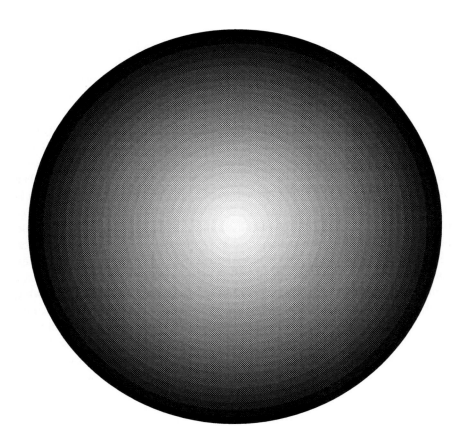

Figure 2.32: Move your head toward the center of this figure, and you should perceive the brightness to "flow" outward towards the edges. It works best on either a computer screen or glossy paper. If the stimulus is probably due to a tunnel with an illumination gradient, then as an observer moves forward the brightness will, indeed, "flow" outward. Thus, the illusion is consistent with latency correction. The illusion was invented by David Widders.

occurred—*A* followed by *B*—subjects took more time to respond that they were different than when the images were presented in the opposite order. It is as if subjects, upon seeing image *A*, forward it a little, so that by the time *B* is displayed, their memory of *A* is already depicted in *B*, and they have difficulty noticing any difference. I introduce the connection here only to note that representational momentum may be another long-known effect that, like the classical illusions, may be due to latency correction.

Other cues to the focus of expansion

Thus far, the cues to the location of the focus of expansion have been projected *z* lines, which, as we have discussed, may be due either to real live *z* line contours, or may be due to optic flow itself (that is, the radial lines may mimic optic blur). Optic flow is not actually necessary, however, to perceive forward movement (Schrater et al., 2001); all that is necessary is that the overall size of image features increases through time. Accordingly, we might expect that we can create a static image such that there is a size gradient, with larger image features (i.e., larger spatial frequency) near the periphery and smaller image features near the center. Such an image would suggest that the peripheral parts of the image are nearby, that the parts nearer to the center are farther away, and that the "flow" is that the smaller image features near the center are becoming the bigger images features on the sides. The probable focus of expansion therefore is the center, and we expect to find the same kinds of illusions as in the radial display stimuli from earlier.

In this light, consider moving down a tubular cave with constant-sized "rocks" along the inside wall. At any given radial angular distance from the focus of expansion, all the rocks at that angular distance will project at the same size. Thus, arguing backward, *if two features in an image have the same size (i.e., the objects causing them are projecting the same size), then they are probably due to rocks at the same angular distance from the focus of expansion.* Consider Figure 2.33 (A) which shows a bunch of similar-sized projected shapes, and the probable location of the focus of expansion is shown at the center. Since this is the probable focus of expansion, the vertical dotted line on the left should increase in angular size more in the next moment than the equal angular size vertical dotted line on the right. And this is what observers perceive. Actually, this is now just like a standard class of variants of the Müller-Lyer illusion, one which is shown in Figure 2.33 (B). These variants may, then, be explained by the fact that the probable focus of expansion for them is a point

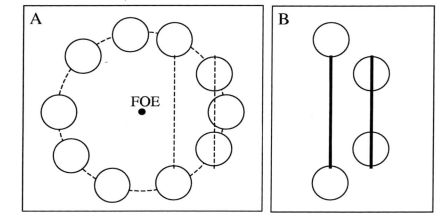

Figure 2.33: *(A) When there is a set of projected shapes all of the same size and projecting roughly equal angular distances from a single point in the visual field, this point is probably the focus of expansion. (This is because the projected shapes are probably due to similar objects at similar positions relative to the observer.) Since the observer is probably, then, moving toward this point, the angular distance of the dotted line on the left is expected, if latency correction is true, to be perceived to have greater angular size than the one on the right. This is consistent with what observers, in fact, perceive. (B) Furthermore, this explains a class of variants of the Müller-Lyer illusion, where there is some object—in this case a circle—outside and inside the vertical lines. The vertical line with the object outside is always perceived to have greater angular size, no matter the object's shape. This is because the probable focus of expansion is the point that is equi-angular-distant from the four objects, and this point must be nearer to the left vertical line, as depicted in (A).*

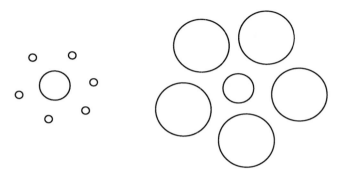

Figure 2.34: *The Ebbinghaus illusion: the middle circles on the left and right are identical, but the one on the left appears to have greater angular size. Since the circle on the left is surrounded by smaller circles, it is probably nearer to the focus of expansion, and should increase in angular size in the next moment (supposing it is not too different in distance from the observer than the other circle).*

nearer to the line with the objects on the outside of the line.

The same observations concerning the tubular cave above can allow us to make another qualitative conclusion. The objects nearer to the focus of expansion project smaller, being farther away. Thus, all things equal, *the part of the visual field with smaller image features is more probably nearer the focus of expansion.* Consider now Figure 2.34 showing a classical illusion called the Ebbinghaus. The circle in the middle on the left and in the middle on the right are identical, and yet the one on the left appears to have greater angular size. This is readily explained by the conclusion just made: since the circle on the left is surrounded by smaller projections, it is probably nearer to the focus of expansion than is the circle on the right. Supposing that the probable distances from the observer for each of the two middle circles is roughly the same, the one on the left will increase in angular size more quickly in the next moment. Thus, latency correction predicts that the left middle circle will be perceived to be larger, which is consistent with actual perception.

These ideas are preliminary at this point, but very promising. They suggest that my latency correction model may be easily extendable beyond x, y and z lines, so that it may be applied to stimuli with many other kinds of projections.

Further discussion of these ideas will have to await further research on my part.

Chapter 3

Induction and Innateness

One of the deepest problems in philosophy concerns how we learn about the world, and whether there are right or wrong ways to go about it. In this chapter I introduce this problem—the "problem of induction"—and describe its relevance to understanding learning in intelligent agents, and brains in particular. One consequence of the problem of induction is that there can be no such thing as a universal learning machine; it is not even possible that brains could enter the world as blank slates equipped with universal learning algorithms. The goal of the chapter is to provide a kind of solution to the problem of induction, and also to put forth something I call a theory of innateness. The latter would be a mathematical framework in which we are able to make sense of the kinds of structures that must be innately generated in a brain in order for that brain to have its own innate way of learning in the world. I present a theory called *Paradigm Theory* (Changizi and Barber, 1998) that purports to do these things.

What is induction?

"John is a man. All men are mortal. Therefore, John is mortal." This argument from two premises to the conclusion is a *deductive* argument. The conclusion logically follows from the premises; equivalently, it is logically impossible for the conclusion not to be true *if* the premises are true. Mathematics is the primary domain of deductive argument, but our everyday lives and scientific lives are filled mostly with another kind of argument.

Not all arguments are deductive, and 'inductive' is the adjective labelling any non-deductive argument. Induction is the kind of argument in which we typically engage. "John is a man. Most men die before their 100th birthday.

151

Therefore John will die before his 100th birthday." The conclusion of *this* argument can, in principle, be false while the premises are true; the premises do not logically entail the conclusion that John will die before his 100th birthday. It nevertheless is a pretty good argument.

It is through inductive arguments that we learn about our world. Any time a claim about infinitely many things is made on the evidence of only finitely many things, this is induction; e.g., when you draw a best-fit line through data points, your line consists of infinitely many points, and thus infinitely many claims. Generalizations are kinds of induction. Even more generally, any time a claim is made about more than what is given in the evidence itself, one is engaging in induction. It is with induction that courtrooms and juries grapple. When simpler hypotheses are favored, or when hypotheses that postulate unnecessary entities are *dis*favored (Occam's Razor), this is induction. When medical doctors diagnose, they are doing induction. Most learning consists of induction: seeing a few examples of some rule and eventually catching on. Children engage in induction when they learn the particular grammatical rules of their language, or when they learn to believe that objects going out of sight do not go out of existence. When rats or pigeons learn, they are acting inductively. On the basis of retinal information, the visual system generates a percept of its guess about what is in the world in front of the observer, despite the fact that there are always infinitely many ways the world could be that would lead to the same retinal information—the visual system thus engages in induction.

If ten bass are pulled from a lake which is known to contain at most two kinds of fish—bass and carp—it is induction when one thinks the next one pulled will be a bass, or that the probability that the next will be a bass is more than $1/2$. Probabilistic conclusions are still inductive conclusions when the premises do not logically entail them, and there is nothing about having fished ten or one million bass that logically entails that a bass is more probable on the next fishing, much less some specific probability that the next will be a bass. It is entirely possible, for example, that the probability of a bass is now *decreased*—"it is about time for a carp."

What the problem is

Although we carry out induction all the time, and although all our knowledge of the world depends crucially on it, there are severe problems in our understanding of it. What we would *like* to have is a theory that can do the following. The theory would take as input (i) a set of hypotheses and (ii) all the evidence known concerning those hypotheses. The theory would then assign each hypothesis a probability value quantifying the degree of confidence one logically *ought* to have in the hypothesis, given all the evidence. This theory would interpret probabilities as *logical probabilities* (Carnap, 1950), and might be called a theory of logical induction, or a theory of logical probability. (Logical probability can be distinguished from other interpretations of probability. For example, the *subjective* interpretation interprets the probability as how confident a person actually is in the hypothesis, as opposed to how confident the person ought to be. In the *frequency* interpretation, a probability is interpreted roughly as the relative frequency at which the hypothesis has been realized in the past.)

Such a theory would tell us the proper method in which to proceed with our inductions, i.e., it would tell us the proper "inductive method." [An *inductive method* is a way by which evidence is utilized to determine *a posteriori* beliefs in the hypotheses. Intuitively, an inductive method is a box with evidence and hypotheses as input, and *a posteriori* beliefs in the hypotheses as output.] When we fish ten bass from the lake, we could use the theory to tell us exactly how confident we should be in the next fish being a bass. The theory could be used to tell us whether and how much we should be more confident in simpler hypotheses. And when presented with data points, the theory would tell us which curve ought to be interpolated through the data.

Notice that the kind of theory we would like to have is a theory about what we *ought* to do in certain circumstances, namely inductive circumstances. It is a *prescriptive* theory we are looking for. In this way it is actually a lot like theories in ethics, which attempt to justify why one ought or ought not do some act.

Now here is the problem: *No one has yet been able to develop a successful such theory!* Given a set of hypotheses and all the known evidence, it sure *seems* as if there is a single right way to inductively proceed. For example, if all your data lie perfectly along a line—and that is *all* the evidence you have to go on—it seems intuitively obvious that you should draw a line through the data, rather than, say, some curvy polynomial passing through each point. And after seeing a million bass in the lake—and assuming these observations are all you have to help you—it has just *got* to be right to start betting on bass, not carp.

Believe it or not, however, we are still not able to defend, or justify, why one really ought to inductively behave in those fashions, as rational as they seem. Instead, there are multiple inductive methods that seem to be just as

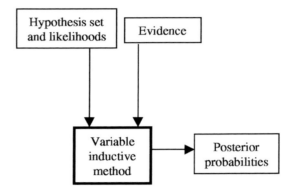

Figure 3.1: *The purpose of an inductive method is to take a set of hypotheses and the evidence as input, and output the degree to which we should believe in each hypothesis in light of the evidence, i.e., output the posterior probability distribution over the set of hypotheses. Inductive methods may, in principle, be any function from the hypotheses and evidence to a posterior probability distribution, but some inductive methods seem better than others. Which one ought we use? That is the riddle of induction. An ideal answer would be a theory of logical probability that tells us, once and for all, which inductive method to use. But there is no such ideal theory.*

good as one another, in terms of justification. Figure 3.1 depicts the problem. The hypothesis set and evidence need to be input into some inductive method in order to obtain beliefs in light of the evidence. But the inductive method is, to this day, left variable. Different people can pick different inductive methods without violating any mathematical laws, and so come to believe different things even though they have the same evidence before them.

But do we not use inductive methods in science, and do we not have justifications for them? Surely we are not picking inductive methods willy nilly! In order to defend inductive methods as we actually use them today, we make *extra* assumptions, assumptions going beyond the data at hand. For example, we sometimes simply assume that lines are more *a priori* probable than parabolas (i.e., more probable *before* any evidence exists), and this helps us conclude that a line through the data should be given greater confidence than the other curves. And for fishing at the lake, we sometimes make an *a priori* assumption that, if we pull n fish from the lake, the probability of getting n bass and no carp is the same as the probability of getting $n - 1$ bass and one carp, which is the same as the probability of getting $n - 2$ bass and two carp, and so on; this assumption makes it possible (as we will see later) to begin to favor bass as more and more bass, and no carp, are pulled from the lake. Making different *a priori* assumptions would, in each case, lead to different inductive methods, i.e., lead to different ways of assigning inductive confidence values, or logical probabilities, to the hypotheses.

But what justifies our making these *a priori* assumptions? That's the problem. If we had a theory of logical probability—the sought-after kind of theory I mentioned earlier—we would not have to make any such undefended assumption. We would know how we logically ought to proceed in learning about our world. By making these *a priori* assumptions, we are just *a priori* choosing an inductive method; we are not bypassing the problem of justifying the inductive method.

I said earlier that the problem is that "no one has yet been able to develop a successful such theory." This radically understates the dilemma. It suggests that there could really *be* a theory of logical probability, and that we have just not found it yet. It is distressing, but true, that there simply *cannot* be a theory of logical probability! At least, not a theory that, given only the evidence and the hypotheses as input, outputs the degrees of confidence one really "should" have. The reason is that to defend any one way of inductively proceeding requires adding constraints of some kind—perhaps in the form of extra assumptions—constraints that lead to a distribution of logical probabilities on the hypothesis set even *before* any evidence is brought to bear. That is, to get induction going, one needs something equivalent to *a priori* assumptions about the logical probabilities of the hypotheses. But how can these hypotheses have degrees of confidence that they, *a priori*, simply *must* have. Any theory of logical probability aiming to once-and-for-all answer how to inductively proceed must essentially make an *a priori* assumption about the hypotheses, and this is just what we were hoping to avoid with our theory of logical probability. That is, the goal of a theory of logical induction is to explain why we are justified in our inductive beliefs, and it does us no good to simply assume inductive beliefs in order to explain other inductive beliefs; inductive beliefs are what we are trying to explain.

Bayesian formulation of the problem

We have mentioned probabilities, but it is important to understand a simple, few-centuries-old theorem of Bayes. Using Bayes' Theorem it will be possible to understand inductive methods more deeply. As set up thus far, and as depicted in Figure 3.1, the inductive method is left entirely variable. Any way of using evidence to come to beliefs about hypotheses can fill the 'inductive method' role. Different inductive methods may utilize evidence in distinct ways to make their conclusions. Bayes' Theorem allows us to lay down a fixed principle dictating how evidence should modify our beliefs in hypotheses. The variability in inductive methods is constrained; inductive methods cannot now differ in regards to how evidence supports hypotheses. As we will see, the Bayesian framework does not dictate a single unique inductive method, however; the variability is pushed back to prior probabilities, or the degrees of confidence in the hypotheses before having seen the evidence. Let me first explain Bayes' Theorem and the framework.

First, the Bayesian framework is a probabilistic framework, where degrees of confidence in hypotheses are probabilities and must conform to the axioms of Probability Theory. The axioms of probability are these: (i) Each probability is in the interval $[0,1]$. (ii) The sum of all the probabilities of the hypotheses in the hypothesis set must add to 1. (iii) The probability of no hypothesis being true is 0. And (iv), the probability of two possibilities A and B is equal to the sum of their individual probabilities minus the probability of their co-occurrence. We will be assuming our hypotheses in our hypothesis sets to be mutually exclusive, and so no two hypotheses can possibly co-occur, making axiom (iv) largely moot, or trivially satisfied for us.

Suppose the probability of event A is $P(A)$, and that for event B is $P(B)$. What is the probability of *both* A and B. We must first consider the probability that A occurs, $P(A)$. Then we can ask, given that A occurs, what is the probability of B; this value is written as $P(B|A)$. The probability of A and B occurring is the product of these two values. That is, we can conclude that

$$P(A\&B) = P(A) \cdot P(B|A).$$

But note that we could just as well have started with the probability that B occurs, and then asked, given that B occurs, what is the probability of A. We would then have concluded that

$$P(A\&B) = P(B) \cdot P(A|B).$$

The right hand sides of these two equations differ, but the left hand sides are the same, so we may set them equal to one another, resulting in

$$P(A) \cdot P(B|A) = P(B) \cdot P(A|B).$$

This is essentially Bayes' theorem, although it is usually manipulated a little.

To see how it is usually stated, let us change from A and B to h and e, where h denotes some hypothesis, and e denotes the evidence. The formula now becomes

$$P(h) \cdot P(e|h) = P(e) \cdot P(h|e).$$

What do these values mean?

- $P(h)$ stands for the probability of hypothesis h *before* any evidence exists. It is called the *prior probability* of h. Each hypothesis might have its own distinct prior probability.

- $P(h|e)$ is the probability of hypothesis h *after* the evidence has been considered; it is the hypothesis' probability given the evidence. Accordingly, it is called the *posterior probability* of h. Each hypothesis might have its own distinct posterior probability.

- $P(e|h)$ is the probability of getting the evidence *if* hypothesis h were true. It is called the *likelihood*. Each hypothesis might have its own distinct likelihood, and its likelihood is usually determinable from the hypothesis.

- $P(e)$ is the probability of getting that evidence. This value does not depend on the hypothesis at issue. It may be computed from other things above as follows:

$$P(e) = \sum_h [P(h)P(e|h)].$$

Ultimately, the value that we care about most of all is $P(h|e)$, the posterior probability. That is, we want to know how much confidence we should have in some hypothesis given the evidence. So, let us solve for this term, and we get a formula that is the traditional way of expressing Bayes' Theorem.

$$P(h|e) = \frac{P(h) \cdot P(e|h)}{P(e)}.$$

Since $P(e)$ does not depend on which hypothesis is at issue, it is useful to simply forget about it, and write Bayes' Theorem as

$$P(h|e) \sim P(h) \cdot P(e|h).$$

That is, the posterior probability is proportional to the prior probability times the likelihood. This makes intuitive sense since how much confidence you have in a hypothesis should depend on both how confident you were in it before the evidence—the prior probability—and on how much that hypothesis is able to account for the evidence—the likelihood.

Using the evidence to obtain posterior probabilities is the aim of induction. Figure 3.2 shows the material needed to obtain posterior probabilities within the Bayesian framework. As in Figure 3.1, the hypothesis set (along with the likelihoods) and the evidence are inputs to the inductive method (which may be of many different kinds, and is thus variable), which outputs posterior probabilities. But now the inductive method box has some boxes within it; inductive methods are now determined by variable prior probability distributions and the fixed Bayes' Theorem.

Consider an example first. I present to you a coin, and tell you it is possibly a trick coin. I tell you that there are three possibilities: it is fair, always-heads, or always-tails. These three possibilities comprise the hypothesis set. Your task is to flip the coin and judge which of these three possibilities is true. Your evidence is thus coin flip outcomes. Your likelihoods are already defined via the decision to consider the three hypotheses. For example, suppose two heads are flipped. The likelihood of getting two heads for the coin-is-fair hypothesis is $(1/2)^2 = 1/4$. The likelihood for the always-heads hypothesis is $1^2 = 1$, and for the always-tails hypothesis it is $0^2 = 0$. What is the posterior probability for these three hypotheses given that the evidence consists of the two heads? To answer this, we still need prior probability values for the hypotheses. This is where things get hairy. In real life, we may have experience with tricksters and coins with which we can make guesses as to the prior (i.e., the prior probability distribution). But the point of this example is to imagine that you have no experience whatsoever with tricksters or coins, and you somehow need to determine prior probabilities for these three hypotheses. Let us suppose you declare the three to be equally probable, *a priori*. Now you can engage in induction, and the posterior probabilities are as follows:

- $P(\text{fair}|\text{two heads}) = \frac{\frac{1}{3} \cdot \frac{1}{4}}{\frac{1}{3} \cdot \frac{1}{4} + \frac{1}{3} \cdot 1 + \frac{1}{3} \cdot 0} = \frac{1}{5}$.

- $P(\text{always-heads}|\text{two heads}) = \frac{\frac{1}{3} \cdot 1}{\frac{1}{3} \cdot \frac{1}{4} + \frac{1}{3} \cdot 1 + \frac{1}{3} \cdot 0} = \frac{4}{5}$.

- $P(\text{always-tails}|\text{two heads}) = \frac{\frac{1}{3} \cdot 0}{\frac{1}{3} \cdot \frac{1}{4} + \frac{1}{3} \cdot 1 + \frac{1}{3} \cdot 0} = 0$.

Different prior probability assignments would have led to different posterior probability assignments; i.e., led to different inductive conclusions.

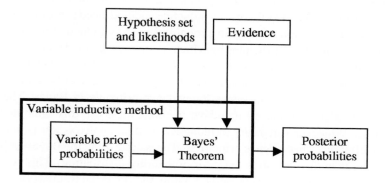

Figure 3.2: To acquire beliefs about the world, evidence and a set of hypotheses must be input into an inductive method, whose job it is to output the degrees of belief about those hypotheses one ought to have given the evidence. In the Bayesian framework, the inductive method is determined by a choice of prior probabilities over the hypothesis set. This variable prior is put, along with the evidence, into Bayes' Theorem, which outputs the posterior probabilities. Now it is not the case that any old inductive method is justified, unlike in Figure 3.1. However, there is still tremendous variability in the possible inductive methods due to the variability in the choice of prior. One of the nice things about this is that the variability no longer concerns how evidence is brought to bear on hypotheses; this is kept constant by the use of Bayes' Theorem. All the variability in inductive methods is reduced to just one kind of thing: one's degrees of belief in the hypotheses before having seen the evidence. Also, note that the Bayesian framework is also a probabilistic framework, which constrains the numerical degrees of confidence in hypotheses to satisfy the axioms of Probability Theory; this constraint is not depicted in the figure.

What does the Bayesian framework for induction buy us? After all, we still have many possible inductive methods to choose from; we have not solved the problem of the variability, or indeterminacy, of inductive methods. For one thing, it rules out whole realms of possible inductive methods; inductive methods must now fit within the framework. Algorithmic learning rules that take evidence and assign probabilities to the hypotheses are not allowable inductive methods if they cannot be obtained by starting with a prior probability distribution and grinding it through Bayes' Theorem. The second nice thing about the Bayesian framework is that it gets inside inductive methods and helps to distinguish between two things an inductive method needs in order to do its job: evidence principles and prior probabilities. Any inductive method needs "evidence principles," principles by which it employs the evidence to affect the degrees of confidence in the hypotheses. For example, if I fish one more bass, is this good or bad for the hypothesis that the next fish will be a bass? The Bayesian framework encapsulates its evidence principle in Bayes' Theorem, effectively declaring that all inductive methods must use this same evidence principle. Whatever variability in inductive method choice is left is not, then, due to differences in evidence principles. The second thing the Bayesian framework serves to distinguish is the prior probability distribution. This is left indeterminate, but any inductive method within the Bayesian framework requires some setting for this variable. All the variability in inductive methods is, then, reduced to one kind: one's *a priori* degrees of confidence in the hypotheses. A final important thing about the Bayesian framework is that the evidence principle is not just any old evidence principle; it is justifiable in the sense that it follows from probability axioms that everyone believes. Not only does "everyone believe" the probability axioms, they are, in a certain clear sense, principles a reasoner ought to hold. This is due to the fact that if someone reasons with numerical confidences in hypotheses that do not satisfy the probability axioms, then it is possible to play betting games with this fellow and eventually take all his money. This is called the *Dutch Book Theorem*, or the Ramsey-de Finetti Theorem (Ramsey, 1931; de Finetti, 1974; see also Howson and Urbach, 1989, pp. 75–89 for discussion). And Bayes' Theorem follows from these axioms, so this evidence principle is rational, since to not obey it would lead one to being duped out of one's money.[1]

---

[1]Things are actually a bit more complicated than this. Using Bayes' Theorem as our principle of evidence (or our "principle of conditionalization," as it is sometimes said) is *the* rational principle of evidence—i.e., in this case because any other will lead you to financial ruin—*if*, upon finding evidence *e*, *e* does not entail that your future degree of confidence in the hypoth-

With this machinery laid before us, the riddle of induction can now be stated more concisely as, "What prior probability distribution ought one use?" By posing induction within the Bayesian framework, one cannot help but see that to have a theory of logical induction would require a determinate "best" choice of prior probabilities. And this would be to make an *a priori* assumption about the world (i.e., an assumption about hypotheses concerning the world). But our original hope was for a theory of logical induction that would tell us what we ought to do *without* making *a priori* assumptions about the world.

What would a theory of logical probability look like?

There is, then, no solution to the riddle of induction, by which we mean there is no theory of logical probability which, given just a set of hypotheses and some evidence, outputs *the* respectable inductive method. There simply *is no* unique respectable inductive method.

If one tries to solve a problem, only to eventually realize that it has no solution, it is a good idea to step back and wonder what was wrong with the way the problem was posed. The problem of induction must be ill posed, since it has no solution of the strong kind for which we were searching. Let us now step back and ask what we want out of a theory of logical probability.

The Bayesian framework serves as a strong step forward. Within it, we may make statements of the form,

> If the prior probability distribution on H is $P(h)$, then, given the evidence, the posterior probability distribution ought to be given by $P(h|e)$, as dictated by Bayes' Theorem.

There are a number of advantages we mentioned earlier, but a principal downside remains. What the inductive method is depends entirely on the prior probability distribution, but the prior probability distribution comprises a set of beliefs about the degrees of confidence in the hypotheses. That is, prior probabilities are judgements about the world. Thus, the Bayesian statement becomes something like,

> If one has certain beliefs about the world before the evidence, then he should have certain other beliefs about the world after the evidence.

esis given *e* will be different from that given by Bayes' Theorem. That is, if, intuitively, the evidence does not somehow logically entail that Bayes' Theorem is inappropriate in the case at issue. This 'if' basically makes sure that some very weird scenarios are not occurring; no weird circumstances... Bayes' Theorem is the rational principle of evidence. See Howson and Urbach (1989, pp. 99–105) for details.

But one of the goals of a logical theory of induction is to tell us which beliefs about the world we ought to have. The Bayesian framework leaves us unsatisfied because it does not tell us which *a priori* beliefs about the world we should have. Instead, it leaves it entirely open for us to believe, *a priori*, anything we want!

In our move from the pre-Bayesian framework (Figure 3.1) to the Bayesian framework (Figure 3.2), we were able to encapsulate a fixed evidence principle, and were left with variable prior probabilities. Now *I submit that the task of a theory of logical probability is to put forth fixed principles of prior probability determination, and to have left over some variable that does* not *possess information about the world (unlike prior probabilities).* Just as the left over variable in the Bayesian framework was non-evidence-based, the variable left over within this new framework will be non-induction-based, or non-inductive, or not-about-the-world. If we had something like this, then we could make statements like,

> If one has non-inductive variable Q, then one ought to have prior probability distribution $P_Q(h)$, as dictated by the principles of prior probability determination.

It should also be the case that the non-inductive variable has some coherent (non-inductive) interpretation, lest one not know how anyone would ever pick any value for it. The principles of prior probability determination would possess a few things that one ought to do when one picks prior probabilities given the non-inductive variable. In this way, we would have reduced all oughts found in induction to a small handful of principles of ought, and no undefended assumptions about the world would need to be made in order to get different inductive methods up and going.

Figure 3.3 is the same as Figure 3.2, but now shows the kind of machinery we need: (i) some fixed, small number of axioms of *a priori* logical probability determination, in the form of rationality principles, and (ii) some variable with a meaningful interpretation, but not with any inductive significance.

The bulk of this chapter consists of the development and application of a theory of logical induction aiming to fill these shoes. The theory is called *Paradigm Theory.* Three abstract symmetry-related principles of rationality are proposed for the determination of prior probabilities, and a kind of non-inductive variable—called a "paradigm"—is introduced which is interpreted as a conceptual framework, capturing the kinds of properties of hypotheses one acknowledges. A paradigm and the principles together entail a prior probability distribution; the theory allows statements of the form,

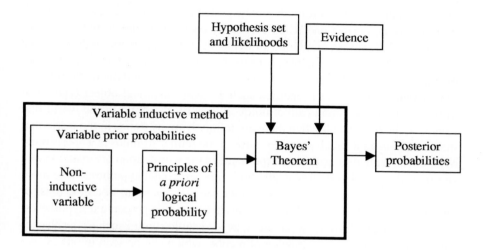

Figure 3.3: The structure of a sought-after theory of logical induction. The prior probability distribution should follow from the combination of a small number of rationality principles—things a rational agent ought to do—and some non-inductive variable with a meaningful interpretation.

If one has paradigm Q, then one ought to have prior probability distribution $P_Q(h)$, as dictated by the symmetry-related principles of prior probability determination.

Innateness

The brain learns. It therefore entertains hypotheses, and implements inductive methods. What do we mean by a hypothesis in regards to the brain? Here are a couple examples. The human brain quickly learns the grammar of natural language, and there are (infinitely) many possible hypotheses concerning which grammar is the correct one. Kids eventually converge to the correct grammar; i.e., after sufficient accumulation of evidence, they impart the highest degree of belief to the correct (or nearly correct) grammatical hypothesis. Another example is vision. The retina is a two-dimensional sheet, and the world is three-dimensional, with many properties such as reflectance and object type. The information on the retina cannot uniquely specify the thing in the world that caused it, the reason being that there are infinitely many things in the world that may have caused it. Each possible cause of the retinal stimulus is a "perceptual hypothesis," and after acquiring experience in the world, upon presentation of a stimulus, the visual system typically finds one perceptual hypothesis to be more probable than the others, which is why we see just one scene at a time most of the time. When the probabilities are tied between two perceptual hypotheses, we jump back and forth between them, as in bistable stimuli such as the Necker Cube (which is just a line drawing of a cube, which can be seen in one of two orientations). These two examples for hypotheses entertained by the brain do not even scratch the surface; the brain indeed is a kind of learning machine, and thus entertains possible hypotheses at every turn.

Not only does the brain learn, but it is thought by many to enter the world a blank slate, and to be endowed with powerful and general learning abilities. One can get the impression that the cortex is some kind of universal learning engine. The relatively homogenous nature of the anatomy and connectivity of the cortex is one reason scientists come away with this impression: the cortex is a few millimeter thick sheet (its exact thickness depending on the brain's size), with six layers, and with statistically characterized connectivity patterns for the neurons within it. Roughly, the cortex seems to be built from many repeating units called "minicolumns." And although the cortex is divided up into distinct areas, connecting to other areas primarily via myelinated white matter axons, and although the areas often have distinguishing anatomical features, they appear to be fairly similar in basic design. The high degree of plasticity of the cortex also suggests that it is a general learning machine, not a prisoner to instinct. When, for example, a limb is lost, somatosensory areas formerly devoted to the limb sometimes become employed by other areas. Also, the basic connectivity features of our cortex do not appear much different than that of monkey or cat, animals leading drastically different lives. The intuitive conclusion sometimes drawn is that we differ from monkeys merely in that our brains are relatively much larger, and that our ecologies and thus experiences are different. Our perception of the world appears to rely on general learning strategies by the visual system. People raised in non-carpentered environments, like Bushmen, do not experience some of the classical geometrical illusions that we find illusory (Segall et al., 1966). Even thirst and hunger, two appetitive states one might imagine would be innate if anything is innate, appear to be learned (Changizi et al., 2002b): rats do not know to orient toward a known water source when cellularly dehydrated unless they have experienced dehydration paired with drinking water, and similarly they do not know to orient toward a known food source when food restricted unless they have experienced food restriction paired with eating.

But being highly homogeneous and plastic does not entail that the brain does not possess innate content, or knowledge. Whatever is innate could well be wrapped up in the detailed connectivity patterns in the brain. And a strong role for experience does not mean the cortex is a universal learning machine. Even those scientists that are fans of a strongly innate brain, such as those that believe that grammar is innate (e.g., Chomsky, 1972; Pinker, 1994), obviously believe in an immense role for learning.

With an understanding of the riddle of induction under our belts, we can say, without knowing anything about the particulars of our brains, that we *must* enter the world with innate knowledge. There is no universal learning machine. There are, instead, just lots and lots of different inductive methods. Whatever our brains are doing when they learn, they are engaging in an inductive method (although perhaps a different inductive method for different kinds of learning). As discussed earlier, the brain must therefore, in effect, be making an assumption about the world in the form of a prior probability distribution over the possible hypotheses. That is, in order to learn, brains must enter the world with something equivalent to preconceptions about the degrees of confidence of all the possible hypotheses. Brains are not blank slates; they are born with what are, in effect, *a priori* assumptions about the world.

What would a theory of innateness be?

Brains, then, come furnished with an inductive method; i.e., they have some way by which they take evidence and determine the posterior probabilities of hypotheses. Different brain types—e.g., human versus cat—may employ different inductive methods, and these differences are innate. We will assume that the principal innate differences between brain types are due to their instantiating different inductive methods. (They may also differ in their choice of what hypotheses to consider in the first place, and they may well differ concerning what things matter to them (i.e., utilities).)

What I wish to consider here is a *theory of innateness*, a theory aimed at characterizing the nature of the information that must be innately generated. How much must innately differ between two kinds of brain (or two parts of the same brain) in order for them to possess distinct inductive methods? The theory of innateness I seek is not one that actually claims that brains conform to the theory. Rather, the aim is to construct a mathematical theory, or framework, within which we can conceptually distinguish among the kinds of structure required for an innate inductive method. With a theory of innateness in hand, we will *then* have the conceptual apparatus to begin to speak about the principles governing how brains—or any intelligent learning agents—have innate inductive methods.

Here is one thing that we would like out of a theory of innateness. I have already mentioned that brains of different kinds have a lot in common. It would accordingly be useful to find a theory of innateness that postulates no greater innate differences than are absolutely necessary to account for the different inductive methods used. We would like to be able to model brains of different types—i.e., brains employing different inductive methods—as following the same underlying principles, principles used in determining their inductive method. All these brains are, after all, brains, and the way they go about their learning should be describable using universal principles. Furthermore, these principles should be rationality principles of some kind, or principles stating what a rational agent would do. We would then be able to model brains having different innatenesses as nevertheless being similar to the extent that they follow the same rationality principles. We would be able to say that all these kinds of brains may be different in *some* regard that specifies what is innate, but that in all other ways we may model these brains identically.

The second aspect of our theory of innateness that requires concern is the distinguishing feature between brains of different types—the feature that is the possessor of the innate information. There must be some variable property of brains, distinct settings of the variable which lead to distinct inductive methods used by the brain. As mentioned earlier, the theory of innateness for which we search would postulate no greater innate differences than are absolutely necessary to account for the different inductive methods used. Accordingly, we want the variable that determines the innate differences to be as weak as possible. Furthermore, innate content is derided by many because it seems absurd that, say, natural language grammar could be encoded into the brain at birth. Surely it is an incredible claim that brains enter the world with *a priori* beliefs, or assumptions. With this in mind, we would also like the "innateness variable" to say as little as possible about the world; i.e., to be non-inductive. Finally, this innateness variable should be interpretable in some plausible fashion; if it has no interpretation, then one begins to suspect that it is just a stand-in for an *a priori* inductive assumption.

In short, we would like a theory of innateness that models brains, or any intelligent agent, as following fixed principles of rationality in their learning, and models the differences with an innateness variable that is weak, non-inductive, and has a meaningful interpretation.

If you recall the earlier subsection on what we want out of a theory of logical probability, you will notice a close connection to that and to what we here want out of a theory of innateness. This is not a coincidence. The discovery of a theory of logical probability of the kind described would state how, through a fixed set of prior probability determination principles, a rational agent with a setting of the non-inductive variable should proceed in assigning his prior probabilities, and consequently what inductive method he ought to follow. On the one hand, the theory would tell us what we ought to do, but on the other hand, the theory tells us what a rational agent will, in fact, do—since this agent will do what he should. If our interest is in modeling innateness in assumed-to-be-rational brains, then the theory of logical probability can be used to describe brains, not just to say how brains ought to perform.

Let us go through the connection between a theory of induction and a theory of innateness more slowly, beginning with the earlier Figure 3.1. One way to treat innateness differences in different brain types is to postulate that they are governed by entirely different principles altogether. Brains of different types are, in regards to learning, just (computable) functions of any old kind taking evidence and outputting posterior probabilities. Each brain would innately make different assumptions about the world, and follow different rules concerning how evidence supports hypotheses (i.e., follow different evidence

principles). But we wish to be able to retain the view that brains and other intelligent agents learn in a rational fashion, and thus are all fundamentally similar,
following identical principles, and differing only in regard to some small, interpretable, weak variable that does not correspond to an assumption about the
world.

Bayesianism provides a great first step toward satisfying these demands,
just as it provided a great first step for a theory of logical induction. Figure
3.2 was our corresponding figure for the problem of induction, and it is apt to
look at it again for innateness. Bayes' Theorem is helpful toward a theory of
innateness and learning because it allows us to treat all agents, or brains, as
following Bayes' Theorem in their modification of their degrees of confidence
in hypotheses in the light of evidence. And the Bayesian evidence principle is
not just any old evidence principle, it seems to be the right principle—it is the
way one should use evidence to modify the degree of confidence in hypotheses.
This is why Bayesian approaches have been so popular in the psychological,
brain and decision sciences. This Bayesian framework is used to model the
visual system, memory, learning, behavior, economic agents and hosts of other
cases where there is some kind of "agent" dealing with an uncertain world.
The argument goes something like this: (a) These agents have probably been
selected to learn in an optimal, or rational, manner. (b) The optimal learning
manner is a Bayesian one. (c) Therefore, we may treat these agents as following Bayesian principles. The Bayesian framework also severely constrains the
space of possible inductive methods, from anything-goes down to only those
using its evidence principle.

As powerful as the Bayesian framework is, it leaves us with some resid-
ual dissatisfaction concerning a theory of innateness. The Bayesian framework
has prior probabilities that differ between agents that follow different inductive methods. A prior probability distribution, then, is the innateness variable.
Brains that differ in innateness would be postulated to enter the world with
different *a priori* beliefs about the degree of confidence in the hypotheses. As
discussed earlier, this is one thing we want to avoid with a theory of innateness. We would like it to be the case that innateness can be much more subtle
than *a priori* beliefs about the world in the head. Perhaps there are further
principles—principles beyond Bayes' Theorem—that an optimally engineered
agent will follow, so that two such agents might innately differ in some non-
inductive fashion, yet by following these fixed principles they come to have
different prior probabilities. Figure 3.3 from earlier is again appropriate, for it
shows what we are looking for. Such a theory would even further constrain the

space of possible inductive methods, from any-prior-probability-distribution-goes down to only those using the fixed principles of prior probability determination.

That is our goal for a theory of innateness. The theory that I will propose in the next section—called *Paradigm Theory*—consists of fixed symmetry and symmetry-like principles of rationality—or principles of non-arbitrariness—which I argue any rational agent will follow. The non-inductive variable is something I call a "paradigm", which is just the kinds of hypotheses the agent acknowledges; for example, you and I might possess the same hypothesis set, but I may carve it up into kinds differently than do you. The intuition is that we have different conceptual frameworks, or belong to different (Kuhnian) paradigms. Innateness differences, then, would be attributable to differences in the conceptual frameworks they are born with. But in all other regards agents, or brains, of different innatenesses would be identical, having been selected to follow fixed optimal, or rational, principles, both of prior probability determination and of evidence.

Are there really innate paradigms in the head? I don't know, and at the moment it is not my primary concern. Similarly, the Bayesian framework is widely considered a success, yet no one appears particularly worried whether there is any part of the developing brain that corresponds to prior probabilities. The Bayesian framework is a success because it allows us to model brains as if they are rational agents, and it gives us the conceptual distinctions needed to talk about evidence principles and *a priori* degrees of belief in hypotheses. Similarly, the importance of Paradigm Theory in regards to innateness will be that it allows us to model brains as if they are rational agents, giving us more conceptual distinctions so that, in addition to evidence principles and *a priori* degrees of belief in hypotheses, we can distinguish between principles of non-arbitrariness for prior probability determination and *a priori* conceptual frameworks. Paradigm Theory gives us the power to make hypotheses we otherwise would not be able to make: that brains and intelligent learners could have their innate inductive methods determined by innate, not-about-the-world paradigms, along with a suite of principles of rationality. Whether or not brains actually utilize these possibilities is another matter.

3.1   Paradigm Theory

In this section I introduce a theory of logical probability (Changizi and Barber, 1998), with the aim of satisfying the criteria I put forth in the previous section. The plan is that it will simultaneously satisfy the demands I put forward for a theory of innateness. The theory's name is "Paradigm Theory," and it replaces prior probabilities with a variable that is interpreted as a conceptual framework, and which we call a "paradigm." A paradigm is roughly the way an agent "carves up the world"; it is the kinds of hypotheses acknowledged by the agent.

[The idea that induction might depend on one's conceptual framework is not new. For example, Harsanyi (1983, p. 363) is sympathetic to a dependency on conceptual frameworks for simplicity-favoring in induction. Salmon (1990) argues for a Kuhnian paradigmatic role for prior probabilities. Earman (1992, p. 187) devotes a chapter to Kuhnian issues including paradigms. Di Maio (1994, especially pp. 148–149) can be interpreted as arguing for a sort of conceptual framework outlook on inductive logic. DeVito (1997) suggests this with respect to the choice of models in curve-fitting. Also, Gärdenfors (1990) develops a conceptual framework approach to address Goodman's riddle, and he attributes a conceptual framework approach to Quine (1960), Carnap (1989) and Stalnaker (1979).]

Having a paradigm, or conceptual framework, cannot, all by itself, tell us how we ought to proceed in our inductions. Oughts do not come from non-oughts. As discussed in the previous section, we are looking for principles of ought telling us how, given a paradigm, we should assign *a priori* degrees of belief in the hypotheses. I will put forward three symmetry-related principles that enable this.

Before moving to the theory, let us ask where the hypothesis set comes from. This is a difficult question, one to which I have no good answer. The difficulty is two-fold. First, what hypotheses should one include in the hypothesis set? And second, once this set is chosen, how is that set parameterized? I make some minimal overtures toward answering this in Changizi and Barber (1998), but it is primarily an unsolved, and possibly an unsolvable problem. I will simply assume here that the hypothesis set—a set of mutually exclusive hypotheses—and some parameterization of it is a given.

3.1.1   A brief first-pass at Paradigm Theory

Before presenting Paradigm Theory in detail, I think it is instructive to give a short introduction to it here, with many of the intricacies missing, but nevertheless capturing the key ideas. Paradigm Theory proposes to replace the variable prior probabilities of the Bayesian framework with variable "paradigms," which are interpreted as comprising the inductive agent's way of looking at the set of hypotheses, or the agent's conceptual framework. For example, you and I might share the same hypothesis set, but I might acknowledge that there are simple and complex hypotheses, and you might, instead, acknowledge that some are universal generalizations and some are not. More generally, a paradigm consists of the kinds of hypotheses one acknowledges. One of the most important aspects of paradigms is that they do not make a claim about the world; they are non-inductive. If, in complete ignorance about the world, I choose some particular paradigm, I cannot be charged with having made an unjustifiable assumption about the world. Paradigms are just a way of carving up the space of hypotheses, so they make no assumption. Prior probabilities, on the other hand, are straightforwardly claims about the world; namely, claims about the *a priori* degree of confidence in the hypotheses. The justification of induction in Paradigm Theory rests not on a variable choice of prior probabilities as it does in the Bayesian framework, but, instead, on a variable choice of a non-inductive paradigm. Paradigm Theory puts forth three principles which prescribe how prior probabilities ought to be assigned given that one possesses a paradigm. Different inductive methods differ only in the setting of the paradigm, not on any *a priori* differences about claims about the world or about how one ought to go about induction.

To understand the principles of prior probability determination, we have to understand that any paradigm naturally partitions the hypothesis set into distinct sets. [A *partition* of a set B is a set of subsets of B, where the subsets do not overlap and their union is B.] The idea is this. From the point of view of the paradigm—i.e., given the properties of hypotheses acknowledged in the paradigm—there are some hypotheses which cannot be distinguished using the properties in the paradigm. Hypotheses indistinguishable from one another are said to be *symmetric*. Each partition consists of hypotheses that are symmetric to one another, and each partition is accordingly called a *symmetry type*. Hypotheses in distinct partitions *can* be distinguished from one another. Since hypotheses cannot be distinguished within a symmetry type, the symmetry types comprise the kinds of hypothesis someone with that paradigm can distinguish.

Note that the symmetry types may well be different than the properties in the paradigm; the properties in the paradigm imply a partition into symmetry types of distinguishable (from the paradigm's viewpoint) types of hypotheses.

With an understanding that paradigms induce a natural partition structure onto the hypothesis set, I can state Paradigm Theory's principles for how one should assign prior probabilities. One principle states that each distinguishable type of hypothesis should, *a priori*, receive the same degree of confidence; this is the *Principle of Type Uniformity*. The intuition is that if one is only able to distinguish between certain types of hypotheses—i.e., they are the kinds of hypotheses one is able to talk about in light of the paradigm—and if there is no apparatus within the paradigm with which some of these distinguished types can *a priori* be favored (and there is no such apparatus), then it would be the height of arbitrariness to give any one type greater prior probability than another. The second principle states that hypotheses that are symmetric to one another—i.e., the paradigm is unable to distinguish them—should receive the same probability; this is the *Principle of Symmetry*. The motivation for this is that it would be entirely arbitrary, or random, to assign different *a priori* degrees of confidence to symmetric hypotheses, given that the paradigm has no way to distinguish between them; the paradigm would be at a loss to explain why one gets a higher prior probability than the other. There is one other principle in the full Paradigm Theory, but it is less central than these first two, and we can skip it in this subsection.

The Principle of Type Uniformity distributes equal shares of prior probability to each symmetry type, and the Principle of Symmetry distributes equal shares of the symmetry type's probability to its members. In this way a prior probability distribution is determined from a paradigm and the principles. Paradigms leading to different symmetry types usually lead to different prior probability distributions. Justifiable inductive methods are, then, all the same, in the sense that they share the Bayesian principle of evidence, and share the same principles of prior probability determination. They differ only in having entered the world with different ways of conceptualizing it. I can now make claims like, "If you conceptualize the world in fashion Q, then you ought to have prior probabilities $P_Q(H)$ determined by the principles of Paradigm Theory. This, in turn, entails a specific inductive method you ought to follow, since you ought to follow Bayes' Theorem in the application of evidence to your probabilities."

The remainder of this subsection, and the next subsection, develop this material in detail, but if you wish to skip the details, and wish to skip example applications of Paradigm Theory (to enumerative induction, simplicity favor-ing, curve-fitting and more), you may jump ahead to Section 3.3.

3.1.2 Paradigms, Symmetry and Arbitrariness

In the next subsection I will present the principles of prior probabilities de-termination, i.e., principles of ought which say what one's prior probabilities should be given that one has a certain paradigm. But first we need to intro-duce paradigms, and to motivate the kinds of symmetry notions on which the principles will rest.

Paradigms

Let us begin by recalling that we are assuming that we somehow are given a hypothesis set, which is a set filled with all the hypotheses we are allowed to consider. The hypotheses could concern the grammar of a language, or the curve generating the data, and so on. The hypothesis set comprises an inductive agent's set of all possible ways the world could be (in the relevant regard).

Now, what is a paradigm? A paradigm is just a "way of thinking" about the set of hypotheses. Alternatively, a paradigm is the kinds of similarities and differences one appreciates among the hypotheses. Or, a paradigm stands for the kinds of hypotheses an inductive agent acknowledges. A paradigm is a kind of conceptual framework; a way of carving up the set of hypotheses into distinct types. It is meant to be one way of fleshing out what a Kuhnian paradigm might be (Kuhn, 1977). If the hypothesis set is the "universe," a paradigm is the properties of that "universe," a kind of ontology for hypothesis sets. When an inductive agent considers there to be certain kinds of hypotheses, I will say that the agent *acknowledges* those kinds, or *acknowledges* the associated properties. I do not mean to suggest that the agent would not be able to discriminate, or notice, other properties of hypotheses; the agent can presumably tell the difference between any pair of hypotheses. The properties in the paradigm, however, are the only properties that are "sanctioned" or endowed as "genuine" properties in the ontology of that universe of hypotheses.

For example, suppose the hypothesis set is the set of six outcomes of a roll of a six-sided die. One possible paradigm is the one that acknowledges being even and being odd; another paradigm is the one that acknowledges being small (three or less) and big (four or more). Or, suppose that the hypothesis set is the set of all points in the interior of a unit circle. One possible paradigm is the one that acknowledges being within distance 0.5 from the center. Another possible paradigm would be the one acknowledging the different distances from the center of the circle; that is, points at the same radius would be of the same acknowledged kind. For a third example, suppose the hypothesis set is the set of all possible physical probabilities p of a possibly biased coin; i.e., the hypothesis set is $H = [0, 1]$, or all the real numbers from 0 to 1 included. One possible paradigm is the one that acknowledges the always-heads ($p = 0$) and always-tails ($p = 1$) hypotheses, and lumps the rest together. Another paradigm on this hypothesis set could be to acknowledge, in addition, the coin-is-fair hypothesis ($p = 1/2$).

For each of these examples, there is more than one way to carve up the hypothesis set. One person, or inductive community, might acknowledge properties that are not acknowledged by another person or community. Where do these properties in the paradigm come from? From Paradigm Theory's viewpoint it does not matter. The properties will usually be interpreted as if they are subjective. There are two kinds of subjective interpretations: in the first kind, the properties in the paradigm have been consciously chosen by the inductive agent, and in the second kind, the properties are in the paradigm because the inductive agent has evolved or been raised to acknowledge certain properties and not others.

Recall that our aim for a theory of logical probability was to have an interpretable, non-inductive variable to replace prior probabilities. In Paradigm Theory, the variable is the paradigm, and we have just seen that paradigms are interpreted as conceptual frameworks. But we also want our variable—namely, paradigms—to also be non-inductive, or not-about-the-world. (And, similarly, for our hoped-for theory of innateness, the innate content was to have some interpretable, non-inductive variable.)

Are paradigms about the world? A paradigm is just the set of properties acknowledged, and there is no way for a paradigm to favor any hypotheses over others, nor is there any way for a paradigm to favor any properties over others—each property is of equal significance. Paradigms cannot, say, favor simpler hypotheses, or disfavor hypotheses inconsistent with current ontological commitments; paradigms can *acknowledge* which hypotheses are simpler than others, and *acknowledge* which hypotheses are inconsistent with current ontological commitments. Paradigms make no mention of degrees of belief, they do not say how inductions ought to proceed, and they do not presume that the world is of any particular nature. Do not confuse a paradigm with information. Being unbiased, the properties in the paradigm give us no information about the success or truth of any hypothesis, and in this sense the paradigm is not information. Therefore, paradigms are non-inductive.

To help drive home that paradigms are non-inductive, suppose that an agent discounts certain hypotheses on the basis of something not measured by the paradigm (e.g., "too complex") or favors some properties over others. Paradigm Theory is not then applicable, because the inductive agent now effectively already has prior probabilities. Paradigm Theory's aim is to attempt to defend inductive beliefs such as priors themselves. If an agent enters the inductive scenario with what are in effect prior probabilities, then Paradigm Theory is moot, as Paradigm Theory is for the determination of the priors one should have. Consider the following example for which Paradigm Theory is inapplicable. A tetrahedral die with sides numbered 1 through 4 is considered to have landed on the side that is face down. Suppose one acknowledges that one of the sides, side 4, is slightly smaller than the others, and acknowledges nothing else. The paradigm here might seem to be the one acknowledging that side 4 is a unique kind, and the others are lumped together. If this were so, Paradigm Theory would (as we will see) say that 4 should be preferred. But side 4 should definitely *not* be preferred! However, Paradigm Theory does not apply to cases where one begins with certain inductive beliefs (e.g., that smaller sides are less likely to land face down). Paradigm Theory is applicable in those kinds of circumstances where one has not yet figured out that smaller sides are less likely to land face down. [There may remain an issue of how we assign a precise prior probability distribution on the basis of an imprecise inductive belief such as "smaller sides are less likely to land face down," but this issue of formalization of imprecise inductive beliefs is a completely different issue than the one we have set for ourselves. It is less interesting, as far as a theory of logical probability goes, because it would only take us from imprecise inductive beliefs to more precise inductive beliefs; it would not touch upon the justification of the original imprecise inductive belief.]

I now have the concept of a paradigm stated, but I have not quite formally defined it. Here is the definition.

Definition 1 A *paradigm* is any set of subsets of the hypothesis set that is closed under complementation. The complements are presumed even when, in defining a paradigm, they are not explicitly mentioned. $\triangle$

Recall that when you have a set of objects of any kind, a property is just a subset of the set: objects satisfying the property are in the set, and objects not satisfying the property are not in the set (i.e., are in the complement of the set). The definition of a paradigm just says that a paradigm is a set of subsets, or properties; and it says that for any property P in the paradigm, the property of not being P is also in the set. And that is all the definition says.

Being Symmetric

We now know what a paradigm is: it is the non-inductive variable in our theory of logical probability that I call Paradigm Theory, and paradigms are interpreted as conceptual frameworks, or ways of conceptualizing the set of hypotheses. Our goal is to present compelling principles of rationality which prescribe how one ought to assign prior probabilities *given* one's paradigm; we would thereby have fixed principles of prior probability determination that all rational agents would follow, and all justifiable differences in inductive methods would be due to differences in the way the inductive agent carved up the world before having known anything about it.

Before we can understand Paradigm Theory's principles of prior probability determination, we must acquire a feel for the intuitive ideas relating to symmetry, and in this and the following subsubsection I try to relate these intuitions.

One of the basic ideas in the rational assignment of prior probabilities will be the motto that *names should not matter*. This motto is, generally, behind every symmetry argument and motivates two notions formally introduced in this subsubsection. The first is that of a *symmetry type*. Informally, two hypotheses are of the same symmetry type if the only thing that distinguishes them is their names or the names of the properties they possess; they are the same type of thing but for the names chosen. One compelling notion is that hypotheses that are members of smaller symmetry types may be chosen with less arbitrariness than hypotheses in larger symmetry types; it takes less arbitrariness to choose more unique hypotheses. The principles of Paradigm Theory in the Subsection 3.1.3, founded on different intuitions, respect this notion in that more unique hypotheses should receive greater prior probability than less unique hypotheses. The second concept motivated by the "names should not matter" motto, and presented in the next subsubsection, is that of a *defensibility hierarchy*, where picking hypotheses higher in the hierarchy is less arbitrary, or more defensible. The level of defensibility of a hypothesis is a measure of how "unique" it is. Subsection 3.1.3 describes how the principles of rationality of Paradigm Theory lead to a prior probability assignment which gives more defensible types of hypotheses greater prior probability. Onward to the intuition pumping.

Imagine having to pick a kitten for a pet from a box of five kittens, numbered 1 through 5. Imagine, furthermore, that you deem no kitten in the litter
to be a better or worse choice for a pet. All these kittens from which to choose,
and you may not wish to pick randomly. You would like to find a reason to
choose one from among them, even if for no other reason but that one is distinguished in some way. As it turns out, you acknowledge some things about
these kittens: the first four are black and the fifth is white. These properties
of kittens comprise your paradigm. Now suppose you were to pick one of the
black kittens, say kitten #1. There is no reason connected with their colors you
can give for choosing #1 that does not equally apply to #2, #3 and #4. "I'll
take the black kitten" does not pick out #1. Saying "I'll take kitten #1" picks
out that first kitten, but these number-names for the kittens are arbitrary, and
had the first four kittens been named #2, #3, #4 and #1 (respectively), "I'll take
kitten #1" would have picked out what is now called the fourth kitten. #1 and
#4 are the same (with respect to the paradigm) save their arbitrary names, and
we will say that they are symmetric; in fact, any pair from the first four are
symmetric.

Imagine that the five kittens, instead of being just black or white, are each a
different color: red, orange, yellow, green and blue, respectively. You acknowledge these colors in your paradigm. Suppose again that you choose kitten #1.
Unlike before, you can at least now say that #1 is "the red one." However,
why is redness any more privileged than the other color properties acknowledged in this modified paradigm? 'red' is just a name for a property, and had
these five properties been named 'orange', 'yellow', 'green', 'blue' and 'red'
(respectively), "the red one" would have picked out what is now called the blue
one. #1 and #5 will be said to be symmetric; in fact, each pair will be said to
be symmetric.

For another example, given an infinite plane with a point above it, consider
the set of all lines passing through the point. If the plane and point "interact" via some force, then along which line do they do so? This question was
asked by a professor of physics to Timothy Barber and myself as undergrad-
uates (we shared the same class), and the moral was supposed to be that by
symmetry considerations the perpendicular line is the only answer, as for every other line there are lines "just as good." In our theoretical development
we need some explicit paradigm (or class of paradigms) before we may make
conclusions. Suppose that you acknowledge the properties of the form "having
angle θ with respect to the plane," where a line parallel to the plane has angle
0. Any pick of, say, a parallel line will be arbitrary, as one can rotate the world

about the perpendicular line and the parallel line picked would become another one. Each parallel line is symmetric to every other. The same is true of each non-perpendicular line; for any such line there are others, infinitely many others, that are the same as far as the paradigm can tell. The perpendicular line is symmetric only with itself, however.

In the remainder of this subsubsection we make the notion of symmetry precise, but there is no real harm now skipping to the next subsubsection if mathematical details bother you. The following defines the notion of being symmetric.

Definition 2 Fix hypothesis set H and paradigm Q. h_1 and h_2 are Q-*symmetric* in H if and only if it is possible to rename the hypotheses respecting the underlying measure such that the paradigm is unchanged but the name for h_1 becomes the name for h_2. Formally, for $p : H \to H$, if $X \subseteq H$ then let $p(X) = \{p(x)|x \in X\}$, and if Q is a paradigm on H, let $p(Q) = \{p(X)|X \in Q\}$. h_1 and h_2 are Q-*symmetric* in H if and only if there is a measure-preserving bijection $p : H \to H$ such that $p(Q) = Q$ and $p(h_1) = h_2$. $\triangle$

In the definition of 'Q-symmetric' each measure-preserving bijection $p :$ $H \to H$ is a renaming of the hypotheses. Q represents the way the hypothesis set H "looks," and the requirement that $p(Q) = Q$ means that the renaming cannot affect the way H looks. For example, if $H = \{h_1, h_2, h_3\}$ with names 'a', 'b', and 'c', respectively, and $Q = \{\{h_1, h_2\}, \{h_2, h_3\}\}$, the renaming $p_1 : (a, b, c) \to (c, b, a)$ preserves Q, but the renaming $p_2 : (a, b, c) \to (c, a, b)$ gives $p_2(Q) = \{\{h_3, h_1\}, \{h_1, h_2\}\} \neq Q$. Suppose we say, "Pick a." We are referring to h_1. But if the hypotheses are renamed via p_1 we see H in exactly the same way yet we are referring now to h_3 instead of h_1; and thus h_1 and h_3 are Q-symmetric. Two hypotheses are Q-symmetric if a renaming that swaps their names can occur that does not affect the way H looks. Only arbitrary names distinguish Q-symmetric hypotheses; and so we say that Q-symmetric hypotheses cannot be distinguished non-arbitrarily. Another way of stating this is that there is no name-independent way of referring to either h_1 or h_3 because they are the same symmetry type. h_1 and h_3 are of the same type in the sense that each has a property shared by just one other hypothesis, and that other hypothesis is the same in each case.

But cannot one distinguish h_1 from h_3 by the fact that they have different properties? The first property of Q is, say, 'being red,' the second 'being short.' h_1 is red and not short, h_3 is short and not red. However, so the intuition goes, just as it is not possible to non-arbitrarily refer to h_1 because of the "names should not matter" motto, it is not possible to non-arbitrarily refer to the red hypotheses since $p_1(Q) = Q$ and $p_1(\{h_1, h_2\}) = \{h_3, h_2\}$ (i.e., $p_1(red) = short$). Our attempt to refer to the red hypotheses by the utterance "the red ones" would actually refer to the short hypotheses if 'red' was the name for short things. The same observation holds for, say, $Q' = \{\{h_\alpha\}, \{h_\beta\}, \{h_\gamma\}\}$. The fact that each has a distinct property does not help us to refer to any given one non-arbitrarily since each pair is Q'-symmetric.

Consider h_2 from above for a moment. It is special in that it has the unique property of being the only hypothesis having both properties. I say that a hypothesis is *Q-invariant* in H if and only if it is Q-symmetric only with itself. h_2 is invariant (the white kitten was invariant as well, as was the perpendicular line). Intuitively, invariant hypotheses can be non-arbitrarily referred to.

Three other notions related to 'symmetric' we use later are the following: First, $I(Q, H)$ is the set of Q-invariant hypotheses in H, and $\neg I(Q, H)$ is its complement in H. Above, $I(Q, H) = \{h_2\}$, and $\neg I(Q, H) = \{h_1, h_3\}$. Second, a paradigm Q is called *totally symmetric* on H if and only if the hypotheses in H are pairwise Q-symmetric. Q' above is totally symmetric (on $\{h_\alpha, h_\beta, h_\gamma\}$). Third, t is a *Q-symmetry type* in H if and only if t is an equivalence class with respect to the relation 'Q-symmetric'. $\{h_2\}$ and $\{h_1, h_3\}$ are the Q-symmetry types. In each of the terms we have defined, we omit Q or H when either is clear from context.

The Q-symmetry types are the most finely grained objects one can speak of or distinguish via the paradigm Q. One can distinguish between *no* hypotheses when the paradigm is totally Q-symmetric. When we say that a property is "acknowledged" we mean that the property is in the paradigm. Acknowledging a property does not mean that it is distinguishable, however, as we saw above with Q'. When we say that a property is "distinguishable" we mean that it is a symmetry type (but not necessarily a set appearing in the paradigm). $\{h_1, h_2\}$ is acknowledged in Q above but is not distinguishable. $\{h_2\}$ is distinguishable but not acknowledged in the paradigm.

Invariant hypotheses, then, can be non-arbitrarily referred to—non-invariant hypotheses cannot. From the point of view of the paradigm, invariant hypotheses can be "picked for a reason," but non-invariant hypotheses cannot. In this sense to pick an invariant hypothesis is to make a non-random choice and to pick a non-invariant hypothesis is to make a random choice; however I will try to avoid using this terminology for there are already many rigorous notions of randomness and this is not one of them. Any "reason" or procedure that picks a non-invariant hypothesis picks, for all the same reasons, any other hypothesis in its symmetry type; where "reasons" cannot depend on names. We say that invariant hypotheses are more *defensible,* or less *arbitrary,* than non-invariant ones. Picking a hypothesis that is not invariant means that had it been named differently you would have chosen something else; this is bad because surely a defensible choice should not depend on the names. Invariant hypotheses would therefore seem, *a priori,* favorable to non-invariant hypotheses. More generally, the intuition is that hypotheses that are members of larger symmetry types are less preferred, as picking one would involve greater arbitrariness. These intuitions are realized by the rationality principles comprising Paradigm Theory (as we will see later).

Consider the following example. $H_a = \{h_0, h_1, h_2, h_3\}$, $Q_a = \{\{h_0\}, \{h_1\}, \{h_2\}, \{h_2, h_3\}\}$. The reader may check that h_0 is symmetrical to h_1, and that h_2 and h_3 are each invariant. Suppose one chooses h_0. Now suppose that the hypotheses h_0, h_1, h_2, h_3 are renamed h_1, h_0, h_2, h_3, respectively, under the action of p. Since $p(Q_a) = Q_a$, the choice of hypotheses is exactly the same. However, this time when one picks h_0, one has really picked h_1. h_3 is invariant because, intuitively, it is the only element that is not in a one-element set. h_2 is invariant because, intuitively, it is the only element occurring in a two-element set with an element that does not come in a one-element set.

One way to visualize paradigms of a certain natural class is as an undirected graph. Hypothesis set H and paradigm Q are *associated with* undirected graph G with vertices V and edges $E \subset V^2$ if and only if there is a bijection $p : V \to H$ such that $Q = \{\{p(v)\}|v \in V\} \cup \{\{p(v_1), p(v_2)\} \,|(v_1, v_2) \in E\}\}$. This just says that a graph can represent certain paradigms, namely those paradigms that (i) acknowledge each element in H and (ii) the other sets in Q are each composed of only two hypotheses. Consider the following graph.

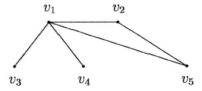

The associated hypothesis set is $H_b = \{v_1, \ldots, v_5\}$ and the associated paradigm is $Q_b = \{\{v_1\}, \ldots, \{v_5\}\} \cup \{\{v_1, v_2\}, \{v_1, v_3\}, \{v_1, v_4\}, \{v_1, v_5\}, \{v_2, v_5\}\}$. Notice that $\{v_1\}$, $\{v_2, v_5\}$, and $\{v_3, v_4\}$ are the Q_b-symmetry types; so only v_1 is Q_b-invariant—informally, it is the vertex that is adjacent to every other vertex. When visualized as graphs, one is able to *see* the symmetry.

Defensibility Hierarchy and Sufficient Reason

In the previous subsubsection I introduced the notion of a symmetry type, and we saw that a paradigm naturally induces a partition on the hypothesis set, where each partitions consists of hypotheses that are symmetric to one another. The symmetry types are the kinds of hypotheses that the inductive agent can distinguish, given his paradigm. Hypotheses that are members of smaller symmetry types can intuitively be chosen with less arbitrariness, as there are fewer hypotheses just like it as far as the paradigm is concerned. An invariant hypothesis—a hypothesis that is all alone in its symmetry type—can be chosen with the least arbitrariness since there are no other hypotheses symmetrical to it. Invariant hypotheses can, intuitively, be picked for a reason.

Although an invariant hypothesis may be able to be picked for a reason and is thus more defensible than non-invariant hypotheses, if there are one hundred other invariant hypotheses that can be picked for one hundred other reasons, how defensible can it be to choose that hypothesis? Why *that* reason and not any one of the others? Among the invariant hypotheses one may wonder if there are gradations of invariance. The way this may naturally be addressed is to restrict the hypothesis set to the invariant hypotheses, consider the induced paradigm on this set (we discuss what this means in a moment), and again ask what is invariant and what is not. Intuitively, concerning those hypotheses that can be picked for a reason, which of these *reasons* is justifiable? That is to say, which of these hypotheses can *now* be picked for a reason?

For the remainder of this subsubsection we say how to make this precise, but if you wish to skip the details, it will serve the purpose to simply know that there is a certain well-motivated, well-defined sense in which a paradigm induces a hierarchy of more and more defensible hypotheses, where being more defensible means that it can, intuitively, be picked with less arbitrariness.

A paradigm Q is just the set of acknowledged properties of the hypotheses in H. If one cares only about some subset H' of H, then the *induced paradigm* is just the one that acknowledges the same properties in H'. Formally, if $H' \subseteq H$, let $Q \sqcap H'$ denote $\{A \cap H' | A \in Q\}$, and call it the *induced paradigm* on H'. $Q \sqcap H'$ is Q after throwing out all of the hypotheses in $H - H'$. For example, let $H_d = \{h_0, h_1, h_2, h_3, h_4\}$ and $Q_d = \{\{h_0, h_2\}, \{h_1, h_2\}, \{h_3\}, \{h_2, h_3, h_4\}\}$. h_0 and h_1 are the non-invariant hypotheses; h_2, h_3 and h_4 are the invariant hypotheses. Now let H'_d be the set of invariant hypotheses, i.e.,

$H'_d = I(Q_d, H_d) = \{h_2, h_3, h_4\}$. The induced paradigm is $Q'_d = Q_d \sqcap H'_d = \{\{h_2\}, \{h_3\}, \{h_2, h_3, h_4\}\}$.

Now we may ask what is invariant at this new level. h_2 and h_3 are together in a symmetry type, and h_4 is invariant. h_4 is the least arbitrary hypothesis among H'_d; and since H'_d consisted of the least arbitrary hypotheses from H_d, h_4 is the least arbitrary hypothesis of all. This hierarchy motivates the following definition.

Definition 3 Fix hypothesis set H and paradigm Q. $H^0 = H$, and for any natural number n, $Q^n = Q \sqcap H^n$. For any natural number n, $H^{n+1} = I(Q^n, H^n)$, which just means that H^{n+1} consists of the invariant hypotheses from H^n. This hierarchy is the *defensibility hierarchy*, or the *invariance hierarchy*. $\triangle$

For instance, for H_d and Q_d above we had:

- $H^0_d = \{h_0, h_1, h_2, h_3, h_4\}$, $Q^0_d = \{\{h_0, h_2\}, \{h_1, h_2\}, \{h_3\}, \{h_2, h_3, h_4\}\}$.

- $H^1_d = \{h_2, h_3, h_4\}$, $Q^1_d = \{\{h_2\}, \{h_3\}, \{h_2, h_3, h_4\}\}$.

- $H^2_d = \{h_4\}$, $Q^2_d = \{\{h_4\}\}$.

- $H^3_d = \{h_4\}$, $Q^3_d = \{\{h_4\}\}$.

- etc.

For any hypothesis set H and paradigm Q there is an ordinal number $\alpha(Q, H)$ such that $H^\alpha = H^{\alpha+1}$; this is the *height* of the defensibility hierarchy of Q on H.[2] We say that a hypothesis h is *at level* m in the defensibility hierarchy if the highest level it gets to $\leq \alpha$ is the mth. For H_d/Q_d, h_2 is at level 1, or the second level of the defensibility hierarchy; h_4 is at level 2, or the third level. We let Δ_m denote the set of hypotheses at level m. Hypotheses at higher levels in the hierarchy are said to be *more defensible*. This defines 'defensibility' respecting our intuition that, other things being equal, the more defensible a hypothesis the less arbitrary it is. h_4 is the lone maximally defensible hypothesis, and the intuition is that it is the most non-arbitrary choice and should, *a priori*, be favored over every other hypothesis.

---

[2]When H is infinite it is possible that the least ordinal number α such that $H^\alpha = H^{\alpha+1}$ is transfinite. To acquire hypothesis sets H^β when β is a limit ordinal we must take the intersection of H^γ for all $\gamma < \beta$. $Q^\beta = Q \sqcap H^\beta$ (as usual).

For H_d/Q_d above, notice that h_2 and h_3 are similar in that, although they are not symmetric with each other at level 0, they *are* symmetric at level 1. We will say that they are Q_d-equivalent. Generally, two hypotheses are *Q-equivalent* in H if and only if at some level H^n they become symmetric (i.e., there is a natural number n such that they are $Q \sqcap H^n$-symmetric). Two invariant hypotheses may therefore be Q-equivalent but not Q-symmetric. d is a *Q-equivalence type* in H if and only if d is an equivalence class of Q-equivalent hypotheses. $\{h_0, h_1\}$, $\{h_2, h_3\}$ and $\{h_4\}$ are the Q_d-equivalence types, whereas $\{h_0, h_1\}$, $\{h_2\}$, $\{h_3\}$ and $\{h_4\}$ are the symmetry types. The equivalence types are therefore coarser grained than the symmetry types. Two members of an equivalence type are equally defensible. For Q-equivalence types d_1 and d_2, we say that d_1 is *less Q-defensible than d_2* if and only if for all $h \in d_1$ and $h' \in d_2$, h is less Q-defensible than h'. Our central intuition was that hypotheses that are more unique are to be preferred, *a priori*. Similarly we are led to the intuition that more defensible types of hypotheses are to be preferred, *a priori*. Paradigm Theory's rationality principles, presented in the next section, result in higher (actually, not lower) prior probability for more defensible equivalence types.

As an example, consider the paradigm represented by the following graph, where $H_f = \{a, \ldots, l\}$.

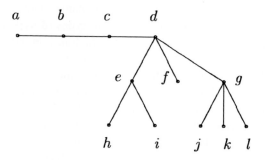

The symmetry types are $\{h, i\}$, $\{j, k, l\}$ and every other vertex is in a singleton symmetry type. The defensibility types are $\{h, i\}$, $\{j, k, l\}$, $\{e, f, g\}$, $\{a, d\}$ and $\{b, c\}$. The defensibility levels are $\Delta^0 = \{h, i, j, k, l\}$, $\Delta^1 = \{e, f, g\}$, and $\Delta^2 = \{a, b, c, d\}$.

We noted earlier that invariant hypotheses can be picked "for a reason," and this is reminiscent of Leibniz's Principle of Sufficient Reason, although not with his metaphysical import,[3] which says, in Leibniz's words, "we can find no true or existent fact, no true assertion, without there being a sufficient reason why it is thus and not otherwise..." (Ariew and Garber, 1989, p. 217.) Rewording our earlier intuition, we can say that invariant hypotheses can be picked "for sufficient reason." The problem with this statement is, as we have seen, that there may be multiple invariant hypotheses, and what sufficient reason can there be to pick from among them? This subsubsection's defensibility hierarchy answers this question. It is perhaps best said that lone maximally defensible hypotheses may be picked "for sufficient reason." More important is that the defensibility hierarchy is a natural formalization and generalization of Leibniz's Principle of Sufficient Reason (interpreted non-metaphysically only), giving a more finely grained breakdown of "how sufficient" a reason is for picking a hypothesis: hypotheses in smaller symmetry types possess more sufficient reason, and hypotheses higher in the hierarchy possess (other things equal) more sufficient reason. Paradigm Theory, further, quantifies the degree of sufficiency of reason with real numbers in [0,1], as we will soon see.

3.1.3   Paradigm Theory's principles

In this subsection I present the guts of Paradigm Theory: its principles of ought. Let us first, though, sum up the previous subsection: I showed how acknowledging *any* set of subsets of a hypothesis set—i.e., a paradigm—naturally determines a complex hierarchical structure. We saw that the "names should not matter" motto leads to a partition of the hypothesis set into types of hypotheses: the symmetry types. Among those hypotheses that are the lone members of their symmetry type—i.e., the invariant (or "unique") hypotheses—there may be some hypotheses that are "more" invariant, and among *these* there may some that are "even more" invariant, etc. This led to the defensibility, or invariance, hierarchy. Hypotheses that "become symmetric" at some level of the hierarchy are equivalent, and are said to be members of the same equivalence type.

We also noted in Subsection 3.1.2 the following related intuitions for which we would like principled ways to quantitatively realize: *a priori*, (i) hypotheses in smaller symmetry types are more favorable; or, more unique hypotheses are to be preferred as it takes less arbitrariness to choose them, (ii) (equivalence)

---

[3]Leibniz believed that Sufficient Reason arguments actually determine the way the world must be. However, he did seem, at least implicitly, to allow the principle to be employed in a purely epistemic fashion, for in a 1716 letter to Newton's friend and translator Samuel Clarke, Leibniz writes, "has not everybody made use of the principle upon a thousand occasions?" (Ariew and Garber, 1989, p. 346).

types of hypotheses that are more defensible are more favorable, and (iii) the lone most defensible hypothesis—if there is one—is most favorable (this follows from (ii)). Each is a variant of the central intuition that less arbitrary hypotheses are, *a priori,* more preferred.

These intuitions follow from the three rationality principles concerning prior probabilities I am about to present. The principles are conceptually distinct from these intuitions, having intuitive motivations of their own. The fact that two unrelated sets of intuitions converge in the way we see below is a sort of argument in favor of Paradigm Theory, much like the way different intuitions on computability leading to the same class of computable functions is an argument for Church's Thesis. The motivations for stating each principle is natural and intuitive, and the resulting prior probability distributions are natural and intuitive since they fit with intuitions (i), (ii) and (iii).

The first subsubsection presents the three principles of rationality, the next discusses the use of "secondary paradigms" to acquire more detailed prior probability distributions, and the final subsubsection sets forth the sort of explanations Paradigm Theory gives.

The Principles

Paradigm Theory consists of three principles of rationality that, from a given paradigm (and a hypothesis set with a finite measure), determine a prior probability distribution. Paradigm Theory as developed in this section is only capable of handling cases where there are finitely many symmetry types.[4] We

---

[4] If one begins with H and Q such that there are infinitely many symmetry types, one needs to restrict oneself to a proper subset H' of H such that there are only finitely many symmetry types with respect to the induced paradigm. There are some compelling rationality constraints on such a restriction that very often suffice: (i) any two members of the same equivalence type in H either both appear in H' or neither, (ii) if an equivalence type from H appears in H', then (a) all more defensible equivalence types appear in H', and (b) all equally defensible equivalence types in H that are the same size or smaller appear in H'. These constraints on hypothesis set reduction connect up with the observation that we do not seriously entertain all logically possible hypotheses. This is thought by F. Suppe (1989, p. 398) "to constitute one of the deepest challenges we know of to the view that science fundamentally does reason and proceed in accordance with inductive logic." These rationality constraints help guide one to focus on the *a priori* more plausible hypotheses, ignoring the rest, and is a first step in addressing this challenge. These constraints give us the ability to begin to break the bonds of a logic of discovery of a prior assessment sort, and claim some ground also as a logic of discovery of a hypothesis generation sort: hypotheses are generated in the first place by "shaving off" most of the other logically possible hypotheses.

will assume from here on that paradigms induce just finitely many symmetry types.[5]

Assume hypothesis set H and paradigm Q are fixed. $P(A)$ denotes the probability of the set A. $P(\{h\})$ is often written as $P(h)$.

Principle of Type Uniformity

Recall that the symmetry types are precisely the types of hypotheses that can be referred to with respect to the paradigm. Nothing more finely grained than symmetry types can be spoken of. *Prima facie,* a paradigm gives us no reason to favor any symmetry type (or "atom") over any other. To favor one over another would be to engage in arbitrariness. These observations motivate the first principle of Paradigm Theory of Induction.

Principle of Type Uniformity: *Every (symmetry) type of hypothesis is equally probable.*

There are other principles in the probability and induction literature that are akin to the Principle of Type Uniformity. For example, if the types are taken to be the complexions (where two strings are of the same complexion if they have the same number of each type of symbol occurring in it), then Johnson's Combination Postulate (Johnson, 1924, p. 183) says to set the probability of the complexions equal to one another. Carnap's m^* amounts to the same thing.

The (claimed) rationality of the Principle of Type Uniformity emanates from the seeming rationality of choosing a non-arbitrary prior; to choose a non-uniform prior over the symmetry types would mean to give some symmetry types higher probability for no good reason. Is favoring some symmetry types over others necessarily arbitrary? Through the eyes of a paradigm the symmetry types are distinguishable, and might not there be aspects of symmetry types that make some, *a priori,* favorable? If any are favorable, it is not because any is distinguished among the symmetry types; each is equally distinguished. Perhaps some could be favorable by virtue of having greater size? Size is, in fact, relevant in determining which sets are the symmetry types. Actually, though, it is size *difference*, not size, that is relevant in symmetry type determination. Paradigms are not capable of recognizing the size of symmetry types; symmetry types *are* the primitive entities, or atoms, in the paradigm's

---

[5]This restriction ensures that the height of the defensibility hierarchy is finite (although having infinitely many symmetry types does not entail a transfinite height).

ontology. From the paradigm's point of view, symmetry types cannot be favored on the basis of their being larger. Given that one possesses a paradigm and nothing else (like particular inductive beliefs), it is plausible that anything but a uniform distribution on the symmetry types would be arbitrary.

Now, perhaps one could argue that the weakness of paradigms—e.g., their inability to acknowledge larger symmetry types—counts against Paradigm Theory. Paradigm Theory aims to be a "blank slate" theory of induction, taking us from innocuous ways of carving the world to particular degrees of belief. Paradigms are innocuous in part because of their weakness. Strengthening paradigms to allow the favoring of symmetry types over others would have the downside of decreasing the explanatory power; the more that is packed into paradigms, the less surprising it is to find that, given them, they justify particular inductive methods. That is my motivation for such a weak notion of paradigm, and given only such a weak paradigm, the Principle of Type Uniformity is rational since to not obey it is to engage in a sort of arbitrariness.

Principle of Symmetry

The second principle of rationality is a general way of asserting that the renaming of objects should not matter (so long as the paradigm Q is unaltered). Recall the convention that the underlying measure on H is finite.

Principle of Symmetry: *Within a symmetry type, the probability distribution is uniform.*

For finite H this is: hypotheses of the same type are equally probable, or, hypotheses that can be distinguished only by their names or the names of their properties are equally probable. Unlike the Principle of Type Uniformity whose intuition is similar to that of the Classical Principle of Indifference (which says that if there is no known reason to prefer one alternative over another, they should receive equal probability), the Principle of Symmetry is truly a symmetry principle. Violating the Principle of Symmetry would result in a prior probability distribution that would not be invariant under renamings that do not alter the paradigm; names would suddenly matter. Violating the Principle of Type Uniformity, on the other hand, would *not* contradict the "names should not matter" motto (and is therefore less compelling).

If one adopts the Principle of Symmetry without the Principle of Type Uniformity, the result is a Generalized Exchangeability Theory. Each paradigm induces a partition of symmetry types, and the Principle of Symmetry, alone, requires only that the probability within a symmetry type be uniform. When the hypothesis set is the set of strings of outcomes (0 or 1) of an experiment and the paradigm is such that the symmetry types are the complexions (see Q_L then the Principle of Symmetry just *is* Johnson's Permutability Postulate (Johnson, 1924, pp. 178–189), perhaps more famously known as de Finetti's Finite Exchangeability.

The Basic Theory

Carnap's m^*-based theory of logical probability (Carnap, 1950, p. 563)—which I will call *Carnap's logical theory*—uses versions of the Principles of Type Uniformity and Symmetry (and leads to the inductive method he calls c^*). His "structure-descriptions," which are analogous to complexions, are given equal probability, which amounts to the use of a sort of Principle of Type Uniformity on the structure-descriptions. Then the probabilities are uniformly distributed to his "state-descriptions," which are analogous to individual outcome strings of experiments, which amounts to a sort of Principle of Symmetry. But whereas Carnap (and Johnson) is confined to the case where the partition over the state-descriptions is given by the structure-descriptions (or for Johnson, the partition over the outcome strings is given by the complexions), Paradigm Theory allows the choice of partition to depend on the choice of paradigm and is therefore a natural, powerful generalization of Carnap's m^*-based Logical Theory. The paradigm determines the symmetry types, and the symmetry types play the role of the structure-descriptions. When the hypothesis set is totally symmetric, one gets something akin to Carnap's $m^\dagger$-based logical theory (which he calls $c^\dagger$).

It is convenient to give a name to the theory comprised by the first two principles alone.

Basic Theory: *Assign probabilities to the hypothesis set satisfying the Principles of Type Uniformity and Symmetry.*

Applying the Basic Theory to H_a and Q_a from Subsection 3.1.2, we get $P(h_0) = P(h_1) = 1/6$ and $P(h_2) = P(h_3) = 1/3$. Applying the Basic Theory to H_b and Q_b from the same subsection, we get $P(v_1) = 1/3$, and the remaining vertices each receive probability $1/6$. Applying it to H_d and Q_d, $P(h_0) = P(h_1) = 1/8$ and $P(h_2) = P(h_3) = P(h_4) = 1/4$.

Notice that since the underlying measure of the hypothesis set is finite, the probability assignment for the Basic Theory is unique. For $h \in H$ let $c(h)$ be the cardinality of the symmetry type of h. Let w denote the number of symmetry types in H. The following theorem is obvious.

Theorem 1 *Fix finite H. The following is true about the Basic Theory. For all $h \in H$, $P(h) = \frac{1}{w \cdot c(h)}$.* $\triangle$

Theorem 1 may be restated more generally to include infinite hypothesis sets: for any measure μ and all $A \subseteq H$ with measure μ that are a subset of the same symmetry type, $P(A) = \frac{1}{w\mu}$.

We see that the probability of a hypothesis is inversely proportional to both the number of symmetry types and the number (or measure) of other hypotheses of the same symmetry type as itself. The fraction $1/w$ is present for every hypothesis, so $c(h)$ is the variable which can change the probabilities of hypotheses relative to one another. The more hypotheses in a type, the less probability we give to each of those hypotheses; this fits with our earlier intuition number (i) from the beginning of this section. The following corollary records that the Basic Theory fits with this intuition and the intuition that invariant hypotheses are more probable. The corollary is true as stated no matter the cardinality of the hypothesis set.

Theorem 2 *The following are true about the Basic Theory.*

1. *Hypotheses in smaller symmetry types acquire greater probability.*

2. *Each invariant hypothesis receives probability $1/w$, which is greater than (in fact, at least twice as great as) that for any non-invariant hypothesis.* $\triangle$

The Basic Theory is not Paradigm Theory, although when the defensibility hierarchy has no more than two levels the two theories are equivalent. The Basic Theory does not notice the hierarchy of more and more defensible hypotheses, and noticing the hierarchy will be key to providing a general explanation for why simpler hypotheses ought to be favored. When I say things like, "only the Basic Theory is needed to determine such and such probabilities," I mean that the probabilities are not changed upon the application of the third principle (to be stated below) of Paradigm Theory.

Principle of Defensibility
The third principle of rationality is, as far as I know, not similar to any previous principle in the induction and probability literature. It encapsulates an intuition

similar to that used when I discussed gradations of invariance in Subsection 3.1.2. I asked: Among the invariant elements, which are more defensible? Now I ask: Among the invariant elements, which are more probable? From the viewpoint of the entire hypothesis set H the invariant hypotheses seem equally and maximally defensible. But when focusing only on the invariant hypotheses we see further gradations of defensibility. Similarly, from the viewpoint of the entire hypothesis set H the invariant hypotheses look equally and maximally probable. But when focusing only on the invariant hypotheses we see further gradations of probability. The third principle of rationality says to refocus attention on the invariant hypotheses.

Principle of Defensibility: *Reapply the Principles of Type Uniformity, Symmetry, and Defensibility to the set of invariant hypotheses* $(H' = I(Q, H))$ *via the induced paradigm* $(Q \sqcap H')$.

Since the Principle of Defensibility is one of the three rationality principles mentioned in its own statement, it applies to itself as well. I have named the principle the Principle of Defensibility because it leads to the satisfaction of intuition (ii) from the beginning of this section, i.e., to more defensible types of hypotheses acquiring higher prior probability. However, neither the intuitive motivation for the principle nor the statement of the principle itself hints at this intuition. The principle only gets at the idea that there is structure among the invariant hypotheses and that it should not be ignored.

Paradigm Theory

With the three principles presented I can state Paradigm Theory.

Paradigm Theory: *Assign probabilities to the hypothesis set satisfying the Principles of Type Uniformity, Symmetry, and Defensibility.*

The proposal for the prior probability assignment is to use the principles in the following order: (i) Type Uniformity, (ii) Symmetry, and (iii) Defensibility (i.e., take the invariant hypotheses and go to (i)). These principles amount to a logical confirmation function, as in the terminology of Carnap, but ours is a function of a hypothesis h, evidence e, *and* paradigm Q; i.e., $c(h, e, Q)$.

Paradigm Theory is superior to the Basic Theory in the sense that it is able to distinguish higher degrees of defensibility. Paradigm Theory on H_a/Q_a and

H_b/Q_b from Section 3.1.2 behaves identically to the Basic Theory. Applying Paradigm Theory to H_d and Q_d is different, however, than the Basic Theory's assignment. First we get, as in the Basic Theory, $P(h_0) = P(h_1) = 1/8$, $P(h_2) = P(h_3) = P(h_4) = 1/4$. Applying the Principle of Defensibility, the probability assignments to h_0 and h_1 remain fixed, but the 3/4 probability assigned to the set of invariant hypotheses is to be redistributed among them. With respect to $\{h_2, h_3, h_4\}$ and the induced paradigm $\{\{h_2, h_3\}, \{h_4\}\}$, the symmetry types are $\{h_2, h_3\}$ and $\{h_4\}$, so each symmetry type receives probability $(3/4)/2 = 3/8$. The probabilities of $h_0, \ldots h_4$ are, respectively, 2/16, 2/16, 3/16, 3/16, 6/16. Recall that h_4 is the lone most defensible element but the Basic Theory gave it the same probability as h_2 and h_3; Paradigm Theory allows richer assignments than the Basic Theory.

It is easy to see that since the underlying measure of the hypothesis set is finite and there are assumed to be only finitely many symmetry types, Paradigm Theory assigns a unique probability distribution to the hypothesis set, and does so in such a way that each hypothesis receives positive prior probability density (i.e., priors are always "open-minded" within Paradigm Theory). Theorem 14 in the appendix at the end of this chapter examines some of its properties. Unlike the Basic Theory, Paradigm Theory respects the intuition (number (ii)) that more defensible (less arbitrary) implies higher probability by giving the more defensible equivalence types not less probability than the less defensible equivalence types. Also, unlike the Basic Theory, Paradigm Theory respects the intuition (number (iii)) that if a hypothesis is lone most defensible (the only least arbitrary one) then it receives higher probability than every other hypothesis. The following theorem states these facts; the proofs along with other properties are given in the appendix to this chapter.

Theorem 3 *The following are true about Paradigm Theory.*

1. *For all equivalence types d_1 and d_2, if d_1 is less defensible than d_2, then $P(d_1) \leq P(d_2)$.*

2. *For all hypotheses h; h is the lone most defensible if and only if for all $h' \neq h$, $P(h') < P(h)$.* $\triangle$

Theorem 3 is an argument for the superiority of Paradigm Theory over the Basic Theory.

Secondary Paradigms

Suppose we have found the prior probability distribution on H given a paradigm Q, and, say, half of the hypotheses end up with the same probability; call this subset H^*. Now what if we acknowledge other properties concerning H^*, properties which are, in some sense, *secondary* to the properties in the original paradigm? May H^*'s probabilities be validly redistributed according to this secondary paradigm? After all, cannot any hypothesis set and paradigm be brought to Paradigm Theory for application, including H^* and this secondary paradigm? The problem is that to do this would be to modify the original, or *primary* probability distribution, and this would violate the principles in the original application of Paradigm Theory.

Here is an example of the sort of thing I mean. Let $H = \{3, \ldots, 9\}$ and Q acknowledge the property of being prime. There are two symmetry types, $\{4, 6, 8, 9\}$ and $\{3, 5, 7\}$, each receiving probability $1/2$. Now suppose that there are secondary paradigms for each symmetry type, in each case acknowledging the property of being odd. The second symmetry type above remains unchanged since all are odd, but the first gets split into $\{4, 6, 8\}$ and $\{9\}$, each receiving probability $1/4$. Notice that this is different than what a primary paradigm that acknowledges both being prime and odd gives; in this case the probability of $\{3, 5, 7\}$, $\{4, 6, 8\}$ and $\{9\}$ are $1/3$, $1/3$, $1/3$ instead of, respectively, $1/2$, $1/4$, $1/4$, as before. The first method treats being prime as more important than being odd in the sense that primality is used to determine the large-scale probability structure, and parity is used to refine the probability structure. The second method treats being prime and being odd on a par. A more Kuhnian case may be where one allows the primary paradigm to acknowledge scope, and allows the secondary paradigm to acknowledge simplicity; this amounts to caring about scope first, simplicity second.

I generalize Paradigm Theory to allow such secondary paradigms in a moment, but I would first like to further motivate it. There is a sense in which Paradigm Theory, as defined thus far, is artificially weak. For simplicity consider only the Principles of Type Uniformity and Symmetry; i.e., the Basic Theory. These two principles are the crux of the probability assignment on the hypothesis set. Together they allow only two "degrees of detail" to probability assignments: one assignment to the symmetry types, and another to the particular hypotheses within the symmetry types. The Principle of Defensibility does allow further degrees of detail *for the invariant hypotheses*, and it accomplishes this without the need for secondary paradigms. But for non-invariant hypotheses there are just two degrees of detail. Why two? This seems to be a somewhat artificial limit.

Allowing secondary paradigms enables Paradigm Theory to break this limit. Paradigm Theory is now generalized in the following way: *Secondary paradigms may modify the primary prior probability distribution by applying the three principles to any subset H^* such that the primary prior in H^* is uniform.* In other words, we are licensed to tinker with the primary prior using secondary paradigms, so long as we tinker only on subsets that were originally equiprobable. When H^* and a secondary paradigm Q^* are brought to Paradigm Theory for application, they can be treated as creating their own primary distribution within H^*. Secondary paradigms with respect to H^* and Q^* are *tertiary* paradigms with respect to the original hypothesis set H and paradigm Q. The point is that any degree of detail in the sense mentioned above is now sanctioned, so long as there are n^{th}-ary paradigms for large enough n.

All this increase in power may make one skeptical that one can create any prior one wants by an ad hoc tuning of the secondary (tertiary, and so on) paradigms. An explanation by Paradigm Theory is only as natural and explanatory as is the paradigm (primary, secondary, and so on) used (see Section 3.1.3). Ad hoc secondary paradigms create ad hoc explanations. The only use of paradigms in this chapter beyond primary ones are secondary ones. I use them later where they are quite explanatory and give Paradigm Theory the ability to generalize a certain logical theory of Hintikka's ($\alpha = 0$). I also note in Subsection 3.2.3 their ability to give a non-uniform prior over the simplest hypotheses. If in any particular application of Paradigm Theory there is no mention of secondary paradigms, then they are presumed not to exist.

The Paradigm Theory Tactic

In the following section Paradigm Theory is used to explain why certain inductive methods we tend to believe are justified are, indeed, justified. The general tactic is two-fold. First, a mathematical statement concerning the power of Paradigm Theory is given (often presented as a theorem). Second, an informal explanatory argument is given. Paradigm Theory's ability to justify induction is often through the latter.

Most commonly, the mathematical statement consists of showing that paradigm Q entails inductive method x. This alone only shows that inductive method x is or is not within the scope of Paradigm Theory; and this is a purely mathematical question. Such a demonstration is not enough to count as an explanation of the justification of inductive method x. Although paradigm Q may determine inductive method x, Q may be artificial or ad hoc and thereby not be very explanatory; "who would carve the world *that* way?" If Q is very unnatural and no natural paradigm entails inductive method x, then this may provide an explanation for why inductive method x is disfavored: one would have to possess a very strange conceptual framework in order to acquire it, and given that we do not possess such strange conceptual frameworks, inductive method x is not justified. Typically, the paradigm Q determining inductive method x is natural, and the conclusion is that inductive method x is justified because we possess Q as a conceptual framework. I do not actually argue that we *do* possess any particular paradigm as a conceptual framework. Rather, "inductive method x is justified because we possess paradigm Q" is meant to indicate the form of a possible explanation in Paradigm Theory. A fuller explanation would provide some evidence that we in fact possess Q as conceptual framework.

A second type of mathematical statement is one stating that every paradigm entails an inductive method in the class Z. The explanatory value of such a statement is straightforward: every conceptual framework leads to such inductive methods, and therefore one cannot be a skeptic about inductive methods in Z; any inductive method not in Z is simply not rational. A sort of mathematical statement that sometimes arises in future sections is slightly weaker: every paradigm Q *of such and such type* entails an inductive method in the class Z. The explanatory value of this is less straightforward, for it depends on the status of the "such and such type." For example, open-mindedness is of this form for the Personalistic (of Subjective) Bayesian on the hypothesis set $H = [0, 1]$: every prior that is open-minded (everywhere positive density) converges in the limit to the observed frequency. If the type of paradigm is extremely broad and natural, and every paradigm not of that type is not natural, then one can conclude that inductive skepticism about inductive methods in Z is not possible, unless one is willing to possess an unnatural paradigm; inductive skepticism about inductive methods in Z is not possible because every non-artificial conceptual framework leads to Z. Similar observations hold for arguments of the form, "no paradigm Q of such and such type entails an inductive method in the class Y."

These claims of the "naturalness" of paradigms emanate from our (often) shared intuitions concerning what properties are natural. The naturalness of a paradigm is *not* judged on the basis of the naturalness of the inductive method to which it leads; this would ruin the claims of explanatoriness.

3.2 Applications

3.2.1 Simple preliminary applications

By way of example we apply the Basic and Paradigm Theories to some preliminary applications, first presented in Changizi and Barber (1998).

Collapsing to the Principle of Indifference

Paradigm Theory (and the Basic Theory) gives the uniform distribution when the paradigm is empty. This is important because, in other words, Paradigm Theory collapses to a uniform prior when no properties are acknowledged, and this is a sort of defense of the Classical Principle of Indifference: be ignorant *and* acknowledge nothing... get a uniform prior. More generally, a uniform distribution occurs whenever the paradigm is totally symmetric. Since being totally symmetric means that there are no distinctions that can be made among the hypotheses, we can say that *Paradigm Theory collapses to a uniform prior when the paradigm does not have any reason to distinguish between any of the hypotheses.* Only the Principle of Symmetry—and not the Principle of Type Uniformity—needs to be used to found the Principle of Indifference as a subcase of Paradigm Theory.

Archimedes' Scale

Given a symmetrical scale and (allegedly) without guidance by prior experiment Archimedes (*De aequilibro*, Book I, Postulate 1) predicts the result of hanging equal weights on its two sides. The hypothesis set in this case is plausibly the set of possible angles of tilt of the scale. Let us take the hypothesis set to include a finite (but possibly large) number, N, of possible tilting angles, including the horizontal, uniformly distributed over the interval $[-90°, 90°]$. Archimedes predicts that the scale will remained balanced, i.e., he settles on $\theta = 0°$ as the hypothesis. He makes this choice explicitly on the basis of the obvious symmetry; that for any $\theta \neq 0°$ there is the hypothesis $-\theta$ which is "just as good" as θ, but $\theta = 0°$ has no symmetric companion.

To bring this into Paradigm Theory, one natural paradigm is the one that acknowledges the amount of tilt but does not acknowledge which way the tilt is; i.e., $Q = \{\{-\theta, \theta\} | 0° \leq \theta \leq 90°\}$. $\theta = 0°$ is the only hypothesis in a single-element set in Q, and it is therefore invariant. Furthermore, every other hypothesis can be permuted with at least its negation, and so $\theta = 0°$ is the only invariant hypothesis. With the paradigm as stated, any pair $-\theta, \theta$ (with $\theta > 0°$) can permute with any other pair, and so there are two symmetry types: $\{0°\}$ and everything else. Thus, $0°$ receives prior probability $1/2$, and every other hypothesis receives the small prior probability $1/(2 \cdot (N - 1))$. Even if N is naturally chosen to be 3—the three tilting angles are $-90°$, $0°$ and $90°$—the prior probabilities are $1/4$, $1/2$ and $1/4$, respectively.

Now let the paradigm be the one acknowledging the property of being within $\theta°$ from horizontal, for every $\theta \in [0°, 90°]$. For each $\theta \in H$, $\{-\theta, \theta\}$ is a symmetry type, and this includes the case when $\theta = 0°$, in which case the symmetry type is just $\{0°\}$. Each symmetry type receives equal prior probability by the Principle of Type Uniformity, and by the Principle of Symmetry each $\theta \neq 0°$ gets half the probability of its symmetry type. $0°$ gets all the probability from its symmetry type, however, as it is invariant. Therefore it is, *a priori*, twice as probable as any other tilting angle. If N is chosen to be 3, the prior probabilities for $-90°$, $0°$ and $90°$ are as before: $1/4$, $1/2$ and $1/4$, respectively.

Explanations for such simple cases of symmetry arguments can sometimes seem to be assumptionless, but certain *a priori* assumptions are essential. Paradigm Theory explains Archimedes' prediction by asserting that he possessed one of the paradigms above as a conceptual framework (or some similar sort of paradigm). He predicts that the scale will remained balanced because, roughly, he acknowledges the angle of tilt but not its direction. Most natural paradigms will entail priors favoring $\theta = 0°$, and I suspect no natural paradigm favors any other.

Leibniz's Triangle

To a second historical example, I noted earlier the connection of Paradigm Theory to Leibniz's Principle of Sufficient Reason (interpreted non-metaphysically), and I stated that Paradigm Theory is a sort of generalization of the principle, giving precise real-valued degrees to which a hypothesis has sufficient reason to be chosen. Let us now apply Paradigm Theory to an example of Leibniz. In a 1680s essay, he discusses the nature of an unknown triangle.

> And so, if we were to imagine the case in which it is agreed that a triangle of given circumference should exist, without there being anything in the givens from which one could determine what kind of triangle, freely, or course, but without a doubt. There is nothing in the givens which prevents another kind of triangle from existing, and so, an equilateral triangle is not necessary. However,

all that it takes for no other triangle to be chosen is the fact that in no triangle
except for the equilateral triangle is there any reason for preferring it to others.
(Ariew and Garber, 1989, p. 101.)

Here the hypothesis set is plausibly $\{\langle\theta_1,\theta_2,\theta_3\rangle|\ \theta_1 + \theta_2 + \theta_3 = 180°\}$,
where each 3-tuple defines a triangle, θ_i being the angle of vertex i of the
triangle. Now consider the paradigm that acknowledges the three angles of a
triangle, but does not acknowledge which vertex of the triangle gets which an-
gle; i.e., $Q = \{\{\langle\theta_1,\theta_2,\theta_3\rangle, \langle\theta_3,\theta_1,\theta_2\rangle, \langle\theta_2,\theta_3,\theta_1\rangle, \langle\theta_3,\theta_2,\theta_1\rangle, \langle\theta_1,\theta_3,\theta_2\rangle,$
$\langle\theta_2,\theta_1,\theta_3\rangle\}\ |\ \theta_1 + \theta_2 + \theta_3 = 180°\}$. This natural paradigm, regardless of the
hypothesis set's underlying measure, results in $\langle60°, 60°, 60°\rangle$ being the only
invariant hypothesis. In fact, every other of the finitely many symmetry types
is of the size continuum, and thus every hypothesis but the 60° one just mentioned receives infinitesimal prior probability. An explanation for why Leibniz
believed the equilateral triangle must be chosen is because he possessed the
conceptual framework that acknowledged the angles but not where they are.

Straight Line

Consider a hypothesis set H consisting of all real-valued functions consistent
with a finite set of data falling on a straight line (and let the underlying measure
be cardinality). It is uncontroversial that the straight line hypothesis is the most
justified hypothesis. Informally, I claim that any natural paradigm favors—if
it favors any function at all—the straight line function over all others, and that
this explains why in such scenarios we all feel that it is rational to choose the
straight line. For example, nothing but the straight line can be invariant if one
acknowledges any combination of the following properties: 'is continuous', 'is
differentiable', 'has curvature κ' (for any real number κ), 'has n zeros' (for
any natural number n), 'has average slope of m' (for any real number m),
'changes sign of slope k times' (for any natural number k). One can extend
this list very far. For specificity, if the curvature properties are acknowledged
for each κ, then the straight line is the only function fitting the data with zero
curvature, and for every other value of curvature there are multiple functions
fitting the data that have that curvature; only the straight line is invariant and
Paradigm Theory gives it highest probability. The same observation holds for
the 'changes sign of slope k times' property. What is important is not any
particular choice of natural properties, but the informal claim that any natural
choice entails that the straight line is favored if any function is. The reader is
challenged to think of a natural paradigm that results in some other function in

H receiving higher prior probability than the straight line.

Reference

For a consistent set of sentences, each interpretation of the language making all the sentences true can be thought of as a hypothesis; that is, each model of the set of sentences is a hypothesis. The question is: Which model is, *a priori*, the most probable? Consider the theorems of arithmetic as our consistent set of sentences. There is one model of arithmetic, called the "standard model," that is considered by most of us to be the most preferred one. That is, if a person having no prior experience with arithmetic were to be presented with a book containing all true sentences of arithmetic (an infinitely long book), and this person were to attempt to determine the author's interpretation of the sentences, we tend to believe that the standard model should receive the greatest prior probability as the hypothesis. Is this preference justified?

Suppose that one's paradigm acknowledges models "fitting inside" other models, where a model M_1 *fits inside* M_2 if the universe of M_1 is a subset (modulo any isomorphism) of that of M_2 and, when restricted to the universe of M_1, both models agree on the truth of all sentences.[6] Intuitively, you can find a copy of M_1 inside M_2 yet both satisfactorily explain the truth of each sentence in the set. As such, M_2 is unnecessarily complex.[7] Does this paradigm justify the standard model? The standard model of arithmetic has the mathematical property that it fits inside any model of arithmetic; it is therefore invariant for this paradigm. We do not know of a proof that there is no other invariant (for this paradigm) model of arithmetic, but it is strongly conjectured that there is no other (M. C. Laskowski, private communication). If this is so, then the standard model is the most probable one (given this paradigm).

Paradigm Theory can be used to put forth a conceptual framework-based probabilistic theory of reference in the philosophy of language: *to members of a conceptual framework represented by paradigm Q, the reference of a symbol in a language is determined by its interpretation in the most probable model, where the prior probabilities emanate from Q and are possibly conditioned via Bayes' Theorem if evidence (say, new sentences) comes to light.* (See Putnam (1980, 1981, p. 33) for some discussion on underdetermination of interpretation and its effect on theories of reference, and Lewis (1984) for some commentary and criticism of it.

---

[6]In logic it is said in this case that M_1 *embeds elementarily* into M_2.
[7]This is a sort of "complexification;" see Subsection 3.2.3.

3.2.2   Enumerative Induction

I consider *enumerative induction* on two types of hypothesis set: (i) the set of strings of the outcomes (0 or 1) of N experiments or observations, and I denote this set H_N; (ii) the set of possible physical probabilities p in $[0, 1]$ of some experiment, with the uniform underlying measure. Three types of enumerative induction are examined: no- , frequency- , and law-inductions. *No-induction* is the sort of inductive method that is completely rationalistic, ignoring the evidence altogether and insisting on making the same prediction no matter what. *Frequency-induction* is the sort of inductive method that converges in the limit to the observed frequency of experimental outcomes (i.e., the ratio of the number of 0s to the total number of experiments). *Law-induction* is the sort of inductive method that is capable of giving high posterior probability to laws. 'all 0s' and 'all 1s' are the laws when $H = H_N$, and '$p = 0$' and '$p = 1$' are the laws when $H = [0, 1]$.

For reference throughout this section, Table 3.1 shows the prior probability assignments for the paradigms used in this section on the hypothesis set H_4.

No-Induction

The sort of no-induction we consider proceeds by predicting with probability .5 that the next experimental outcome will be 0, regardless of the previous outcomes.

$H = H_N$

First we consider no-induction on the hypothesis set H_N, the set of outcome strings for N binary experiments. Table 3.1 shows the sixteen possible outcome strings for four binary experiments. The first column of prior probabilities is the uniform assignment, and despite its elegance and simplicity, it does not allow learning from experience. For example, suppose one has seen three 0s so far and must guess what the next experimental outcome will be. The reader may easily verify that $P(0|000) = P(1|000) = 1/2$; having seen three 0s does not affect one's prediction that the next will be 0. The same is true even if one has seen one million 0s in a row and no 1s. This assignment is the one Wittgenstein proposes (1961, 5.15–5.154), and it is essentially Carnap's $m^\dagger$ (Carnap, 1950) (or $\lambda = \infty$).

Recall that a totally symmetric paradigm is one in which every pair of hypotheses is symmetric. Any totally symmetric paradigm entails the uniform

Table 3.1: The prior probability assignments for various paradigms over the hypothesis set H_4 (the set of possible outcome strings for four experiments) are shown. Q_{law_L} is shorthand for Q_{law} with Q_L as secondary paradigm. The table does not indicate that in the Q_{law} cases the 'all 0s' and 'all 1s' acquire probability $1/4$ no matter the value of N (in this case, $N = 4$); for the other paradigms this is not the case.

string	Q_s	Q_L	Q_{rep}	Q_{law}	Q_{law_L}
0000	1/16	1/5	1/8	1/4	1/4
0001	1/16	1/20	1/24	1/28	1/24
0010	1/16	1/20	1/24	1/28	1/24
0100	1/16	1/20	1/24	1/28	1/24
1000	1/16	1/20	1/24	1/28	1/24
0011	1/16	1/30	1/24	1/28	1/36
0101	1/16	1/30	1/8	1/28	1/36
0110	1/16	1/30	1/24	1/28	1/36
1001	1/16	1/30	1/24	1/28	1/36
1010	1/16	1/30	1/8	1/28	1/36
1100	1/16	1/30	1/24	1/28	1/36
0111	1/16	1/20	1/24	1/28	1/24
1011	1/16	1/20	1/24	1/28	1/24
1101	1/16	1/20	1/24	1/28	1/24
1110	1/16	1/20	1/24	1/28	1/24
1111	1/16	1/5	1/8	1/4	1/4

assignment on H_N. Therefore, any totally symmetric paradigm results in no-induction on H_N. This is true because there is just one symmetry type for a totally symmetric paradigm, and so the Principle of Symmetry gives each string the same prior probability. I have let Q_s denote a generic totally symmetric paradigm in Table 3.1.

The uniform assignment on H_N is usually considered to be inadequate on the grounds that the resulting inductive method is not able to learn from experience. There is a problem with this sort of criticism: it attributes the inadequacy of a particular prior probability assignment to the inadequacy of the inductive method to which it leads. If prior probabilities are chosen simply in order to give the inductive method one wants, then much of the point of prior probabilities is missed. Why not just skip the priors altogether and declare the desired inductive method straightaway? In order to be explanatory, prior probabilities must be chosen for reasons independent of the resulting inductive method. We want to explain the lack of allure of the uniform prior on H_N *without* referring to the resulting inductive method.

One very important totally symmetric paradigm is the empty one, i.e., the paradigm that acknowledges nothing. If one considers H_N to be the hypothesis set, and one possesses the paradigm that acknowledges no properties of the hypotheses at all, then one ends up believing that each outcome string is equally likely. I believe that for H_N the paradigm that acknowledges nothing is far from natural, and this helps to explain why no-induction is treated with disrepute. To acknowledge nothing is to not distinguish between the 'all 0s' string and any "random" string; for example, 0000000000 and 1101000110. To acknowledge nothing is also to not acknowledge the relative frequency. More generally, any totally symmetric paradigm, no matter how complicated the properties in the paradigm, does not differentiate between any of the outcome strings and is similarly unnatural. For example, the paradigm that acknowledges every outcome string is totally symmetric, the paradigm that acknowledges every pair of outcome strings is totally symmetric, and the paradigm that acknowledges every property is also totally symmetric. No-induction is unjustified because we do not possess a conceptual framework that makes no distinctions on H_N. On the other hand, if one really does possess a conceptual framework that makes no distinctions among the outcome strings, then no-induction *is* justified.

There are some ad hoc paradigms that do make distinctions but still entail a uniform distribution over H_N. For example, let paradigm Q acknowledge $\{1\}, \{1, 2\}, \{1, 2, 3\}, \ldots, \{1, \ldots, 16\}$, where these numbers denote the corresponding strings in Table 3.1. Each string is then invariant, and therefore can be distinguished from every other, yet the probability assignment is uniform by the Principle of Type Uniformity. For another example, let the paradigm consist of $\{1, \ldots, 8\}$ and $\{1, \ldots, 16\}$. There are two symmetry types, $\{1, \ldots, 8\}$ and $\{9, \ldots, 16\}$, each subset can be distinguished from the other, but the resulting prior probability assignment is still uniform. These sorts of paradigms are artificial—we have not been able to fathom any natural paradigm of this sort. The explanation for why no-induction is unjustified is, then, because we neither possess conceptual frameworks that make no distinctions nor possess conceptual frameworks of the unnatural sort that make distinctions but still give a uniform distribution.

$H = [0, 1]$

Now we take up no-induction on the hypothesis set $H = [0, 1]$, the set of physical probabilities p of a repeatable experiment. In no-induction it is as if one believes with probability 1 that the physical probability of the experiment (say, a coin flip) is .5, and therefore one is incapable of changing this opinion no matter the evidence. In fact this is *exactly* what the uniform probability assignment over H_N is equivalent to. That is, the prior on $[0, 1]$ leading to no-induction gives $p = .5$ probability 1, and the probability density over the continuum of other hypotheses is zero. What was an elegant, uniform distribution on H_N has as its corresponding prior on $[0, 1]$ an extremely inelegant Dirac delta prior. With $[0, 1]$ as the hypothesis set instead of H_N, there is the sense in which no-induction is *even more* unjustified, since the prior is so clearly arbitrary. The reason for this emanates from the fact that $[0, 1]$ is a "less general" hypothesis set than H_N, for, informally, $[0, 1]$ lumps all of the outcome strings in a single complexion into a single hypothesis (recall, two strings are in the same complexion if they have the same number of 0s and 1s); H_N is capable of noticing the order of experiments, $[0, 1]$ is not. This property of $[0, 1]$, that it presumes exchangeability, severely constrains the sort of inductive methods that are possible and makes frequency-induction "easier" to achieve in the sense that any open-minded prior converges asymptotically to the observed frequency; no-induction is correspondingly "harder" to achieve in $[0, 1]$.

In fact, within Paradigm Theory no-induction on [0,1] is impossible to achieve for the simple reason that paradigms always result in open-minded priors. The reason we believe no-induction is unjustified on [0,1] is because no paradigm leads to no-induction.

Frequency-Induction

If an experiment is repeated many times, and thus far 70% of the time the outcome has been 0, then in very many inductive scenarios most of us would infer that there is a roughly 70% chance that the next experiment will result in 0. This is frequency-induction, and is one of the most basic ways in which we learn from experience, but is this method justifiable? Laplace argued that such an inference is justified on the basis of his Rule of Succession. It states that out of $n + 1$ experiments, if 0 occurs r times out of the first n, then the probability that 0 will occur in the next experiment is $\frac{r+1}{n+2}$. As $n \to \infty$, this very quickly approaches $\frac{r}{n}$; and when $r = n$, it very quickly approaches 1. Derivations of this rule depend (of course) on the prior probability distribution; see Zabell (1989) for a variety of historical proofs of the rule. In this section we demonstrate how Paradigm Theory naturally leads to the Rule of Succession when $H = H_N$ and $H = [0, 1]$.

$H = H_N$

The second column of probabilities in Table 3.1, headed "Q_L," shows the probability assignment on H_4 needed to lead to Laplace's Rule of Succession.[8] Notice, in contrast to Q_s, that for this column $P(0|000) = (1/5)/(1/5 + 1/20) = 4/5$, and so $P(1|000) = 1/5$; it learns from experience. Laplace's derivation was via a uniform prior on the hypothesis set $H = [0, 1]$ (with uniform underlying prior), but on H_N something else is required. Johnson's Combination Postulate and Permutability Postulate (Johnson, 1924, pp. 178–189) together give the needed assignment. The Combination Postulate—which states that it is *a priori* no more likely that 0 occurs i times than j times in n experiments—assigns equal probability to each complexion, and the Permutability Postulate—which states that the order of the experiments does not matter—distributes the probability uniformly within each complexion. Carnap's logical theory with m^* (Carnap, 1950, p. 563) does the same by assigning equal probability to each structure-description (analogous to the complexions), and distributing the probability uniformly to the state-descriptions (analogous to the individual outcome strings) within each structure-description (see the earlier discussion of the "Basic Theory").

In order for Paradigm Theory to give this prior probability assignment it suffices to find a paradigm whose induced symmetry types are the complex-

---

[8]A discussion on the difference between Q_s and Q_L can be found in Carnap (1989).

ions. If a paradigm satisfies this, the Principle of Type Uniformity assigns each complexion the same prior probability, and the Principle of Symmetry uniformly distributes the probability among the outcome strings within each complexion. In other words, if one's conceptual framework distinguishes the complexions, then one engages in frequency-induction via the Rule of Succession. Explanatorily, the Rule of Succession is justified because we possess paradigms that distinguish the complexions.

For distinguishing the complexions it is not sufficient to simply acknowledge the complexions; if the paradigm consists of *just* the complexions, then there are three symmetry types in H_4 as in Table 3.1: $\{0000, 1111\}$, $\{0001,$ $0010, 0100, 1000, 1110, 1101, 1011, 0111\}$, and the "middle" complexion. There *are* very natural paradigms that do induce symmetry types equal to the complexions. One such paradigm is employed in the following theorem whose proof may be found in the appendix to this chapter.

Theorem 4 *Let Q_L ('L' for 'Laplace') be the paradigm containing each complexion and the set of all sequences with more 0s than 1s. The probability assignment of Q_L on H_N via Paradigm Theory is identical to that of Johnson, and so Q_L results in Laplace's Rule of Succession.* $\triangle$

Note that Q_L is quite natural. It is the paradigm that acknowledges the complexions, and in addition acknowledges the difference between having more 0s than 1s and not more 0s than 1s. An explanation for the intuitive appeal of the Rule of Succession is that we often acknowledge exactly those properties in Q_L, and from this the Rule of Succession follows.

Since there are only finitely many inductive methods that may result given H_N via Paradigm Theory, the theory is not capable of handling a continuum of frequency-inductive methods as in Johnson and Carnap's λ-continuum, which says if r of n outcomes have been 1 in a binary experiment, the probability of the next outcome being a 1 is $\frac{r+\lambda/2}{n+\lambda}$. I have not attempted to determine the class of all λ such that there exists a paradigm that entails the λ-rule, but it seems that the only two natural sorts of paradigms that lead to an inductive method in the λ-continuum with $H = H_N$ are totally symmetric paradigms and those that have the complexions as the symmetry types. The first corresponds to $\lambda = \infty$, and the second corresponds to $\lambda = 2$. Reichenbach's Straight Rule (where, after seeing r of n outcomes of 1 in a binary experiment, the probability that the next will be 1 is r/n), or $\lambda = 0$, does not, therefore, seem to be justifiable within Paradigm Theory.

Laplace's Rule of Succession needs the assumption on H_N that, *a priori*, it is no more likely that 1 is the outcome i times than j times in n experiments. Call a *repetition* the event where two consecutive experiments are either both 1 or both 0; two strings are in the same *repetition set* if they have the same number of repetitions. Why, for example, should we not modify Johnson's Combination Postulate (or Principle of Indifference on the complexions) to say that, *a priori*, it is no more likely that a repetition occurs i times than j times in n experiments? The prior probability assignment resulting from this does not lead to Laplace's Rule of Succession, but instead to the "Repetition" Rule of Succession. 'REP' denotes the assignment of equal probabilities to each repetition set, with the probability uniformly distributed among the strings in each repetition set; this is shown for H_4 in Table 3.1 under the heading Q_{rep}. If one has seen r repetitions of 1 thus far with n experiments, the probability the outcome of the next experiment will be the same as the last outcome, via REP, is $\frac{r+1}{n+1}$. The proof is derivable from Laplace's Rule of Succession once one notices that the number of ways of getting r repetitions in a length n binary sequence is $2C_r^{n-1}$; the proof is omitted. This result can be naturally accommodated within Paradigm Theory.

Theorem 5 *Let Q_{rep} be the paradigm that acknowledges the number of repetitions in a sequence as well as acknowledging the sequences with less than half the total possible number of repetitions. The probability assignment of Q_{rep} is identical to that of REP, and so Q_{rep} results in the Repetition Rule of Succession.* $\triangle$

Whereas all of the previously mentioned paradigms on H_N entail prior probability assignments that are de Finetti exchangeable, Q_{rep} does not. It *is* Markov exchangeable, however: where strings with both the same initial outcome and the same number of repetitions have identical prior probability. A conceptual framework that acknowledges both the number of repetitions and which (0 or 1) has the greater number of repetitions results in the Repetition Rule of Succession. When our inductive behavior is like the Repetition Rule, it is because we possess Q_{rep} (or something like it) as our conceptual framework.

Q_L and Q_{rep} generally give very different predictions. However, they nearly agree on the intuitively clear case where one has seen all of the experiments give the same result. For example, Laplace had calculated the probability that the sun will rise tomorrow with his Rule of Succession; "It is a bet of 1,826,214 to one that it will rise again tomorrow" (Laplace, 1820). The Repetition Rule of Succession says that the odds are 1,826,213 to one that tomorrow will be the same as the past with respect to the sun rising or not, and since we know it came up today, those are the odds of the sun rising tomorrow.

$$H = [0, 1]$$

Now we consider frequency-induction on the hypothesis set $H = [0, 1]$ with the natural uniform underlying measure. We noted earlier that $[0, 1]$ "more easily" leads to frequency-induction than H_N; disregarding the order of experiments puts one well on the path toward frequency-induction. We should suspect, then, that it should be easier to acquire frequency-inductions with $[0, 1]$ as the hypothesis set than H_N via Paradigm Theory. In fact, frequency-induction is guaranteed on $[0, 1]$ since paradigms lead to open-minded priors which, in turn, lead to frequency-induction. One cannot be a skeptic about frequency-induction in $[0, 1]$. Frequency-induction on $[0,1]$ is justified because every conceptual framework leads to it.

For Laplace's Rule of Succession, Laplace assigned the uniform prior probability distribution over the underlying measure, from which the Rule follows. Here is the associated result for Paradigm Theory.

Theorem 6 *Any totally symmetric paradigm entails the uniform assignment on* $[0, 1]$. *Therefore, any totally symmetric paradigm results in Laplace's Rule of Succession.* $\triangle$

If one acknowledges nothing on $[0, 1]$, or more generally one makes no distinctions, Paradigm Theory collapses to a sort of Principle of Indifference (see Subsection 3.2.1) and one engages in frequency-induction via Laplace's Rule of Succession. Laplace's Rule of Succession is justified because when presented with hypothesis set $[0, 1]$ we possess a conceptual framework that does not distinguish between any hypotheses.

Law-Induction

Frequency-induction allows instance confirmation, the ability to place a probability on the outcome of the very next experiment. C. D. Broad (1918) challenged whether frequency-induction, Laplace's Rule of Succession in particular, is ever an adequate description of learning. The premises that lead to the Rule of Succession also entail that if there will be N experiments total and one has conducted n so far, all of which are found to be 1 (i.e., $r = n$), then the probability that *all* outcomes will be 1 is $(n + 1)/(N + 1)$. If N is large compared to n, $(n + 1)/(N + 1)$ is small; and this is the origin of Broad's complaint. In real situations N, if not infinite, is very large. Yet we regularly acquire high degree of belief in the general law that all outcomes will be 1 with only a handful of experiments (small n). For example, we all conclude that all crows are black on the basis of only a small (say 100) sample of black crows. If, by 'crow,' we mean those alive now, then N is the total number of living crows, which is in the millions. In this case, after seeing 100 black crows, or even thousands, the probability via the Rule of Succession premises of the law 'all crows are black' is miniscule. The probability that all crows are black becomes high only as n approaches N—only after we have examined nearly every crow! Therefore, the premises assumed for the Rule of Succession cannot be adequate to describe some of our inductive methods.

Carnap (1950, pp. 571–572) makes some attempts to argue that instance confirmation is sufficient for science, but it is certain that we (even scientists) do in fact acquire high probability in universal generalizations, and the question is whether (and why) we are justified in doing so.

Jeffreys (1955) takes Broad's charge very seriously. "The answer is obvious. The uniform assessment of initial probability says that before we have any observations there are odds of $N - 1$ to 2 against any general law holding. This expresses a violent prejudice against a general law in a large class" (ibid., p. 278). He suggests that the prior probability that a general law holds be a constant > 0, independent of N. This allows learning of general laws. For example, fix a probability of .1 that a general law holds, .05 for the 'all 0s' law, .05 for the 'all 1s' law, the probability uniformly distributed over the rest. After seeing just five black crows the probability of the 'all 0s' law is .64, and after seeing ten black crows the probability becomes .98; and this is largely independent of the total number of crows N.

The problem with this sort of explanation, which is the sort a Personalistic Bayesian is capable of giving, is that there seems to be no principled reason for why the general laws should receive the probability assignments they do; why not .06 each instead of .05, or why not .4 each? Paradigm Theory determines exact inductive methods capable of giving high posterior probability to laws, and it does so with very natural paradigms.

$$H = H_N$$

Beginning with H_N as the hypothesis set, suppose one acknowledges only two properties: being a general law and not being a general law. With this comprising the paradigm Q_{law} the induced symmetry types are the same as the acknowledged properties. Paradigm Theory gives probability .5 to a general law holding—.25 to 'all 0s', .25 to 'all ones'—and .5 uniformly distributed to the rest; see the "Q_{law}" column in Table 3.1. Largely independent of the total number of crows, after seeing just one black crow the probability that all crows are black is .5. After seeing 5 and 10 black crows the probability becomes .94 and .998, respectively—near certainty that all crows are black after just a handful of observations. I record this in the following theorem whose proof may be found in the appendix to this chapter.

Theorem 7 *If there will be N experiments and $1 \leq n < N$ have been conducted so far, all which resulted in* 1, *then the probability that all N experiments will result in* 1, *with respect to the paradigm Q_{law} on the hypothesis set H_N, is approximately*

$$\frac{2^{n-1}}{1 + 2^{n-1}} \cdot \Delta$$

One is open to the confirmation of universal generalizations if one acknowledges being a law and acknowledges no other properties. Of course, the theorem holds for any paradigm that induces the same symmetry types as Q_{law}. For example, suppose that a paradigm Q_{const} acknowledges the *constituents*, from Hintikka (1966), where a constituent is one possible way the world can be in the following sense: either all things are 0, some things are 0 and some are 1, or all things are 1. The induced symmetry types are the same as those induced by Q_{law}.

Similar results to Theorem 7 follow from any paradigm that (i) has {'all 0s', 'all 1s'} as a symmetry type (or each is alone a symmetry type), and (ii) there is some natural number k such that for all N the total number of symmetry types is k. Q_{law} and Q_{const} are special cases of this, with $k = 2$. Each paradigm satisfying (i) and (ii) entails an inductive method that is capable of giving high posterior probability to universal generalizations. This is because the two laws each receive the probability $1/(2k)$ (or $1/k$ if each is invariant) no matter how large is the number of "crows in the world" N.

There is a problem with paradigms satisfying (i) and (ii). Paradigms satisfying (i) and (ii) are not able to engage in frequency-induction when some but not all experiments have resulted in 1. This is because frequency-induction on H_N requires that one distinguish among the $N + 1$ complexions, and this grows with N, and so (ii) does not hold. Specifically considering Q_{law} and Q_{const}, the most natural paradigms satisfying (i) and (ii), when some but not all experiments have resulted in 1 the Q_{law} and Q_{const} assignment does not learn at all. This is because the probabilities are uniformly distributed over the outcome strings between the 'all 0s' and 'all 1s' strings, just like when the paradigm is Q_s from earlier.

To "fix" this problem it is necessary to employ a secondary paradigm. We concentrate only on fixing Q_{law} for the remainder of this subsection, but the same goes for Q_{const} as well. What we need is a secondary paradigm on the set of strings between 'all 0s' and 'all 1s' that distinguishes the complexions, i.e., has them as symmetry types. Let the secondary paradigm be the one acknowledging the complexions and the property of having more 0s than 1s, which is like the earlier Q_L, and let the hypothesis set be $H_N - \{$'all 0s','all 1s'$\}$ instead of H_N. The resulting inductive behavior is like Laplace's Rule of Succession for strings that are neither 'all zeros' nor 'all ones', and similar to that of Q_{law} described in Theorem 7 for the 'all 0s' and 'all 1s' strings. We denote this paradigm and secondary paradigm duo by Q_{law_L}, and one can see the resulting prior probability on H_4 in Table 3.1. The proof of part (a) in the following theorem emanates, through de Finetti's Representation Theorem, from part (a) of Theorem 9; (b) is proved as in Theorem 4.

Theorem 8 Q_{law_L} assigns prior probabilities to H_N $(n < N)$ such that if 1 occurs r times out of n total, then (a) if $r = n > 0$ the probability that all outcomes will be 1 is approximately $\frac{n+1}{n+3}$, and (b) if $0 < r < n$ the probability that the next outcome will be a 1 is $\frac{r+1}{n+2}$ (i.e., the inductive method is like that of Q_L). $\triangle$

After seeing 5 and 10 black crows, the probability that all crows are black is approximately .75 and .85, respectively.

How natural is the primary/secondary paradigm pair Q_{law_L}? It acknowledges being a law (or in Q_{const}'s case, acknowledges the constituents), acknowledges the complexions, and acknowledges having more 0s than 1s. But it also believes that the laws (or constituents) are more important (or "more serious" parts of the ontology) than the latter two properties. "Primarily, the members of our paradigm acknowledge laws; we acknowledge whether or not all things are 0, and whether or not all things are 1. Only secondarily do we acknowledge the number of 0s and 1s and whether there is a greater number of 0s than 1s." Having such a conceptual framework would explain why one's inductive behavior allows both frequency-induction and law-induction. Note that if Q_L were to be primary and Q_{law} secondarily applied to each symmetry type induced by Q_L, then the result would be no different than Q_L alone. The same is true if we take as primary paradigm the union of both these paradigms. Thus, if being a law is to be acknowledged independently of the other two properties at all, it must be via relegating the other two properties to secondary status.

The above results on universal generalization are related to one inductive method in Hintikka's two-dimensional continuum (Hintikka, 1966). Q_{law_L} (and Q_{const_L}) corresponds closely to Hintikka's logical theory with $\alpha = 0$ (ibid., p. 128), except that Hintikka (primarily) assigns probability $1/3$ to each constituent: $1/3$ to 'all 0s', $1/3$ to 'all 1s', and $1/3$ to the set of strings in between. In Q_{law} (and Q_{const}) 'all 0s' and 'all 1s' are members of the same symmetry type, and so the probabilities were, respectively, $1/4$, $1/4$, $1/2$. Then (secondarily) Hintikka divides the probability of a constituent evenly among the structure-descriptions, which are analogous to our complexions. Finally, the probability of a structure-description is evenly divided among the state-descriptions, which are analogous to our outcome strings. Q_{law_L}, then, acknowledges the same properties as does Hintikka's "$\alpha = 0$"-logical theory, and in the same order.

It *is* possible for Paradigm Theory to get *exactly* Hintikka's $\alpha = 0$ assignment, but the only paradigms I have found that can do this are artificial. For example, a paradigm that does the job is the one that acknowledges 'all 0s' and the pairs {'all 1s', σ} such that σ is a non-law string. 'all 0s' and 'all 1s' are now separate symmetry types, and the non-law strings in between comprise the third. Each thus receives prior probability $1/3$ as in Hintikka's "$\alpha = 0$"-Logical Theory. This paradigm is indeed artificial, and I do not believe Paradigm Theory can give any natural justification for the $\alpha = 0$ inductive method.

With Q_{law_L} in hand we can appreciate more fully something Paradigm Theory can accomplish with secondary paradigms: a principled defense and natural generalization of Hintikka's "$\alpha = 0$"-logical theory. Well, not exactly, since as just mentioned the nearest Paradigm Theory can naturally get to $\alpha = 0$ is with Q_{law_L} (or Q_{const_L}). Ignoring this, Paradigm Theory gives us a principled reason for why one should engage in law-induction of the $\alpha = 0$ sort: because one holds Q_{law} (or Q_{const}) as the conceptual framework, and Q_L secondarily. Paradigm Theory also allows different notions of what it is to be a law, and allows different properties to replace that of being a law. The $\alpha = 0$ tactic can be applied now in any way one pleases.

$H = [0, 1]$

We have seen in Subsection 3.2.2 that $[0, 1]$ as the hypothesis set makes frequency-induction easier to obtain than when the hypothesis set is H_N. Informally, one must expend energy when given H_N so as to treat the complexions as the primitive objects upon which probabilities are assigned, whereas this work is already done when given $[0, 1]$ instead. To do this job on H_N for law-induction we required secondary paradigms in order to have frequency-induction as well, but it should be no surprise that on $[0, 1]$ having both comes more easily.

As in the previous subsubsection we begin with the paradigm that acknowledges being a law and not. We call it by the same name, Q_{law}, although this is strictly a different paradigm than the old one since it is now over a different hypothesis set. There are two symmetry types, $\{0, 1\}$ and $(0, 1)$. Thus, $p = 0$ and $p = 1$ each receives probability .25, and the remaining .5 is spread uniformly over $(0, 1)$. This is a universal-generalization (UG) open-minded prior probability distribution, where not only is the prior probability density always positive, but the $p = 0$ and $p = 1$ hypotheses are given positive probability; this entails an inductive method capable of learning laws. It is also open-minded, and so is an example of frequency-induction as well; we do not need secondary paradigms here to get this. In fact, because the prior is uniform between the two endpoints the inductive behavior follows Laplace's Rule of Succession when the evidence consists of some 0s and some 1s. The following theorem records this; the proof of (a) is in the appendix to this chapter, and (b) is derived directly from Laplace's derivation of the Rule of Succession.

Theorem 9 Q_{law} on $[0, 1]$ *entails the prior probability distribution such that if 1 occurs r times out of n total, then (a) if $r = n > 0$ the probability that $p = 1$ is $\frac{n+1}{n+3}$, and (b) if $0 < r < n$ the probability that the next outcome will be a 1 is $\frac{r+1}{n+2}$.* $\triangle$

If one holds $[0, 1]$ as the hypothesis set and acknowledges being a law and nothing else, one is both able to give high probability to laws and converge to the relative frequency. Turned around, we should engage in law- and frequency-induction (of the sort of the previous theorem) because our conceptual framework acknowledges the property of being a law. One need make no primitive assumption concerning personal probabilities as in Personalistic Bayesianism, one need only the extremely simple and natural Q_{law}.

Similar results to Theorem 9 can be stated for any paradigm such that the two laws appear in symmetry types that are finite (the laws are distinguished, at least weakly). For any such paradigm the two laws are learnable because they acquire positive prior probability, and frequency-induction proceeds (asymptotically, at least) because the prior is open-minded. In an informal sense, "any" natural paradigm acknowledging the laws results in both law- and frequency-induction.

3.2.3 Simplicity-Favoring

Occam's Razor says that one should not postulate unnecessary entities, and this is roughly the sort of simplicity to which I refer (although any notion of simplicity that has the same formal structure as that described below does as well). Paradigm Theory is able to provide a novel justification for simplicity: *when the paradigm acknowledges simplicity, it is "usually" the case that simpler hypotheses are less arbitrary and therefore receive higher prior probability.* This explanation for the preferability of simpler hypotheses does *not* assume that we must favor simpler hypotheses in the paradigm (something the paradigm does not have the power to do anyway). The paradigm need only *acknowledge* which hypotheses are simpler than which others.[9] In a sentence, Paradigm Theory gives us the following explanation for why simpler hypotheses are preferred: *simpler hypotheses are less arbitrary.*

For any hypothesis there are usually multiple ways in which it may be "complexified"—i.e., unnecessary entities added—to obtain new hypotheses. Each complexification itself may usually be complexified in multiple ways, and so may each of its complexifications, and so on. A *complexification tree* is induced by this complexification structure, starting from a given hypothesis as the root, its complexifications as the children, their complexifications as the grandchildren, etc.[10]

Recall from Subsection 3.1.2 that certain paradigms are representable as graphs. Consider the following two special cases of trees whose associated paradigms result in the root being the lone maximally defensible element; the proof is found in the appendix to this chapter. A tree is *full* if every leaf is at the same depth in the tree.

Theorem 10 *The paradigm associated with any full tree or finite-depth binary tree places the root as the lone maximally defensible element. But not every paradigm associated with a tree does so, and these two cases do not exhaust the trees that do so.* △

---

[9]In fact, it suffices to acknowledge the two-element subsets for which one element is simpler than the other; after all, paradigms as defined for the purposes of this chapter do not allow relations.

[10]I am ignoring the possibility that two hypotheses may "complexify" to the same hypothesis, in which case the structure is not a tree.

If a hypothesis set H consists of h, all of its complexifications and all of their complexifications and so on, and the paradigm on H is the complexification tree with root h—i.e., the paradigm acknowledges the pairs of hypotheses for which one is a complexification of the other—then the paradigm puts h alone at the top of the hierarchy if the tree is full or finite binary.[11] Informally, "most" natural notions of hypothesis and complexification imply complexification trees that are full. Such paradigms naturally accommodate Occam's Razor; acknowledging simplicity results in setting the lone most defensible element to what Occam's Razor chooses for many natural (at least finite binary and full) complexification trees. The hypothesis that posits the least unnecessary entities is, in these cases, the lone most defensible hypothesis, and thus acquires the greatest prior probability (via Theorem 3).

Full Complexification Trees

Let Q_{full} be the paradigm represented by the full tree below.

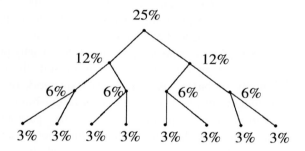

There are four symmetry types (one for each level), so each receives probability 1/4. The approximate probability for each hypothesis is shown in the figure. Only the Basic Theory is needed here—i.e., the Principles of Type Uniformity and Symmetry—the Principle of Defensibility does not apply. If there are m such trees, the m roots each receive probability $\frac{1}{4m}$, the $2m$ children each receive $\frac{1}{8m}$, the $4m$ grandchildren each receive $\frac{1}{16m}$, and the $8m$ leaves

---

[11]We are assuming that the paradigm acknowledges only those pairs of hypotheses such that one is an "immediate" complexification of the other, i.e., there being no intermediate complexification in between. Without this assumption the complexification trees would not be trees at all, and the resulting graphs would be difficult to illustrate. However, the results in this section do not depend on this. If the paradigm acknowledges every pair such that one is simpler than the other, then all of the analogous observations are still true.

each receive $\frac{1}{32m}$. The following theorem generalizes this example. Recall that the depth of the root of a tree is zero.

Theorem 11 *Suppose the paradigm's associated graph consists of m full b-ary ($b \geq 2$) trees of depth n, and that hypothesis h is at depth i in one of them. Then $P(h) = \frac{1}{m(n+1)b^i}$.* $\triangle$

This tells us that the prior probability of a hypothesis drops exponentially the more one complexifies it, i.e., the greater i becomes. For example, consider base 10 numbers as in the hypothesis set $H = \{2; 2.0, \ldots, 2.9; 2.00, \ldots, 2.09; \ldots; 2.90, \ldots 2.99\}$, and suppose the paradigm is the one corresponding to the complexification tree. Here we have a 10-ary tree of depth two; 2 is the root, the two-significant-digit hypotheses are at depth one, and the three-significant-digit hypotheses are at depth two. $P(2) = 1/3$, the probability of a hypothesis at depth one is $1/30$, and the probability of a hypothesis at depth two is $1/300$.

When there are multiple trees, the roots may be interpreted as the "serious" hypotheses, and the complexifications the "ridiculous" ones. Theorem 11 tells us that when one acknowledges simplicity and the resulting paradigm is represented by multiple b-ary trees of identical depth, one favors the serious hypotheses over all others. This is a pleasing explanation for why prior probabilities tend to accrue to the simplest hypotheses, but it results in each of these hypotheses being equally probable. A conceptual framework may be more complicated, acknowledging properties capable of distinguishing between the different complexification trees. In particular, a secondary paradigm may be applied to the set of roots, with the understanding that the properties in the secondary paradigm are acknowledged secondarily to simplicity.

Asymmetrical Complexification Trees

We saw in Theorem 10 that any—even an asymmetrical—finite binary tree results in the root being the lone most defensible element. The Principle of Defensibility tends to apply non-trivially when trees are asymmetrical, unlike when trees are full where it makes no difference. The next example shows an asymmetrical tree where Paradigm Theory "outperforms" the Basic Theory. To demonstrate the last sentence of Theorem 10, that 'full' and 'finite binary' do not exhaust the trees that result in a lone most defensible root, we have chosen the tree to be non-binary. We leave it as an exercise to find the probabilities for a similar asymmetrical binary tree.

Let $H_{asymm} = \{a, \ldots, p\}$, and Q_{asymm} be as pictured.

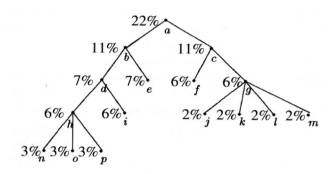

With semicolons between the equivalence types, the invariance levels are Δ^0 $= \{j, k, l, m; n, o, p\}$, $\Delta^1 = \{f, g; h, i\}$, $\Delta^2 = \{d, e\}$, $\Delta^3 = \{b, c\}$ and $\Delta^4 = \{a\}$. Paradigm Theory assigns the probabilities as follows: $P(n) = P(o) = P(p) = 7/231 \approx 3\%$. $P(j) = \ldots = P(m) = 1/44 \approx 2\%$. $P(f) = \ldots = P(i) = 9/154 \approx 6\%$. $P(d) = P(e) = 45/616 \approx 7\%$. $P(b) = P(c) = 135/1232 \approx 11\%$. $P(a) = 135/616 \approx 22\%$. Notice how the Principle of Defensibility is critical to achieve this assignment. The Basic Theory alone agrees with this assignment on the leaves, but on the others it assigns each a probability of $1/11 \approx 9\%$ instead. The Basic Theory does not notice the structure of the invariant hypotheses and so gives them each the same probability.

This example brings out the importance of the Principle of Defensibility. The Basic Theory can be viewed as a natural generalization of Carnap's m^*-logical theory. Except for cases where the tree is full, the Basic Theory is inadequate, ignoring all the structure that we *know* is there. The Basic Theory's weakness is, as discussed in Subsection 3.1.3, that it is capable of seeing only two degrees of detail. The Principle of Defensibility simply says that among the invariant hypotheses there are, from the point of view of the paradigm already before you (i.e., no secondary paradigm is needed), those that are more and less defensible—notice this. It is this principle that allows Paradigm Theory to break the bonds of a simple generalization of Carnap's m^*-logical theory and secure a full explanation and justification for simplicity-favoring.

Discussion

I am in no way elucidating the difficult question of *What is simplicity?* or *What counts as fewer entities?*; if 'grue' is considered simpler than 'green', then it may well end up with greater prior probability. In this subsection we have discussed why simpler hypotheses, supposing we agree on what this means, should be favored. When one acknowledges—*not favors*—simplicity in the paradigm and the paradigm can be represented as a (full or finite binary, among others) tree, the simpler hypotheses receive higher prior probability. This occurs not because they are simpler, but because they are less arbitrary.

Let me address what could be a criticism of this explanation of the justification of simplicity-favoring. This explanation depends on the resulting graph associated with the paradigm being a tree, with the simpler hypotheses near the root. This occurs because, so I asserted, there are usually multiple ways of complexifying any given hypothesis; and these complexifications are a hypothesis' daughters in the tree. What if this is not true? For example, what if one is presented with a hypothesis set consisting of one simple hypothesis and just one of its complexifications? Acknowledging simplicity here does not entail simplicity-favoring; each hypothesis is equally probable. I claim that holding such a hypothesis set is uncommon and unnatural. Most of the time, if we consider a complexification of a hypothesis and notice that it is a complexification, then we also realize that there are other complexifications as well. Choosing to leave the others out of the hypothesis set and allowing only the one to remain is ad hoc. Worse than this example, suppose for each hypothesis there are multiple *simplifications* rather than multiple complexifications for each hypothesis? If this is so, Paradigm Theory ends up favoring more complex hypotheses instead. While certainly one can concoct hypothesis sets where acknowledging simplicity results in a simplification tree instead of a complexification tree, I do not believe there to be very many (if any) natural examples. And if such a hypothesis set *is* presented to one acknowledging simplicity, the most complex hypothesis is indeed the most favorable. These observations are not unexpected: unusual conceptual frameworks may well entail unusual inductive behavior.

For the sake of contrast it is helpful to look at the reasons I have given in this section for favoring simpler hypotheses compared to those of other theorists: (i) they are more susceptible to falsification (Popper, 1959), (ii) they are more susceptible to confirmation (Quine, 1963), (iii) they are practically easier to apply (Russell, 1918; Pearson, 1992; Mach, 1976), (iv) they have greater *a priori* likelihood of being true (Jeffreys, 1948), (v) they have been found in the past to be more successful (Reichenbach, 1938), (vi) following the rule 'pick the simplest hypothesis' leads with high probability to true hypotheses (Kemeny, 1953), (vii) they are more informative (Sober, 1975), (viii) they are more stable (Turney, 1990), and (ix) they have higher estimated predictive accuracy (Forster and Sober (1994)). Paradigm Theory's reason for favoring simpler hypotheses is that we acknowledge simplicity and, since for each hypothesis there tends to be multiple complexifications (and not multiple simplifications), simpler hypotheses are less arbitrary.

3.2.4   Curve-Fitting

In curve-fitting the problem is to determine the best curve given the data points. The phenomenon that needs to be explained is that a curve that is a member of an n parameter family, or model,[12] is typically favored over curves that require $n + 1$ parameters, even when the latter fits the data better than the former. I derive within Paradigm Theory a class of information criteria dictating the degree to which a simpler curve (say, a linear one) is favored over a more complex one.

In curve-fitting generally, the data are presumed to be inaccurate, and no hypothesis can be excluded *a priori*. I concentrate only on the hypothesis set of polynomials, and consider only those up to some finite degree. For definiteness I presume that each dimension is bounded to a finite range, and that the underlying measure is uniform in each $M' - M$ (where M' is a model with one more dimension than M). The first of these conditions on the hypothesis set is perhaps the only questionable one. The parameter bounds may be set arbitrarily high, however; so high that it is difficult to complain that the bound is too restrictive.

Suppose we have models M_0 and M_1, M_0 with parameter a_0 and M_1 with parameters a_0 and a_1, where the parameters range over the reals within some bound and the models are such that for some value of a_1, M_1 makes the same predictions as M_0. In cases such as this Jeffreys (1948) (see also Howson, 1987, pp. 210–211] proposes that M_0 and $M_1 - M_0$ each receive prior probability $1/2$. We shall denote M_0 and $M_1 - M_0$ as, respectively, S_0 and S_1 ("S" for symmetry type). Paradigm Theory gives a principled defense for Jeffreys' prior probability assignment: if the conceptual framework acknowledges the

---

[12]Do not confuse this notion of model with that discussed in Subsection 3.2.1. There is no relation.

two models, then there are two symmetry types—M_0 and $M_1 - M_0$—each receiving prior probability $1/2$ via the Principle of Type Uniformity, and the probability density is uniform over each symmetry type via the Principle of Symmetry.[13]

How the prior probability *density* compares in S_0 and S_1 depends on the choice of underlying measure. Let us first suppose that the measure is the Euclidean one, where length is always smaller than area, area always smaller than volume, etc. Because M_0 is one dimension smaller than M_1, the prior probability density on S_0 is infinitely greater than that on S_1. Thus, any specific curve in S_1 receives prior probability density that is vanishingly small compared to the prior probability of a curve in S_0. More generally, consider M_0, M_1, ..., M_l, where each model is the superset of the previous one resulting from adding one parameter, ranging over the reals within some bound, to allow polynomials of one higher degree. Each subset M_0 and $M_{k+1} - M_k$ for $0 \leq k < l$ is a symmetry type—denoted, respectively, by S_0 and S_{k+1} for $0 \leq k < l$—and receives prior probability $1/(l + 1)$. With the Euclidean underlying measure, the probability density over the symmetry types decreases infinitely as the number of extra parameters is increased. Generally, then, curves that are members of models with fewer parameters are *a priori* favored because we possess a conceptual framework that acknowledges the models (and nothing else).

The Euclidean underlying measure is very strong, resulting in simpler curves having greater posterior probability density *no matter the data.* Since each curve in S_0 has a prior probability density that is infinitely greater than each in S_k for $k > 0$, this effectively means that one restricts oneself to the polynomials of least degree. Perhaps less radical underlying measures should be used, ones that agree that higher degrees have greater underlying measure (intuitively, more polynomials), but not *infinitely* greater (intuitively, not infinitely more polynomials). Suppose, instead, that the underlying measure is s in S_0, and m times greater in each successive degree of greater dimension; i.e., S_k has as underlying measure sm^k for some positive real number m. One may find it convenient to act as if the hypothesis set is finite, and that there are (the truncation of) sm^k curves in S_k. Then one can say that a curve in S_k has prior *probability* equal to so and so, rather than probability *density* equal to so and so. At any rate, the important supposition behind the discussion below is that the underlying measure is m times greater as the degree is increased, not whether the hypothesis set is finite or not. Under these conditions, individual curves

---

[13]Because M_0 is a subset of M_1, elements inside M_0 cannot be interchanged with those outside without affecting the paradigm. This is true regardless of the measure of the two regions.

have probability as stated in the following theorem.

Theorem 12 *Let the hypothesis set $H_{l,s,m}$ be as just described above, i.e., the set of polynomials such that (i) each has degree less than or equal to l, and (ii) $M_k - M_{k-1}$ has a uniform underlying measure equal to sm^k within some finite range. Let Q_{model} be the paradigm that acknowledges the models over $H_{l,s,m}$. If curve h is in S_k for some $0 \le k \le l$, then its prior probability density is $\frac{1}{(l+1)sm^k}$.* $\triangle$

The symmetry types S_k each receive prior probability $1/(l+1)$ by the Principle of Type Uniformity. A hypothesis in S_k must share its probability with a measure of sm^k hypotheses, and by the Principle of Symmetry the theorem follows. If one imagines that $H_{l,s,m}$ is finite, then a curve h in S_k receives prior probability equal to $\frac{1}{(l+1)\lfloor sm^k \rfloor}$ (where $\lfloor x \rfloor$ stands for the truncation of x).

One can see from the m^{-k} term that curves requiring a greater number of parameters receive exponentially lower prior probability density. Acknowledge the natural models...exponentially favor polynomials of lower degree. This observation holds regardless of the value of l and s. As for m, larger values mean that curves requiring a greater number of parameters are more disfavored.

There are a class of curve-fitting techniques called "information criteria" which prescribe picking the model that has the largest value for $\log L_k - \gamma k$, where k is the number of parameters of the model, log is the natural logarithm, L_k is the likelihood ($P(e|h)$) of the maximum likely hypothesis in the model of kth dimension M_k, and γ depends on the specific information criterion. [See Smith and Spiegelhalter (1980, pp. 218) for many of the information criteria (my γ is their $m/2$) and references to the original papers defending them; see also Aitkin (1991).] Once this model is determined, the maximum likely hypothesis in it is chosen, even though it may well not be the maximum likely hypothesis in the entire hypothesis set. Paradigm Theory natural leads to a class of information criteria emanating from the supposition that the paradigm is Q_{model} and the underlying measure of S_{k+1} is m times greater than that of S_k.

Our task is now to find the curve, or hypothesis, with the greatest posterior probability density given that models M_0 through M_l are acknowledged in the paradigm (i.e., Q_{model} is the paradigm). For simplicity, I will for the moment treat $H_{l,s,m}$ as if it is finite, with (the truncation of) sm^k curves in S_k. We want to find h such that it maximizes, via Bayes' Theorem, $P(e|h)P(h)/P(e)$ (e represents the data); that is, we wish to find h with maximum posterior probability. It suffices to maximize the natural logarithm of the posterior probability, or

$$\log P(e|h) + \log P(h) - \log P(e).$$

$P(e)$ is the same for every hypothesis, and we may ignore it. Theorem 12 informs us of the $P(h)$ term, which is the prior probability of h given Q_{model}, and we have

$$\log P(e|h) + \log[\frac{1}{(l+1)sm^k}]$$

if h is in S_k. This manipulates easily to

$$\log P(e|h) - (\log m)k - \log(l+1) - \log s.$$

l and s are the same for each hypothesis, and so they may also be ignored. This allows l, the maximum degree of polynomials allowed in the hypothesis set, to be set arbitrarily high. When the hypothesis set is treated as finite, s can be set arbitrarily high, thereby allowing the set to approximate an infinite one. Thus, the hypothesis with the maximal posterior probability is the one that maximizes

$$\log P(e|h) - (\log m)k.$$

This may be restated in the information criterion form by saying that one should choose the model that has the largest value for

$$\log L_k - (\log m)k,$$

and then choose the maximum likely hypothesis in that model. I have just proven the following theorem, which I state for records sake, and retranslate into its corresponding infinite hypothesis set form.

Theorem 13 *Let the hypothesis set be $H_{l,s,m}$ and the paradigm be Q_{model}; let the prior probability distribution be determined by Paradigm Theory. The hypothesis with the greatest posterior probability density is determined by choosing the model with the largest value for $\log L_k - (\log m)k$ and then picking the maximum likely hypothesis in that model.* $\triangle$

Notice that $\log m$ is filling the role of the γ in the information criteria equation. As m increases, goodness of fit is sacrificed more to the simplicity of the curves requiring fewer parameters since the number of parameters k gets weighed more heavily.

Consider some particular values of m. $m < 1$ means that the underlying measure of S_{k+1} is *less* than that of S_k; that there are, informally, fewer polynomials of the next higher degree. This is very unnatural, and the corresponding information criterion unnaturally favors more complex curves over simpler ones. $m = 1$ implies that moving to higher dimensions does not increase the underlying measure at all. In this case, the second term in the information criterion equation becomes zero, collapsing to the Maximum Likelihood Principle. When moving up in degree and dimension, it is only natural to suppose that there are, informally, more polynomials of that degree. With this in mind, it seems plausible that one chooses $m > 1$. $m = 2$ implies that moving to the next higher dimension doubles the underlying measure, which intuitively means that the number of hypotheses in S_{k+1} is twice as much as in S_k. The value of γ for $m = 2$ is $\gamma = \log m \approx .69$. Smith and Spiegelhalter (1980, pp. 219) observe that when $\gamma < .5$ more complex models still tend to be favored, and this does not fit our curve-fitting behavior and intuition; it is pleasing that one of the first natural values of m behaves well. (My γ is Smith and Spiegelhalters' $m/2$. Their m is not the same as mine.) When $m = e$, the resulting information criterion is precisely Akaike's Information Criterion. This amounts to a sort of answer to Forster and Sobers' (1994, p. 25) charge, "But we do not see how a Bayesian can justify assigning *priors* in accordance with this scheme," where by this they mean that they do not see how a prior probability distribution can be given over the curves such that the resulting information criterion has $\gamma = 1$. Paradigm Theory's answer is that if one acknowledges the natural models, and one assigns underlying measures to degrees in such a way that the next higher degree has e times the underlying measure of the lower degree, then one curve-fits according to Akaike's Information Criterion. When $m = 3, \gamma \approx 1.10$, and the resulting inductive method favors simpler curves just slightly more than in Akaike's. Finally, as $m \to \infty$, the underlying measure on $M_1 - M_0$ becomes larger and larger compared to that of M_0, and all curves requiring more than the least allowable number of dimensions acquire vanishingly small prior probability density; i.e., it approaches the situation in Jeffreys' prior discussed above. (There is also a type of Bayesian Information Criterion, called a "global" one (Smith and Spiegelhalter, 1980), where $\gamma = (\log n)/2$ and n is the number of data (Schwarz, 1978).)

The question that needs to be answered when choosing a value for m is, "How many times larger is the underlying measure of the next higher degree?," or intuitively, "How many times more polynomials of the next higher degree are to be considered?" Values for m below 2 seem to postulate too few polynomials of higher degree, and values above, say, 10 seem to postulate too many. The corresponding range for γ is .69 to 2.30, which is roughly the range of values for γ emanating from the information criteria (Smith and Spiegelhalter, 1980). For these "non-extreme" choices of m, curves requiring fewer parameters quickly acquire maximal posterior probability so long as their fit is moderately good.

Paradigm Theory's explanation for curve-fitting comes down to the following: We favor (and ought to favor) lines over parabolas because we acknowledge lines and parabolas. The reasonable supposition that the hypothesis set includes more curves of degree $k + 1$ than k is also required for this explanation.

Paradigm Theory's class of information criteria avoids at least one difficulty with the Bayesian Information Criteria. The Personalistic Bayesian does not seem to have a principled reason for supposing that the prior probabilities of M_0, $M_1 - M_0$, etc., are equal (or are any particular values). Why not give M_0 much more or less prior probability than the others? Or perhaps just a little more or less? In Paradigm Theory the models induce M_0, $M_1 - M_0$, etc., as the symmetry types, and the Principle of Type Uniformity sets the priors of each equal.

Another advantage to Paradigm Theory approach is that the dependence on the models is explicitly built in through the paradigm. *Any* choice of subsets is an allowable model choice for Paradigm Theory.

3.2.5   Bertrand's Paradox

Suppose a long straw is thrown randomly onto the ground where a circle is drawn. Given that the straw intersects the circle, what is the probability that the resulting chord is longer than the side of an inscribed equilateral triangle (call this event B). This is Bertrand's question (Bertrand, 1889, pp. 4–5). The Principle of Indifference leads to very different answers depending on how one defines the hypothesis set H.

H_0 If the hypothesis set is the set of distances between the center of the chord and the center of the circle, then the uniform distribution gives $P(B) = 1/2$.

H_1 If the hypothesis set is the set of positions of the center of the chord, then the uniform distribution gives $P(B) = 1/4$.

H_2 If the hypothesis set is the set of points where the chord intersects the circle, then the uniform distribution gives $P(B) = 1/3$.

Kneale (1949, pp. 184–188) argues that the solution presents itself once the actual physical method of determining the chord is stated, and a critique can be found in Mellor (1971, pp. 136–146). Jaynes (1973) presents a solution which I discuss more below. Marinoff (1994) catalogues a variety of solutions in a recent article. I approach Bertrand's Paradox in two fashions.

Generalized Invariance Theory

In the first Paradigm Theory treatment of Bertrand's Paradox I take the hypothesis set to be the set of all possible prior probability distributions over the points in the interior of the circle—each prior probability distribution just *is* a hypothesis. To alleviate confusion, when a hypothesis set is a set of prior probability distributions over some other hypothesis set, I call it a *prior set*; I denote the elements of this set by ρ rather than h, and denote the set H_ρ.

I wish to determine a prior probability assignment on H_ρ. What "should" the paradigm be? Jaynes (1973) argues that the problem statement can often hold information that can be used to determine a unique distribution. In the case of Bertrand's Problem, Jaynes argues that because the statement of the problem does not mention the angle, size, or position of the circle, the solution must be invariant under rotations, scale transformations, and translations. Jaynes shows that there is only one such solution (in fact, translational invariance alone determines the solution), and it corresponds to the H_0 case above, with $P(B) = 1/2$: the probability density in polar coordinates is

$$\mathcal{P}(r, \theta) = \frac{1}{2\pi R r}, \ \ 0 \le r \le R, \ \ 0 \le \theta \le 2\pi$$

where R is the radius of the circle. The theory sanctioning this sort of determination of priors I call the *Invariance Theory* (see Changizi and Barber, 1998).

I will interpret the information contained in the problem statement more weakly. Instead of picking the prior distribution that has the properties of rotation, scale, and translational invariance as Jaynes prescribes, suppose one merely *acknowledges* the invariance properties. That is, the paradigm is comprised of the subsets of prior probability distributions that are rotation, scale, and translation invariant, respectively. For every non-empty logical combination of the three properties besides their mutual intersection there are continuum many hypotheses. Supposing that each subset of the prior set corresponding to a logical combination of the three properties has a different measure, Paradigm Theory induces five symmetry types: $T \cap R \cap S$, $\neg T \cap R \cap S$, $R \cap \neg S$, $\neg R \cap S$ and $\neg R \cap \neg S$ (three logical combinations are empty), where T, R and S

denote the set of translation-, rotation- and scale-invariant priors, respectively. Each receives prior probability $1/5$, and since $T \cap R \cap S = \{\frac{1}{2\pi Rr}\}$ and the other symmetry types are infinite, $P(\frac{1}{2\pi Rr}) = 1/5$ and every other prior receives negligible prior probability; $1/(2\pi Rr)$ is the clear choice. In as much as the properties of this paradigm are objective, being implicitly suggested by the problem, this solution is objective.[14]

This "trick" of using Paradigm Theory parasitically on the Invariance Theory can be employed nearly whenever the latter theory determines a unique invariant distribution; and in all but some contrived cases the unique distribution is maximally probable. Some contrived cases may have it that, say, in the prior set ρ_1 is the unique prior that is scale and rotation invariant (where I suppose now that these are the only two properties in the paradigm), but that there is exactly one other prior ρ_2 that is neither scale nor rotation invariant (and there are infinitely many priors for the other two logical combinations). Here there are at most four symmetry types, $\{\rho_1\}$, $\{\rho_2\}$, $R \cap \neg S$ and $\neg R \cap S$. Each of these two priors receives prior probability $1/4$, and so ρ_1 is no longer the maximally probable prior.

Now, as a matter of fact, the invariance properties people tend to be interested in, along with the prior sets that are typically considered, have it that there are infinitely many priors that are not invariant under any of the invariance properties. And, if the Invariance Theory manages to uniquely determine a prior, there are almost always going to be multiple priors falling in every logical combination of the invariance properties except their mutual intersection. If this is true, then Paradigm Theory's induced symmetry types have the unique prior as the only prior alone in a symmetry type, i.e., it is the only *invariant* prior in Paradigm Theory's definition as well. Given that this is so, by Theorem 2 this prior has the greatest prior probability.

Paradigm Theory need not, as in the treatment of Bertrand's Problem above, give infinitely higher prior probability to the unique invariant prior than the others, however. Suppose, for example, that the Invariance Theory "works" in that there is exactly one prior ρ_0 that is both scale and rotation invariant, but that there are exactly two priors ρ_1 and ρ_2 that are scale invariant and not rotation invariant, exactly three priors ρ_3, ρ_4 and ρ_5 that are rotation and not scale invariant, and infinitely many priors that are neither (again, where only rotation and scale invariance are the properties in the paradigm). There are now four

---

[14]And the solution seems to be correct, supposing the frequentist decides such things, for Jaynes claims to have repeated the experiment and verified that $P(B) \approx 1/2$, although see Marinoff's comments on this (Marinoff, 1994, pp. 7–8).

symmetry types, each receiving prior probability 1/4. The probability of the unique invariant prior is 1/4, that of each of the pair is 1/8, and that of each of the triplet is 1/12. The point I mean to convey is that *Paradigm Theory not only agrees with the Invariance Theory on a very wide variety of cases, but it tells us the degree to which the Invariance Theory determines any particular prior.* In this sense Paradigm Theory brings more refinement to the Invariance Theory. In the cases where Paradigm Theory does not agree with the Invariance Theory, as in the "contrived" example above, there is a principled reason for coming down on the side of Paradigm Theory *if* the invariance properties are just *acknowledged* and not favored. Also, not only can Paradigm Theory be applied when the Invariance Theory works, it can be applied when the Invariance Theory fails to determine a unique prior; in this sense, Paradigm Theory allows not only a refinement, but a sort of generalization of the Invariance Theory.

H is the Sample Space

The second way of naturally approaching Bertrand's Paradox within Paradigm Theory takes the hypothesis set to be the set of possible outcomes of a straw toss. In determining the hypothesis set more precisely, one informal guide is that one choose the "most general" hypothesis set. This policy immediately excludes H_0 (see the beginning of this subsection) since it does not uniquely identify each chord in the circle. H_1 and H_2 are each maximally general and are just different parametrizations of the same set. I choose H_1 as, in my opinion, the more natural parametrization, with the underlying measure being the obvious Euclidean area.

What "should" the paradigm be? The problem has a clear rotational symmetry and it would seem very natural to acknowledge the distance between the center of the chord and the center of the circle; this set of distances just *is* H_0 and we will be "packing in" H_0 into the paradigm. Rather than acknowledging *all* of the distances, suppose that one acknowledges n of them (n equally spaced concentric rings within the circle); we will see what the probability distribution looks like as n approaches infinity. Each ring has a different area, and so each is its own symmetry type. Therefore each has a probability of $1/n$. The probability density is

$$\mathcal{P}(r, \theta) = \frac{1/n}{A_i} = \frac{1/n}{(2i-1)\pi R^2/n^2} = \frac{n}{(2i-1)\pi R^2}, \quad r \in [iR/n, (i+1)R/n],$$

where, $i = 1, \ldots, n$, A_i is the area of the ith concentric ring from the center.

As n gets large, $iR/n \approx r$, so $i \approx rn/R$. Thus

$$\mathcal{P}(r, \theta) = \frac{n}{(2rn/R - 1)\pi R^2} = \frac{n}{(rn - R/2)2\pi R}$$

and since n is large, $rn - R/2 \approx rn$, giving

$$\mathcal{P}(r, \theta) = \frac{1}{2\pi Rr}$$

which is exactly what Jaynes concludes. Acknowledge how far chords are from the center of the circle and accept one of the more natural parametrizations... get the "right" prior probability density function.

If instead of acknowledging the distance from the center of the circle one acknowledges the property of being *within* a certain radius, then the sets in the paradigm are nested and the resulting symmetry types are the same as before, regardless of the underlying measure.

3.3   "Solution" to riddle and theory of innateness

The intuition underlying Paradigm Theory is that *more unique is better*, or *arbitrariness is bad*, and this is related to the idea that *names should not matter*, which is just a notion of symmetry. The more ways there are to change a hypothesis' name without changing the structure of the inductive scenario (i.e., without changing the paradigm), the more hypotheses there are that are just like that hypothesis (i.e., it is less unique), which means that there is less "sufficient reason" to choose it. The principles of Paradigm Theory link with this intuition. The Principle of Type Uniformity and Principle of Symmetry give more unique hypotheses greater prior probability, and the Principle of Defensibility entails that among the more unique hypotheses, those that are more unique should receive greater prior probability. Recall (from Subsection 3.1.3) that these are the *links* of the principles to the "more unique is better" motto— the principles do not actually *say* anything about the uniqueness of hypotheses, but are motivated for completely different, compelling reasons of their own. Nevertheless, it is a convenient one-liner to say that Paradigm Theory favors more unique hypotheses, and not just qualitatively, but in a precise quantitative fashion. In this sense the theory is a quantitative formalization of Leibniz's Principle of Sufficient Reason, interpreted nonmetaphysically only.

The favoring of more unique hypotheses, despite its crudeness, is surprisingly powerful, for it is a natural, radical generalization of both Carnap's

m^*-logical theory and (through the use of secondary paradigms) Hintikka's
"$\alpha = 0$"-logical theory, arguably the most natural and pleasing inductive methods from each continuum. Besides these achievements, Paradigm Theory gives
explanations for a large variety of inductive phenomena:

- it "correctly" collapses to the Classical Theory's Principle of Indifference when no distinctions are made among the hypotheses,

- it suggests a conceptual framework-based solution to the problem of the underdetermination of interpretation for language,

- it explains why no-inductions are rarely considered rational,

- it explains why frequency-inductions and law-inductions *are* usually considered rational,

- it gives a foundation for Occam's Razor by putting forth the notion that simpler hypotheses are favored because one acknowledges simplicity, and simpler hypotheses are
  (usually) less arbitrary,

- it accommodates curve-fitting by supposing only that one acknowledges the usual models—constants, lines, parabolas, etc.,

- it allows a sort of generalization of the Invariance Theory for choosing unique prior
  probability distributions, and this is used to solve Bertrand's Paradox,

- and it accounts for Bertrand's Paradox in another fashion by acknowledging the distance
  from the center of the circle.

In the first section of this chapter I laid out the goals of a theory of logical
induction, and the related goals of a theory of innateness. How does Paradigm
Theory fare in regard to these goals?

How Paradigm Theory "solves" the riddle of induction

Let us briefly recall our basic aim for a logical theory of induction. Ultimately,
we would like to reduce all oughts in induction and inference—you *ought* to
choose the simplest hypothesis, you *should* believe the next fish caught will
be a bass, it is *wrong* to draw a parabola through three collinear points, and
so on—to just a small handful of basic, axiomatic, or primitive oughts. The
hope is that all oughts we find in induction can be derived from these primitive
oughts. We would then know, given just a set of hypotheses and the evidence,
exactly what degrees of confidence we should have in each hypothesis. If we
had this, we would have a solution to the riddle of induction.

Alas, as discussed at the start of this chapter, this is impossible; there is no solution to the riddle of induction. There are, instead, multiple inductive methods, and although some may well be irrational or wrong, it is not the case that there is a single right inductive method. This was because any inductive method makes what is, in effect, an assumption about the world, an assumption which is left hanging without defense or justification for why *it* should be believed.

If we are to have a theory of logical induction, we must lower the bar. We would *still*, however, like the theory to consist of a small handful of primitive oughts. But we are going to have to resign ourselves to the persistence of a leftover variable of some kind, such that different settings of the variable lead to different inductive methods. A theory of logical induction would, at best, allow statements of the form

> If the variable is X, then the primitive oughts entail that one should proceed with inductive method M.

But it would defeat the whole purpose of our theory if this variable stood for variable *a priori* beliefs about the world, because the theory would then only be able to say that if you started out believing X, then after seeing the evidence you should believe Y. We want to know why you should have started out believing X in the first place. How did you get *those* beliefs in ignorance about the world?

And this was the problem with the Bayesian approach. The Bayesian approach was good in that it declares a primitive ought: one should use Bayes' Theorem to update probabilities in light of the evidence. And to this extent, Paradigm Theory also utilizes Bayes' Theorem. But the Bayesian approach leaves prior probabilities left over as a free-for-the-picking variable, and priors are just claims about the world.

With this in mind, we required that the variable in any successful theory of logical induction not stand for beliefs or claims about the world. Because any choice of the variable leads, via the primitive principles of ought, to an inductive method, any choice of variable ends up entailing a claim about the world. But that must be distinguished from the variable itself being a claim about the world. We required that the variable have some meaningful, non-inductive interpretation, so that it would be meaningful to say that an inductive agent entered the world with a setting of the variable but nevertheless without any *a priori* beliefs about the world. We would then say that the agent, being rational, should follow the primitive principles of ought and thereby end up

with what are claims about the world. But the claims about the world were
not inherent to the variable, they only come from joining the variable with the
principles of ought.

In this chapter I introduced a kind of variable called a "paradigm," which
is central to Paradigm Theory. Paradigms are not about the world. Instead,
they are *conceptualizations* of the world, and more exactly, conceptualizations
of the space of hypotheses. Paradigms say which hypotheses are deemed to be
similar to one another, and which are not. More precisely, a paradigm is the
set of properties of hypotheses the inductive agent acknowledges. The set of
properties of hypotheses acknowledged does not comprise a claim about the
world, nor does it possess any 'ought's. It is just a way of looking at the set
of hypotheses, and no more than that. Paradigms, then, are non-inductive and
have a meaningful interpretation. This is the kind of variable we wanted in a
theory of logical induction.

But we also needed principles capable of taking us from the variable—
the paradigm in Paradigm Theory—to an inductive method. Paradigm Theory
achieves this via three primitive principles of ought, along with the Bayesian
principle of evidence (Bayes' Theorem). The three principles concern non-
arbitrariness in the assignment of prior probabilities, and given a paradigm the
principles entail a unique prior probability distribution. The Bayesian principle
of evidence finishes the job by stating how one ought to modify prior probabilities to posterior probabilities as evidence accumulates. In sum, Paradigm
Theory allows statements like this:

> If, before knowing anything about the world, you conceptualize the space of
> hypotheses in a fashion described by paradigm Q, then via the three primitive
> principles of prior probability determination you should have certain prior proba-
> bilities $P_Q(h)$ on those hypotheses. And, furthermore, when evidence is brought
> to bear on the logical probabilities of the hypotheses, one should obtain posterior
> probabilities by using Bayes' Theorem.

The most important thing to notice about this is that the statement begins with
the inductive agent *not* making any claim about the world. The statement does
not simply say that if you have certain beliefs you ought to have certain others.
It requires only that the completely-ignorant-about-the-world inductive agent
enter the world with a way of looking at it. Without any preconceptions about
the world (although he has preconceptions about the properties of hypotheses),
the theory nevertheless tells the agent how he ought to proceed with induction.
The theory thereby reduces all inductive oughts to a few primitive principles of

ought, and these primitive oughts are the *only* inductive primitives one needs for a theory of induction. *At base, to justifiably follow an inductive method is to have a paradigm and to follow certain abstract principles of non-arbitrariness and principles of evidence.*

Some readers might say that this is all well and good, but does it really get us anywhere? We are still stuck with paradigms, and there is no way to justify why an inductive agent has the paradigm he has. We have simply pushed the indeterminacy of inductive methods downward, to prior probabilities, and then further downward to paradigms. We still have not answered the question of which inductive method we should use, because we have not given any reason to pick any one paradigm over another. That is, suppose that—poof—a rational, intelligent agent suddenly enters a universe. We still do not know what he should do in regards to learning, and so Paradigm Theory is useless for him.

The response to this kind of criticism is that it is essentially taking Paradigm Theory to task for not being a solution to the riddle of induction. To see this, note that the criticism can be restated as, "If Paradigm Theory is so great, why isn't it telling us what one should believe given just the hypothesis set and the evidence?" But this is just to ask why Paradigm Theory does not solve the riddle of induction. The answer, of course, is that there is no solution to the riddle of induction; i.e., there is no single way that one ought to take a set of hypotheses and evidence and output posterior probabilities in the hypotheses. This kind of criticism has forgotten to lower the bar on what we should be looking for in a theory of logical probability. At best, we can only expect of a theory of logical probability that it reduce inductive oughts to a small number of primitive ones, and to some variable that is not about the world. We cannot expect to have no variable left over.

It should be recognized that it was not *prima facie* obvious, to me at least, that it would even be possible to obtain this lowered-bar theory of logical induction. *Prima facie*, it seemed possible that there would be no way, even in principle, to reduce inductive oughts to a few primitive oughts and some meaningful, non-inductive variable. Paradigm Theory is an existence proof: a lowered-bar theory of logical probability exists. I have presented no argument that no other theory of logical probability could not also satisfy these requirements I have imposed; there probably exist other such theories, perhaps others with superiorities over Paradigm Theory.

How Paradigm Theory serves as a theory of innateness

Paradigm Theory provides a kind of best-we-can-hope-for solution to the riddle of induction. But I had also stated at the start of this chapter that we were simultaneously looking for a theory that would serve as a theory of innateness, and I had put forth requirements we demanded of such a theory. The requirements were that we be able to model rational intelligent agents as following certain fixed learning principles, and that any innate differences in their resultant inductive method would be due to some setting of a variable with a weak, non-inductive, meaningful interpretation. Under the working assumption that the brain is rational, the theory would apply to the brain as well. The theory of innateness would provide a way of economically explaining why different agents—or different kinds of brain—innately engage in different inductive methods. We would not have to commit ourselves to a belief that the principles of learning may be innately chosen willy nilly; there is a single set of learning principles that anyone ought to follow. We would also not be committed to a view that brains enter the world with *a priori* beliefs about it, a view that seems a little preposterous. Instead, brains would only have to innately be equipped with some other kind of difference, although what that difference might be will depend on the kind of theory of innateness that is developed.

Recall that the Bayesian framework is a nice step forward in this regard, and has accordingly been taken up in psychology, neuroscience, computer and the decision sciences to study learning and interactions with an uncertain world. All innate differences in inductive methods will be due *not* to innate differences in how evidence is to be used to modify the degrees of confidence in hypotheses. All innate differences stem from innate differences in prior probabilities, and here lies the problem with the Bayesian framework as a theory of innateness: prior probabilities are *a priori* beliefs about the world, and thus they are not non-inductive, as we require for a theory of innateness.

The Bayesian framework should not, however, be abandoned: it gets the evidence principle right. What we would like is to dig deeper into prior probabilities and find principles of prior probability determination that any agent should follow, so that from some non-inductive innate variable comes prior probabilities via these principles. And this is where Paradigm Theory enters. Objects called "paradigms" were introduced which were interpreted as conceptual frameworks, or ways of conceptualizing the space of hypotheses. Paradigms were not about the world. Paradigm Theory introduced principles of prior probability determination saying how, given a paradigm, one ought to assign prior probabilities. Paradigm Theory, then, appears to satisfy the requirements of a theory of innateness.

But, someone might criticize, we are still left with innate paradigms, or innate ways of conceptualizing the set of hypotheses, or innate ways of lumping some hypotheses together as similar or of the same type. Is this any better than innate prior probabilities? Perhaps it is strange to hypothesize that brains have *a priori* beliefs about the world, but is it not also strange to hypothesize that brains have *a priori* ways of carving up the space of hypotheses?

As a response, let me first admit that it *is, prima facie*, a bit strange. However, one has to recognize that if a brain engages in an inductive method, then it *must* have entered the world with *some* innate structure that is sufficient to entail the inductive method. Such innate "structure" will either be learning algorithms of some kind unique to that kind of brain, or perhaps the Bayesian evidence principle along with prior probabilities unique to that kind of brain, or perhaps the Bayesian evidence principle and principles of prior probability determination along with a paradigm unique to that kind of brain, etc. It may seem *prima facie* odd to believe that *any* of these kinds of "structures" could be innate. One reason for this first reaction to innate structures is that there is, I believe, a tendency to revert to thinking of brains as blank slate, universal learning machines: brains enter the world completely unstructured, and shape themselves by employing universal learning algorithms to figure out the world. But as we have discussed, there is no universal learning algorithm, and so there cannot be brains that enter the world without innate learning-oriented structures. We are, then, stuck with innate learning-oriented structure, no matter how strange that might seem. Thus, the fact that innate paradigms strike us as strange is not, alone, an argument that Paradigm Theory is supposing something outlandish.

But, one may ask, are paradigms any less outlandish than prior probabilities? What have we gained by moving from innate structure in the form of prior probabilities to innate structure in the form of paradigms? We have gained in two ways. First, we have isolated further principles of rationality that inductive agents ought to follow; namely, the non-arbitrariness principles of prior probability determination (the Principles of Type Uniformity, Symmetry and Defensibility). Second, paradigms are a much weaker innate structure, being only about the kinds of hypotheses there are, rather than about the degree of confidence in the hypotheses.

Note that Paradigm Theory as a theory of innateness is not necessarily committed to actual innate paradigms in the head, whatever that might mean. It is commonplace for researchers to hypothesize that different kinds of organisms have what are, in effect, different innate prior probabilities, but such researchers do not commit themselves to any view of what mechanisms may instantiate this. Prior probabilities are primarily a theoretical construct, and allow us to understand brains and learning agents within the Bayesian framework. Similarly, Paradigm Theory is not committed to any particular mechanism for implementing innate paradigms. Rather, paradigms are a theoretical construct, allowing us to describe and explain the behaviors of inductive agents and brains in an economical fashion.

Paradigm Theory is *a* theory of innateness satisfying the requirements we set forth, but there is no reason to believe there are not others also satisfying the requirements, perhaps better theories in many ways.

Appendix to chapter: Some proofs

This section consists of some proofs referred to in this chapter.

Here are some definitions. $\delta(h) = \gamma$ (h is γ-Q-invariant in H) if and only if $h \in \Delta^\gamma$. $\delta(h)$ is the ordinal number indicating the invariance level of h. Say that t is a Q^γ-*symmetry type* in H if and only if t is a $Q \sqcap H^\gamma$-symmetry type in H^γ. Let κ_n be the cardinality of H^n (which is also the number of singleton Q^{n-1}-symmetry types), let s_n be the number of non-singleton Q^n-symmetry types, and let $e(h)$ be the cardinality of the Q-equivalence type of h. Notice that $\kappa_{n+1} = card(I(Q^n, H^n))$ ('*card(A)*' denotes the cardinality of set A). We denote $\frac{\kappa_{i+1}}{s_i + \kappa_{i+1}}$ by r_i and call it the *singleton symmetry type ratio at level i*. The following theorem states some of the basic properties of Paradigm Theory.

Theorem 14 *The following are true concerning Paradigm Theory.*

1. $P(H^{n+1}) = r_n P(H^n) \ (P(H^0) = 1).$

2. $P(H^{n+1}) = r_0 r_1 \cdots r_n.$

3. $P(\Delta^n) = (1 - r_n) P(H^n).$

4. $P(h) = \frac{r_{\delta(h)}}{e(h)\kappa_{\delta(h)+1}} P(H^{\delta(h)}).$

5. $P(h) = \frac{r_0 \cdots r_{\delta(h)}}{e(h)\kappa_{\delta(h)+1}}.$

Proof. Proving 1, there are $s_n + \kappa_{n+1}$ Q^n-symmetry types, and κ_{n+1} of them are singletons which "move up" to the $n + 1^{\text{th}}$ level. Since each Q^n-symmetry type gets the same probability, H^{n+1} gets the fraction

$$\frac{\kappa_{n+1}}{s_n + \kappa_{n+1}}$$

of the probability of H^n. 2 is proved by solving the recurrence in 1. 3 follows from 1 by recalling that $P(\Delta^n) = P(H^n) - P(H^{n+1})$. To prove 4, notice that the probability of a hypothesis h is

$$\frac{P(\Delta^{\delta(h)})}{s_{\delta(h)} e(h)}.$$

Substituting $P(\Delta^{\delta(h)})$ with the formula for it from 3 and some algebraic manipulation gives the result. Finally, 5 follows from 2 and 4. $\triangle$

Proof of Theorem 3. To prove 1, it suffices to show that for all i, $\frac{P(\Delta^i)}{s_i} \leq \frac{P(\Delta^{i+1})}{s_{i+1}}$. By Theorem 14,

$$P(\Delta^i) = \frac{s_i}{s_i + \kappa_{i+1}} P(H^i)$$

and

$$P(\Delta^{i+1}) = \frac{s_{i+1}}{s_{i+1} + \kappa_{i+2}} P(H^{i+1}) = \frac{s_{i+1}}{s_{i+1} + \kappa_{i+2}} \frac{\kappa_{i+1}}{s_i + \kappa_{i+1}} P(H^i)$$

By substitution we get

$$\frac{P(\Delta^{i+1})}{s_{i+1}} = \frac{P(\Delta^i)}{s_i} \frac{\kappa_{i+1}}{s_{i+1} + \kappa_{i+2}}$$

It therefore suffices to show that

$$1 \leq \frac{\kappa_{i+1}}{s_{i+1} + \kappa_{i+2}},$$

and this is true because the denominator is the total number of Q^{i+1} symmetry types, which must be less than or equal to the numerator, which is the total number of hypotheses in H^{i+1}. 2 follows easily from 1. $\triangle$

It is not the case that less defensible equivalence types always have less probability. It is also not the case that more defensible hypotheses never have lower probability than less defensible hypotheses. A more defensible hypothesis h_1 can have less probability than a less defensible hypothesis h_2 if the equivalence type of h_1 is large enough compared to the equivalence type of h_2. The following theorem states these facts.

Theorem 15 *The following are true about Paradigm Theory.*

1. *There are equivalence types d_1 less defensible than d_2 such that $P(d_1) = P(d_2)$.*

2. *There are hypotheses h_1 not more defensible than h_2 such that $P(h_1) \not\leq P(h_2)$.*

Proof. To prove 1, consider a paradigm represented by a two-leaf binary tree. The root comprises one equivalence type, and the pair of leaves is the other. Each equivalence type is also a symmetry type here, and so each gets probability $1/2$.

Proving 2, consider the tree on H_f from Section 3.1.2. The reader may verify that h and i receive probability $\frac{1}{18}$, but e, f, and g receive probability $\frac{14}{15}\frac{1}{18} < \frac{1}{18}$. $\triangle$

Proof of Theorem 4. When n is even there are $2n$ complexions and Laplace's method gives each a probability of $1/2n$. For each complexion there is a symmetrical one with respect to Q_L with which it may be permuted (without changing Q_L), so there are n symmetry types, each receiving via Q_L a probability of $1/n$. Each symmetry type contains exactly two complexions of equal size, and so each complexion gets a probability assigned of $1/2n$. (The non-complexion set in Q_L does not come into play when n is even.)

When n is odd there are $2n - 1$ complexions and Laplace's method gives each a probability of $1/(2n-1)$. Now there are an odd number of complexions, and the "middle" one is not symmetrical with any other complexion. Furthermore, because Q_L contains the set of all sequences with more 0s than 1s, and this set is asymmetrical, none of the complexions are symmetrical with any others. Thus, each complexion is a symmetry type, and each complexion receives a probability of $1/(2n - 1)$. $\triangle$

Proof of Theorem 7. There are $2^N - 2$ sequences that are not predicted by the 'all 1s' or 'all 0s' laws, and these must share the .5 prior probability assignment.

There are $2^{N-n}-1$ sequences of length N with the first n experiments resulting in 1 but not all the remaining $N - n$ experiments resulting in 1; the total prior probability assigned to these strings is therefore

$$q = \frac{1}{2} \frac{2^{N-n} - 1}{2^N - 2}.$$

The probability that after seeing n 1s there will be a counterexample is

$$\frac{q}{.25 + q}.$$

With some algebra, the probability that after seeing n 1s the remaining will all be 1 is

$$\frac{1}{2} \frac{2^N - 2}{2^N(2^{-n} + 2^{-1}) - 2},$$

which, for any moderately sized N becomes, with some algebra, approximately

$$\frac{2^{n-1}}{1 + 2^{n-1}}. \triangle$$

Proof of (a) in Theorem 9. We want the probability that $p = 1$ given that we have seen n 1s and no 0s $(n > 0)$; i.e., $P(p = 1|1^n)$, where 1^n denotes the string with n 1s. By Bayes' Theorem

$$P(p = 1|1^n) =$$

$$\frac{P(p = 1)P(1^n|p = 1)}{P(p = 1)P(1^n|p = 1) + P(p \in (0,1))P(1^n|p \in (0,1)) + P(p = 0)P(1^n|p = 0)}.$$

The only term that is not immediately obvious is $P(1^n|p \in (0,1))$, which is $\int_0^1 p^n dp = 1/(n+1)$. Thus we have

$$\frac{.25(1)}{.25(1) + .5\frac{1}{n+1} + .25(0)},$$

and with a little manipulation this becomes $\frac{n+1}{n+3}$. $\triangle$

Proof Sketch of Theorem 10. 1 is simple and 2 is proved by induction on the depth of the binary tree. 1 and 2 do not exhaust the types of trees that result in the root being the lone maximally defensible element; see the H_{asymm}/Q_{asymm} example in Section 3.2.3 for a non-binary non-full tree that puts the root alone at the top. Informally, most trees put the root at the top. We have made no attempt to characterize the class of trees that put the root at the top. See the following tree for 3.

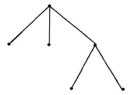

△

Proof of Theorem 11. There are $n + 1$ symmetry types (one for each level), each receiving probability $1/(n + 1)$. The symmetry type at depth i has b^i elements. △

Chapter 4

Consequences of a Finite Brain

Do the ultimate limits on what it is possible to compute have a role to play in explaining our world? Can the ultimate limits on computers tell us anything about ourselves? Can recursion theory, the most theoretical discipline of computer science, be applied to the brain? One major point of this chapter is to suggest that the answers are "Yes": Studying the limits of computation can give one a handle on principles governing any sufficiently intelligent agent, whether the agent is meaty or metal. If you are a finite machine, as we are, then there are certain necessary consequences. *Any* brain will have certain similarities. The similarity that I concentrate on is the phenomenon called *vagueness*, and I show why any finite-brained agent will have to deal with vagueness.

As a student, one of my interests was to understand the ultimate limits of thought, learning, understanding and intelligence. My interest was not really about brains, per se; my interest concerned the ultimate limits of *any* thinking machine. Now, it is primarily brains that think, learn, understand and have intelligence, and so one might expect that studying neuroscience might help one to discover these limits of thought. Not so, at least not with the state of neuroscience today. The reasons are that

(a) brains are not well-understood,

(b) brain activity is not easily describable, and

(c) brains are not susceptible to proving theorems about them and their limits.

And even if someday (a), (b) and (c) are averted with a mature neuroscience, it may be that

(d) the brain, being just one of presumably infinitely many possible kinds of thinking machine, barely scratches the surface as to what the ultimate limits are.

This reasoning led me to learn computer science, and, more specifically, logic, theoretical computer science, complexity theory and recursion theory. Why? Because

(a) computers *are* well-understood,

(b) their activities *are* nicely describable, and

(c) computers *are* susceptible to proving theorems about them and their limits.

Furthermore,

(d) in theoretical computer science one may study what is arguably the class of *all* thinking machines, not just one specific kind of thinking machine.

Understanding theoretical computer science and logic not only illuminates the limits of machines, I believe it can tell us interesting things about one particular machine: the brain. A computer is, alas, not a brain. The fine details of how the brain works will probably not be illuminated merely by understanding computers and their limitations; those interested in the fine details need to study real brains in addition to computers. But I have never been interested in fine details—I care only about general, foundational principles—and this goes for the brain as well. Given the plausible notion that the brain is a kind of computer, understanding computers and their limits may provide us with insights into the brain. That is, using computers as a model of the brain may have a payoff.

There is a danger that one might take me to mean that "using computers to model the brain may have a payoff." Now, *this* is uncontroversially true; computational models in neuroscience and psychology are widespread. Generally, the idea is to concoct a program that captures relevant aspects of the system of interest. But this is not what I mean. When I say that the brain is to be modeled as a computer, I do not mean that I have devised some particular kind of program that seems to capture this or that aspect of the brain or behavior. Instead, I mean that the brain is to be treated as a computational device—whether a neural network machine, a random access architecture, or other—and is thus subject to the same computational limits as computers.

This, too, is almost entirely uncontroversial: nearly everyone considers the brain a computing machine. (Although Penrose (1994) is one counterexample.) What has not been given sufficient attention, however, is that the brain-as-computer hypothesis—all by itself and without reference to any more detailed hypothesis about the program it computes or what its specific limitations are—has certain empirical implications. Rational, computer-brained agents—although they may be radically different in behavior, personality, intelligence, likes, and so on—are going to have certain similarities. These similarities are things true of any rational computationally bound agent; they are "behavioral invariants" for such machines.

Our limitations

Consider the informal plots in Figure 4.1. Each plot has human "activities," or behaviors, along the x axis. Along the y axis in each plot is the human "ability"; $f(x)$ is the degree of ability humans possess in doing activity x. For example, perhaps we are interested in the activity x of running, in which case $f(x)$ represents the human ability to run so and so fast. Or, consider the activity x of holding items in working memory. Humans have the ability to hold around seven (plus or minus two) items in working memory, and $f(x)$ represents this ability.

The top horizontal line in each graph represents the line at which greater ability become logically impossible. Graph A is what I will call the "usual conception," where our human abilities always fall short of the logical limits of what is possible. For example, having a working memory space of seven is presumably not explained by reference to any logical limits on what is possible. The explanation for it is likely to be brain-specific and historical in nature. Graph B, on the other hand, depicts an activity (the arrow) for which the limits of logical possibility are the active constraint in explaining the human ability in that activity.

For example, consider the activity x of determining truths of Peano Arithmetic. First, what is "Peano Arithmetic"? For now, you just need to know that it is a short list of obviously true sentences of arithmetic, from which one may derive infinitely many other truths of arithmetic (but not all the truths of arithmetic). And, note that "arithmetic" just refers to what you might think it does: mathematical sentences concerning addition and multiplication on the natural numbers $0, 1, 2 \ldots$. Thus, I am asking you to consider the activity of determining whether or not a sentence in arithmetic follows from Peano's axioms of

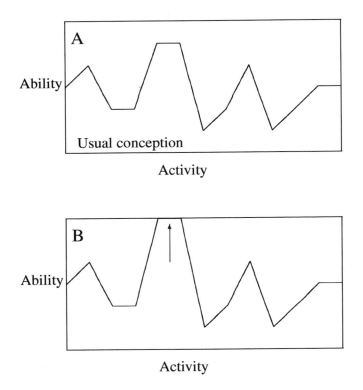

Figure 4.1: Ability to do a given activity. The top horizontal line in each plot demarcates the boundary between logical possibility and, above the line, logical impossibility. The usual conception is to view our ability as falling short of the line of logical impossibility. That is, the explanation for our ability being what it is usually refers to hosts of contingent physical or biological details about us. On the other conception, perhaps there are abilities of ours (the arrow) that are, in some sense, as good as they possibly can be, where "possibly" refers to logical possibility. The explanation for our ability (or inability) being what it is refers entirely (or almost entirely) to the limits of what is logically possible.

arithmetic. It turns out that we humans are not perfect at this; we are not perfect determiners of which sentences follow from Peano's axioms and which do not. This is not the kind of thing that anyone has actually experimentally tested, mind you, but no savant has come forth able to, without error, tell whether or not an arithmetic sentence follows from Peano's axioms. We humans seem to have a limited ability in this regard.

Another place we have a similar inability concerns the problem of determining whether or not some program is going to halt on some given input. That is, I hand you some software for your computer, and I give you a copy of the programming instructions, or code, underlying the software. I then give you something to input into the program. Two things can possibly happen. First, the program could take the input, grind it through some instructions and so on, and eventually terminate. This is what we almost always *want* our software to do: to eventually stop running without us having to reboot the system to force it to stop. The other possible result of placing the input into the program is that the program may never stop running; it is said to never halt. It just keeps carrying out more and more computations, going on and on and on without end. This is when one must definitely resort to rebooting the computer, or breaking out of the program somehow. Your task, with the program code in one hand and the input in the other, is to determine whether the program will or will not halt when that input is entered. This problem is called the *halting problem*, and people are known to be notoriously bad at solving it. Again, it is not as if there have been psychological experiments in this regard (not that I know of); it is just known within computer science circles, for example, that we are *all* prone to error in solving this problem.

One possible source of our limitations: logic itself

What is the source of our limited abilities in determining the truths of Peano arithmetic and in determining whether a program halts on some given input? It is thought that our limited abilities in these activities are explained by the undecidability of the problems. In particular, the set of truths of Peano Arithmetic is undecidable, as is the set of pairs of programs and inputs for which the program halts on the input. What do we mean by "undecidability"? We say that a set is *undecidable* if there is no computer program that exists, even in principle, that can take elements as input and always correctly output whether or not the element is a member of the set. If one did have such a program, the program would be said to *decide* the set; and the set would be decidable since there exists at least one program that can decide it. (We will talk more about this later.) Thus, when we say that the set of truths of Peano arithmetic is undecidable, we mean that there is no program that can be run on a computer that will take as input a sentence of arithmetic and output whether or not it is true. And when we say that the halting set—i.e., the set of pairs of programs and inputs for which the program halts on that input—is undecidable, we mean that there is no program Q that can be implemented on a computer that will take a program code P along with an input x together as input (i.e., the pair $\langle P, x \rangle$ is the input to the program Q) and output YES if program P halts on input x, and outputs NO if program P does not halt on input x.

What does the undecidability of these problems have to do with our limited ability in solving them? Since they are undecidable, no computing machine can solve them perfectly. And since we are just computing machines, we, too, cannot solve them perfectly. This argument depends on something called Church's Thesis, which states that if something is intuitively computable—i.e., if it seems in some sense as if one is able to compute it—then it is computable, in principle, by today's computers. In other words, it says that there is no other notion of computing something that we have not already captured in our understanding of computers. (We'll be discussing this at more depth at the appropriate time later.) With Church's Thesis in hand, it is argued that we can compute nothing a computer cannot also compute, and since a computer has limited ability with the Peano arithmetic and Halting problems, so must we.

Such an explanation, if true, utilizes in an essential way the logical limits of what is possible for finite agents (by which I will mean for now computationally bound agents, although I shall mean something more precise in Section 4.2.3), and thus $f(x)$ in the plot from earlier would be depicted as reaching the top horizontal line.

Vagueness and logical limits

Vagueness—the phenomenon that, roughly, natural language words have borderline regions (see Section 4.1)—is a phenomenon, not an activity, but phenomena and activities are sometimes related. The activity of human running is associated with the phenomenon that humans run only so and so fast. The activity of remembering via working memory is associated with the phenomenon that humans can only hold seven items. The activity of determining truths of Peano Arithmetic is associated with the phenomenon that humans are incapable of acting as perfect Peano Arithmetic truth-determiners. And the activity of discovering if a program halts on a given input is associated with the phenomenon that humans are not very good halting-determiners, or "bug-checkers."

In this vein, the phenomenon of vagueness will be seen to be associated with two certain logical limits on the ability of finite, sufficiently powerful, rational agents:

1. their inability to determine whether or not a program halts on a given input, and

2. their inability to generally acquire algorithms—programs that eventually halt on every input—for their concepts.

Thus, in explaining vagueness in this chapter, I will be arguing for a picture of explanations of human behavior as in graph B of Figure 4.1, where vagueness is connected with an activity for which the human ability is bound by the ultimate limits on what is possible; mathematics shapes us. We will see that vagueness is not due to any particularly human weakness, but due to a weakness that any computationally bound agent possesses; even HAL from *2001: A Space Odyssey* and Data from *Star Trek* will probably experience vagueness.[1]

4.1 Vagueness, the phenomenon

One of the main points of this chapter is to show why one particular, very important, all-pervasive, long-known, and not well-understood phenomenon is explained primarily by the fact that the human brain is finite. That phenomenon is *vagueness*. Vagueness applies to *predicates* of natural language. A *predicate* is a word that applies to a subset of the set of all possible objects or events. For example, the predicate 'dog' applies to all possible dogs, the predicate 'bald' applies to all possible bald heads, and the predicate 'eat' applies to all possible occasions of eating. Nouns, adjectives and verbs are predicates, but many other kinds of words are not predicates, such as logical connectives like 'and', 'or', and 'therefore', or other "function" words (as they are called in linguistics) such as 'after', 'each', 'in', 'must', 'he', 'is', 'the', 'too', and 'what'.

A predicate is vague if it has borderline cases. Yul Brynner (the lead in *The King and I*) is definitely bald, I am (at the time of this writing) definitely not,

---

[1] Very early ideas of mine along the lines presented here appeared in Changizi (1995). A paper on my theory was presented at the 1998 vagueness conference in Bled, Slovenia (Changizi, 1999a), and at the 1998 Irish Conference on Formal Methods (Changizi, 1999b). The latter concentrates on logics and semantics of vagueness motivated by my theory of vagueness. The main ideas were published in Changizi (1999c).

and there are many people who seem to be neither. These people are in the "borderline region" of the predicate 'bald', and this phenomenon is central to vagueness. Nearly every predicate in natural language is vague. From 'person' and 'coercion' in ethics, 'object' and 'red' in physical science, 'dog' and 'male' in biology, to 'chair' and 'plaid' in interior decorating; vagueness is the rule not the exception. Pick any natural language predicate you like, and you will almost surely be able to concoct a case—perhaps an imaginary case—where it is unclear to you whether or not the case falls under the predicate. Take the predicate 'book', for example. The object from which you are reading this is definitely a book, your light source is definitely not a book. Is a pamphlet a book? If you dipped this book in acid and burned off all the ink, would it still be a book? If I write this book in tiny script on the back of a turtle, is the turtle's back a book? We have no idea how to answer such questions. The fact that such questions appear to have no determinate answer is roughly what we mean when we say that 'book' is vague.

And, by 'vague' we do *not* include conundrums such as whether redness is or is not bald; the word 'bald' does not apply within the domain of colors, and so some might say that neither 'bald' nor 'not bald' "nicely applies" to redness. If 'bald' is vague—and it is—it is *not* because of the colors that it is vague. In cases of vagueness, the inability to nicely apply the word and its negation is *not* due to the word not applying to objects in that domain. 'bald' is vague because there are heads which do not nicely fall under 'bald' or 'not bald', even though there are lots of other heads which *do* fall nicely under one or the other of 'bald' or 'not bald'.

For the uninitiated, why should we care about vagueness? There are a number of reasons.

Most importantly for my interests here, the fact that natural language is vague needs to be *explained*. Why are natural language users unable to draw a single sharp line between definitely bald and definitely not bald? Why do natural language users seem to find cases that are borderline bald? Why cannot natural language users determine the boundaries of the borderline region? Is it possible to have a non-vague language? If so, under what conditions?

The most central phenomenon of vagueness is the borderline region, where for a vague predicate P there are objects which are not clearly classifiable as either P or 'not P'. This borderline region phenomenon seems to stab at the very heart of the idea that our concepts divide the world into two parts: those objects to which the predicate applies—the predicate's *extension*—and those objects to which the predicate does not apply—the complement of the predicate's extension. Although some would like to argue that the true meanings of P and 'not P' are as in classical two-valued logic where the extension of 'not P' is the complement of the extension of P (such a semantics is called *determinate*), the concepts as we actually use them do not *seem* to be like this, lest there be no phenomenon of vagueness at all. What are our concepts like, if they are not as in classical logic? If we are not carving up the world a la classical logic, how *do* we carve up the world?

A third reason the study of vagueness is important is related to the previous one, but now the concern is one of logic rather than the question of what is a concept. The issue is that classical two-valued logic seems to be at stake, for classical two-valued logic says (i) that the meaning of a predicate is a precise set and (ii) that the meaning of the negation of the predicate is the complement of that precise set. Prima facie, (i) and (ii) do not seem to be consistent with the existence of the phenomenon of vagueness. This threatens the usefulness of over seventy years of technical work in classical logic. What *is* the proper model of our interpretations and use of natural language predicates? And what *is* the logic of our inferences if we are not "classical logic machines"? Given that we humans are the paradigm example of rationality, it would serve us well to understand the logic we engage in for the purposes of (a) better understanding what is rationality and (b) building artificial machines that can mimic the intelligence and humanity of our inferences and utterances—how better to do this than to first understand how we do this?

Now that we care about vagueness, it is necessary to become clearer about what exactly the phenomenon of vagueness is? What is it that is in need of explanation? Some names of phenomena comprising the phenomenon of vagueness are 'borderline region', 'higher-order vagueness', 'sorites paradox', 'ineliminability', and 'essentialness'; the first two are central. All things equal, a theory that satisfies more of the phenomena is more favorable. But what are these phenomena? It is dangerous to attempt to precisely and formally define them since we have no clear pre-theoretic agreement on what exactly are the data, and any such definition is likely to be theory-laden to some extent. Accordingly I want to remain open-minded. I will give the rough idea for each phenomenon with the understanding that I am in no way defining what it is exactly. The best I hope for is an independently motivated theory that results in certain plausible phenomena that seem to match closely with the rough definitions of those named above. On to the phenomena.

What is the borderline region?

The *borderline region* phenomenon is roughly the phenomenon that for a vague predicate P we find ourselves with objects for which P neither clearly applies nor clearly does not apply; these objects are in the borderline region. Or, an object is borderline if it does not fit neatly into just one category. Alternatively, an object is borderline P if when we are given the choice "Which is it, P or not, and not both?" we do not know quite how to respond, and our non-response is seemingly not because we simply do not know, but because it seems fantastic to suppose there is exactly one correct response; this is partially what distinguishes vagueness from other sorts of unknowabilities. I say "seemingly" above because otherwise I exclude the possibility of an epistemic determinist theory of vagueness being correct, i.e., a theory where every object either falls into the extension of the predicate or the complement of the extension, and vagueness is due to our problems in seeing the boundary. The borderline region is also connected with the phenomenon that we are incapable of drawing a single sharp line distinguishing things P from things not P, and more than this, it is that any line drawn would seem ad hoc, arbitrary and wrong. Sometimes the borderline region is defined more epistemically as that region for which knowledge concerning membership in P is unattainable. The phenomenon is probably best communicated by example: wolves are borderline dog, violet is borderline blue, and so on.

What is higher-order vagueness?

Higher-order vagueness, second-order vagueness in particular, is the phenomenon that we find ourselves incapable of determining boundaries of the borderline region. Alternatively, imagine pulling hairs out of the head of a man who is definitely not bald. Second-order vagueness is exemplified by the fact that we do not find ourselves pulling out a single hair for which we are able to determine that the man suddenly becomes borderline bald. We find objects, or states of this man's head, which are borderline borderline bald. More explicitly epistemically, knowledge of the boundaries—if there are any—of the borderline region is unattainable. Higher-order vagueness, more generally, is the phenomenon that we find ourselves incapable of determining any semantically distinct boundaries at all between definitely bald and definitely not bald.

The "no boundaries" dogma

On these first, most central, two phenomena of vagueness—the borderline region and higher-order vagueness—what needs to be explained is not necessarily the real existence of a borderline region and higher-order vagueness, but rather why there seems to be a borderline region and higher-order borderline regions. The borderline region could be defined as the region which is semantically distinct from the definite regions, but a less theory-laden and more scientific approach would be to say that any adequate theory of vagueness must explain why there is a region which *seems* to be semantically distinct; this still leaves it open as to whether there is semantic indeterminacy. Also, sometimes (very often, in fact) higher-order vagueness is taken to be the phenomenon that there is no sharp semantic boundary between the definite regions and the borderline region, and even more radically it is sometimes taken that there are no semantic lines at all to be drawn, no matter how small the semantic difference or how impossible it is to see the lines. To have any lines posited by one's theory is, so it is reiterated, "not to take vagueness seriously." This seems straightforwardly bad science. There are our experiences of vagueness, and there are our theories about them; only the former can possibly count as data. Theories may well explain the data by positing that there is a semantically distinct borderline region, or that there are no sharp semantic lines, and perhaps such theories can in the end be victorious over epistemic theories like Sorensen's (1988), Williamson's (1994), a version of Koons (1994) and mine. What one cannot do is criticize epistemic theories on the basis that they do not posit a semantically distinct borderline region, or that they do posit sharp lines, for to do so is effectively to criticize epistemic theories for not being non-epistemic theories. In fact, if we are to be biased by our metaphysical prejudices, we have a long history of success in the drawing-sharp-lines business (e.g., classical logic) and should therefore be biased toward the existence of sharp lines.

Not only are epistemic theories unfairly criticized, so are many-valued theories (truth-valuational or probabilistic). I have never understood the criticism of fuzzy logic, for example, that it does not adequately handle higher-order vagueness. The charge is that there is a sharp, and let me suppose knowable, line between 'a is P' having truth value 0, where a is definitely not P, and having truth value > 0 (and < 1), where a is borderline P. This is criticized as being fantastic just as is a semantics with a sharp line between 'a is P' having truth value 0 and having, say, indeterminate truth value. Furthermore, it is argued, fuzzy logic is full of sharp lines everywhere, and this is just crazy.

Edgington (1992), who proposes a probabilistic theory, nicely states the feeling
I have always had on this when she writes,

> A twofold classification into the true and the false is inadequate where things
> hover on the edge of this great divide. A threefold classification is little improve-
> ment, I agree, with things hovering on the edge of the still-substantial divide
> between the true and the indeterminate. But in a manyfold classification, the
> difference between close degrees of truth, *including the difference between clear
> truth and its near neighbors,* is (almost always) insignificant, and a good theory
> must preserve this fact—must not deliver significantly different verdicts for in-
> significantly different cases. It does not matter if things still hover on the edges
> of inconsequential divides. (Edgington, 1993, p. 198.)

The motto "no boundaries" has become a dogma when theories postulating
inconsequential divides are dismissed out of hand.

Why does so much of the vagueness community strongly believe that there
are no sharp lines? I understand the intuition of there being no semantic lines
of any kind drawn since we "feel" like there are no such lines. But we also
feel a lot of ways counter to our current physics, say, but we can explain why
we feel that way, and we allow ourselves to conclude that it is just a feeling.
For example, we do not feel like we are spinning around as the Earth rotates,
but we know now that we are, and we can explain why we feel the way we do.
What is it about vagueness that so many—indeed most—in the field are so bent
on not only explaining the feeling, but making it a real part of the semantics?

And it is not even the case that the "no boundaries" dogma has proved fruit-
ful. I do not believe there is even one extant theory satisfying the "no boundaries" constraint that is remotely adequate as a description, much less an explanation. One wonders whether the dogma is really just a skeptical "no possible
theory of vagueness" dogma, given that the "no boundaries" constraint seems
simply impossible to satisfy in a coherent fashion. Horgan (1994) goes so far as
arguing that one should accept the incoherence of the "no boundaries" motto—
that the incoherence is not vicious. Not all "no boundaries" theorists are so
ready to take this route of Horgan; they hold out hope, presumably, for some
theory satisfying their constraint. For example, Sainsbury (1990) proposes an
attempted such theory that Edgington (1993) shows does have boundaries.

The sorites paradox

Moving on, there is a third phenomenon linked to vagueness: the *sorites paradox.* Its standard form is exemplified by the following two-premise argument

and conclusion: (i) 1 grain of sand cannot make a heap, (ii) for all n, if n grains of sand cannot make a heap, then $n + 1$ grains of sand cannot make a heap, (iii) there are no heaps of sand. (i) is obviously true.[2] (ii) is very compelling, since to deny it means that there is some n such that n grains of sand cannot make a heap but $n + 1$ grains can make a heap—that there is a to-the-grain distinction between being a heap and not—and this seems fantastic. (i) and (ii), though, imply (iii), which is obviously false. The sorites paradox is part of the phenomenon of vagueness in that it may be built using any vague predicate; and it may not be built with non-vague predicates.

There are two related constraints on a theory of vagueness. The first is that it locate the fault in the sorites argument, and do so without having to make fantastic claims. I do not mean to imply that the classical negation of the induction step cannot possibly be the solution to the sorites paradox. Epistemic, determinist theories do exactly this, but it is incumbent upon the theory to say why it is not so fantastic; for example, that we cannot ever determine or know the boundary, and this is why the suggestion that there is a boundary seems incredible. The second constraint is that a theory's post-theoretic notion of the phenomenon of vagueness should be such that a sorites argument built around a predicate displaying the phenomenon is paradoxical; i.e., denying the induction step must seem paradoxical. If the argument loses its paradoxical aspect, then the phenomenon claimed to be vagueness has a lesser claim to vagueness. For example, if we cannot ever determine or know the boundary but still believe quite reasonably that there is one, then there is nothing paradoxical in the sorites argument since the induction step can be (classically) denied readily without intuitive difficulties.

Ineliminability and essentialness

The final two phenomena are less central to vagueness.

The first of these is *ineliminability*: it is often felt that vagueness is not something we can simply eliminate from natural language. For example, it is sometimes said that any attempt to eliminate vagueness through precisification (i.e., making predicates precise) would, at best, radically undermine the meanings of natural language concepts. Also, restricting oneself to some delimited context is also thought to be unhelpful in eliminating vagueness—

---

[2] Although some have sought to save us from paradox by denying the base case. Unger (1979) and Wheeler (1979) deny that there are non-heaps by denying that there is a concept heapness at all.

vagueness occurs *within* contexts. The Undecidability Theory—i.e., my theory of vagueness—explains and accommodates a variety of ways in which vagueness is ineliminable (Subsection 4.3.5). I am less confident about this phenomenon being a necessary constraint on a theory of vagueness, and accordingly I do not criticize other theories on the basis that they do not explain or accommodate ineliminability. I do think that some degree of ineliminability must be addressed, however, lest we be left to wonder why we have not cured ourselves of vagueness.

Or a theory may alleviate this worry just mentioned by giving a good reason for why we should not want to "cure ourselves" of vagueness. This is the final phenomenon of vagueness: its seemingly *essential* nature, in the sense that it is often felt that even if we could eliminate vagueness, we would not want to because it fills an important and essential role for us. Perhaps it quickens our communication, or makes our utterances more informative, or gives us more power to correctly carve up the world, etc. There need not be any such reason, but if a theory has no reason, then it ought to say why vagueness is ineliminable lest, as mentioned above, we wonder why we have not cured ourselves long ago.

In addition to our shared pretheoretic notions which serve as our guide to saying roughly what is vagueness, we have stronger shared intuitions concerning what predicates are vague. If a theory says that a vague predicate is not vague, or that a non-vague predicate is vague, then this counts against the theory.

Explanatoriness

The phenomena from the previous subsections need to be explained, not just modeled. That is, a logic devised just to accommodate these phenomena is not sufficient to have the status of an explanatory theory of vagueness. I want to know why there is vagueness, not just how to describe it. In this section I mention a handful of accounts of vagueness that are not explanatory.

Take many-valued theories such as multiple valued logic—fuzzy logic (Zadeh, 1965) in particular—and probabilistic degrees such as Edgington's (1992). Multiple valued logics allow sentences to have truth values besides simply true and false, or 1 and 0. They allow truth values in between true and false, e.g., a truth value of 1/2, say. For vague predicates R there will be objects c falling in the borderline region of R, and the sentence 'c is R' accordingly has truth value in between 0 and 1. Probabilistic models of vagueness, on the other hand, accommodate vagueness by saying that the probability that 'c is R' is true is somewhere in between 0 and 1. Probabilistic degrees have superiorities over many-valued logics with respect to modeling natural language (Edgington, 1992), many-valued logics whose deficiencies are well catalogued [for starters see Williamson (1994, Chapter 4), and Chierchia and McConnell-Ginet (1990, pp. 389 ff.)]. One problem with many-valued descriptions of vagueness is that it is not clear that they describe vagueness. To be sure, many-values are a good description in many cases: many vague properties come in degrees, like baldness or redness. But even some non-vague mathematical concepts have been shown to come in degrees to subjects, like 'even' (Armstrong et al., 1983), so degrees are not uniquely connected with vagueness. My principal problem with many-valued theories is that even if we agree that they provide a satisfactory model, or description, of natural language semantics—and allow useful applications in computer science and engineering—they do not make for an explanation for vagueness. Many-valued theories are silent on explanatory questions, and, in fairness, description and not explanation is their aim. They do not tell us why natural language is vague, and they do not even tell us why natural language predicates tend to come in degrees.

Consider briefly supervaluations (see Fine (1975) and Kamp (1975); see also Williamson (1994) for some history), which is the model of vagueness wherein, roughly, a sentence 'c is R' is "super-true," or definitely true, if it comes out true on every precisification of the borderline region of R, "super-false" if it comes out false on every precisification, and borderline, or indeterminate, truth value otherwise. Despite its problems concerning whether it is an adequate description of higher-order vagueness and more generally natural language, my problem is that I want to know why the non-linguistic facts do not determine a single precise extension for natural language predicates. What is it about us or the world that makes meanings incomplete? As in many-valued theories, description is the main task, not explanation; supervaluation aims to be a logic of vagueness, not a reason for why vagueness exists.

Sorensen (1988, pp. 199–216) puts forth an epistemic, determinist account in which vagueness is due to ignorance of the sharp line separating the positive and negative extension, but his theory is not aimed at explanation. Vagueness is, he argues, identical to a phenomenon he calls blurriness. Let $N_1, \ldots, N_{100}$ be mystery natural numbers, and consider the mystery sentences 'N_i is even' for every i from 1 to 100. Now say that an integer is *miny* if and only if it is less than or equal to the number of true mystery sentences. 'miny' is blurry. 0 is definitely miny. 1 is almost certainly miny (i.e., its probability of being miny, presuming some natural assumptions, is $1 - (.5)^{100} \approx 1$), 101 is certainly not miny, and somewhere in between things are difficult to say. Sorensen pushes the idea that vague predicates possess the phenomena they do for the same reasons 'miny' possesses the phenomenon of blurriness; i.e., he pushes the idea that vagueness is blurriness. Even if I were to agree that blurriness is a perfect description of the phenomenon of vagueness, I still would want to understand why natural language predicates are blurry, i.e., why predicates are built as if from mystery numbers, and so on.

My point in this subsection is not to seriously entertain these theories on which I have touched, but to emphasize that much of the work on vagueness has concentrated on describing vagueness rather than explaining it; Hyde's (1997) defense of a "subvaluational" logic and Putnam's (1983) intuitionistic solution are two others.

4.2   Unseeable holes in our concepts

In this section I present the guts of my theory of why there is vagueness. I will become precise about what is a "finite, sufficiently powerful, rational agent" and I will argue that any such agent will have concepts with "unseeable holes" in them. What this all has to do with vagueness will not be discussed until Section 4.3. I believe it is useful to separate out this "unseeable holes" thesis from the explanation of vagueness, because, for me at least, that there are unseeable holes in our concepts—and also for any rational, computationally bound agent—is just as interesting and important as that there is vagueness.

We will see that, in my view, vagueness is definitely *not* a good thing for us, in the sense that it is not as if language has evolved to be vague because it was *independently* useful. Vagueness is something we are stuck with because we are rational, finite entities. If you were a computationally bound, rational alien agent given the task of figuring out what our natural language predicates mean, you would very probably end up with vagueness. I will explain how vagueness could, in principle, be avoided: it would require that we either have inaccessible meanings for our predicates, or that we radically confine our possible meanings for predicates to a very reduced set. Given that we do *not* want inaccessible meanings and *do* want to have a rich choice of meanings for predicates, vagueness is thrust upon us. Vagueness is a cost, but it is worth it because of the benefits it brings. In *this* sense, vagueness is good for you, since without vagueness you would be worse off.

My theory for why there is vagueness is simple, and in order that you not miss the overall point in all the details (not technical details, just details), I first give you a quick introduction to my theory.

4.2.1 Brief introductions to the theory of vagueness

Very brief introduction

Here is the theory: When you or I judge whether or not a predicate P applies to an object a, we are running a program in the head for P on input a. This program for determining the meaning of predicate P we may call C_P. For every one of our natural language predicates P, we have a program for it, C_P, in the head. The job of program C_P is to output YES when input with objects that are P, and NO when input with objects that are not P.

The central problem, though, is that we are unable to always have these programs give an answer to every input; these programs will often take an input, but never respond with an answer of YES or NO. Instead of responding, the program will just keep running on and on, until eventually you must give up on it and conclude that the object does not seem to clearly fit under either P or 'not P'.

The reason we are susceptible to this difficulty is that it is a general difficulty for *any* finite-brained entity. And the reason this is true is because the problem of determining if a program halts on every input is undecidable. Thus, generally speaking, there will be, for each predicate P, objects x for which your program in the head for P, C_P, does not ever halt and output YES or NO. These objects will appear to be in the borderline region of P. In sum, we have borderline regions because we are not computationally powerful enough—no finite-brained agent is—to make sure our programs for our predicates always halt. We have holes in our concepts.

The other major feature of vagueness is that it is not generally possible to see the boundary between the definite regions and the borderline regions; this is called higher-order vagueness. This falls out easily from the above, and goes as follows. The borderline region for a predicate P is the set of all x such that the program $C_P(x)$ does not halt. How accessible is this set? For computational agents, the answer is, "not very." To determine that an object a is borderline P, one must determine that $C_P(a)$ does not halt. This, though, is the halting problem which we discussed a little bit in the introduction to this chapter. We mentioned there that the halting problem is undecidable, and so it is not generally possible for computational entities to solve. In sum, then, you will not generally be able to see the boundary of the borderline region because

it is too hard for you to determine which things are and are not borderline P—
too hard, in fact, for any finite-brained entity. Not only, then, do we have holes
in our concepts, we have *unseeable* holes!

This is the explanation of vagueness, as simple as I can make it. There
are more intricacies to the story. For example, being finite-brained is not completely sufficient for the conclusion; one actually needs finite-brained and rational. But it gets across the main idea, which is, I think, embarrassingly simple.

A less brief introduction

I now give a little less brief primer on my theory of vagueness. The "vagueness
is good for you" arguments will still not appear in this introduction. I will take
you to be my example natural language user.

There are three hypotheses.

(1) The first hypothesis is the *Church-Bound Hypothesis*, and it states that
you can compute no more and no less than what a computer can compute.

(2) The second hypothesis is the *Programs-in-Head Hypothesis*, and it
states that what natural language predicates extensionally mean to you is determined by programs in your head. For example, an object is a dog to you if and
only if your program in the head for 'dog' outputs YES when the (name of the)
object is input into the program. It is much less plausible that many scientific,
mathematical and technical predicates get their meaning to you via programs
in the head, and this difference is what prevents my theory from concluding
that such predicates, many which are not vague, are vague.

(3) The third and last hypothesis is the *Any-Algorithm Hypothesis*, and it
states that you allow yourself the choice of any algorithm when choosing programs in the head for determining your natural language predicate meanings.
(An *algorithm* is a program that halts on every input; programs sometimes do
not halt on some inputs.)

Informally and crudely, the three hypotheses are that (1) you are a computer, (2) you have programs in the head determining what natural language
predicates mean to you, and (3) you allow yourself the fullest range of possible
meanings for natural language predicates.

If these three hypotheses are true, what follows? The Programs-in-Head
Hypothesis says you choose programs to determine your meanings of natural
language predicates. The Any-Algorithm Hypothesis says that the set of programs from which you are choosing is a superset of the set of all algorithms.
But here is the catch: one of the basic undecidability results implies that any

such set of programs is undecidable. (A set is *decidable* if and only if there is program that outputs YES whenever input with an object from the set and NO whenever input with an object not in the set.) Because of the Church-Bound Hypothesis, this undecidability is a difficulty for you: in choosing from the set of programs you cannot always obtain algorithms. In fact, because picking algorithms is computationally more difficult than picking non-algorithms, you will "usually" pick non-algorithms; "most" of your programs determining the meanings of natural language predicates will not be algorithms. So, in an attempt to acquire a meaning for 'dog' via a program in the head that outputs YES when something is a dog to you and NO when something is not a dog to you, *there will be objects on which your program does not halt at all.* This does not mean that you will actually run into an infinite loop; it just means that you will eventually give up when running the program on such inputs.

What does this have to do with vagueness? Consider the set of objects for which the program for 'dog' does not halt. For any object in this set the program will neither say YES nor NO; the object will neither be a dog to you nor not a dog to you. My first theoretical claim is that this is the set of borderline cases for the predicate.

What about higher-order vagueness, the phenomenon that the boundaries of the borderline region are vague? Consider trying to determine exactly which objects are part of the borderline region. To determine that some object is in the borderline region of 'dog' requires that you determine that your program for 'dog' does not halt on that object. But now we have another catch: possibly the most well-known undecidability result is the "halting problem," which says that whether or not a program will halt on a given input is undecidable. This undecidability is a difficulty for you because of the Church-Bound Hypothesis: objects in the borderline region are generally difficult to determine as such, and where the boundaries of the borderline region are is not generally possible for you to determine. Imagine moving from 'dog' cases to borderline cases. Your program for 'dog' will no longer output YES, and will, in fact, never halt; but you will not know it will never halt. You will be unable to see the boundary. My second theoretical claim is that this inability is the phenomenon of higher-order vagueness. Here is a simple representation of the behavior of your program for 'dog', where 'Y' denotes YES, 'N' denotes NO, and '↑' denotes "does not halt".

Y Y Y Y Y Y ↑ ↑ ↑ ↑ ↑ ↑ ↑ ↑ ↑ ↑ ↑ N N N N N N N

{ - - 'dog' - - }{ - - borderline - - }{ - - 'not dog' - - }

So, the three hypotheses entail that for "most" of your natural language predicate meanings there are objects for which your program for that predicate does not halt. Add to this my two theoretical claims just mentioned and it follows that "most" natural language predicates are vague. That is the Undecidability Theory of Vagueness in a nutshell. Now to develop and defend it in more detail. The remainder of this section presents the Undecidability Theory and Section 4.3 discusses how it explains vagueness.

4.2.2   Theory

In this subsection I discuss the three hypotheses comprising the Undecidability Theory of Vagueness and show how they lead to what I call the "Thesis," which is central to the Undecidability Theory of Vagueness's characterization of vagueness. Here is the Thesis, followed by an explanation of the terminology used.

Thesis: For "most" natural language predicates P

1. your interpretation of P is determined by a program in the head that is capable of semideciding but not deciding it,

2. your interpretation of 'not P' is determined by a program in your head that is capable of semideciding but not deciding it, and

3. there are objects neither in your interpretation of P nor in your interpretation of 'not P'.

I will explain the scare quotes around 'most' later. By a "program in the head" I mean the method used by you to determine whether or not a given object is in your interpretation of P. One may usefully and informally think of the program as your intension of the predicate P. The "interpretation of P ('not P') determined by a program" is the set of objects on which the program in the head for P ('not P') outputs YES. A set is *decidable* by a program C if and only if for all x, C on input x outputs YES if x is in the set, and outputs NO otherwise. A set is *semidecidable* by a program C if and only if for all x, C on input x outputs YES exactly when it is in the set; if x is not in the set then C may well not halt at all, though. Do not confuse the notion of a set being semidecidable but not decidable by the program for it with the notion of an underdefined or incompletely specified set. The former, which appears in my theory, is a precise set that happens to be computationally difficult for the program to identify nonmembers, whereas the latter is not a well-defined set at all. Also, do not confuse a set's being semidecidable but not decidable by a program C with a set's being semidecidable but not decidable simpliciter. The latter means the set is computationally complex (in fact, it means it is recursively enumerable but not recursive), but the former, which appears in my theory, only means that the set is complex as far as the program C is concerned; C is unable to decide it, even though it may well be decidable.

There is a simpler, equivalent way of stating the Thesis, one I implicitly used in the brief introduction to the theory. I had written there about a single program in the head, call it $C_{P/nonP}$, doing the work for both a predicate and its natural language negation: the interpretation of P was the set of objects on which $C_{P/nonP}$ outputs YES, and the interpretation of 'not P' was the set of objects on which the *same* program outputs NO. In the Thesis and throughout the remainder of the section the single program is treated as two distinct programs: one program, C_P, for P; and another, C_{nonP}, for 'not P'. The interpretation of P is the set of objects on which C_P outputs YES, and the interpretation of 'not P' is the set of objects on which C_{nonP} outputs YES. Each of these two programs can output only a YES, if they halt at all; they do not ever output NO. Realize that there is no difference in these approaches: running $C_{P/nonP}$ on an input is equivalent to simultaneously running both C_P and C_{nonP} on the input and seeing who halts first (if any); if C_P halts first then $C_{P/nonP}$ would have output YES, but if C_{nonP} halts first then $C_{P/nonP}$ would have output NO. In terms of a single program, the Thesis would be the following: *for "most" natural language predicates P there are objects for which $C_{P/nonP}$ does not halt.* Although this is simpler than the statement at the start of this section, the two-programs version helps to clearly identify distinct aspects of the Thesis.

In the subsections that follow I indulge in a sort of fantasy. I imagine that you are a rational, computationally bound agent who has entered into our culture. Your task is to learn language for the first time and to determine what our natural language predicates mean. We will see that the Thesis is very likely to be true of any such agent. Such an agent will likely choose to have vagueness because its costs are less than the costs of avoiding it. I will also show that it is plausible that the Thesis does, in reality, apply to you.

As part of the fantasy I suppose that the true extensions (as opposed to your interpretations) of natural language predicates are determinate; that is, every object is either in the extension of the predicate or its complement. [For defenses of a determinate semantics within the vagueness literature see Campbell (1974), Cargile (1979), Sorensen (1988, 1994) and Williamson (1994).] I use 'extension' to refer to the true meaning of a predicate, and 'interpretation' to refer to whatever you mean by the predicate. All capital letters will be used to signify the extension of a predicate; e.g., 'bald' has the set BALD as its extension, and 'not bald' has the complement of BALD as its extension. In trying to figure out your interpretations for the language in the fantasy scenario, I suppose that you are presented with examples, somehow, from the true extensions. What I wish to communicate by this is that even if the true semantics of natural language predicates were determinate (via, perhaps, a semantic externalist account), you would still very likely end up with interpretations as specified in the Thesis (and thereby end up with vagueness). Thus, while standard classical two-valued logic would be a correct model of natural language true semantics, we will see that it is not a correct model of the way we actually interpret natural language predicates. On the question what *really* is the true semantics of natural language predicates my theory can remain agnostic.

4.2.3   Church-bound

The Undecidability Theory of Vagueness applies only to those agents that are computationally bound. Specifically, it applies only to those agents that are "finite" and "sufficiently powerful."

By a *finite agent* I mean an agent (i) that has a finite but possibly unbounded memory, (ii) that has an upper bound on the speed at which it can compute, (iii) whose primitive computations are simple (e.g., adding 1 to a number), and (iv) who cannot (or at least does not) utilize in its computing any aspects of the universe allowing it to achieve supertasks (i.e., to achieve infinitely many steps in a finite period of time finite "brain"). To defend my use of the term 'finite' in this way, I informally rephrase these requirements as follows: by a finite agent I mean an agent that is finite in (i) memory, (ii) speed, (iii) degree of complexity of primitive computations and (iv) resourcefulness in utilizing nature to achieve supertasks.

Without (i), an agent could have infinitely large look-up tables in the head. Such an agent could compute any function at all by simply storing the entire function (i.e., storing every pair $\langle x, f(x) \rangle$) in its head, so long as the function has domain and range with cardinality no greater than the cardinality of the look-up table. The agent could merely check his look-up table to see what $f(x)$ is. Without (ii), an agent could compute the first step of a computation in half a second, the next in a fourth, the next in an eighth, etc., thereby computing infinitely many steps in one second (such an agent is called a Plato Machine). Without (iii), an agent may have primitive computations that are themselves as mathematically computationally difficult as one pleases; of course, from such an agent's point of view these computations would *seem* utterly simple, requiring only the least amount of "thinking" to compute (see, e.g., Copeland 1998). Finally, without (iv), it is logically possible that the laws of physics might make it possible to compute supertasks (despite (ii)) (see Earman and Norton (1993, 1996) and Hogarth (1994)). Being a finite agent severely constrains what an agent can compute, as I now describe.

We have an informal, pre-theoretic notion of what it is to compute something. Such an intuitive notion of a computation typically connotes that there be only a finite number of steps involved, that the amount of memory (and scratch paper) required also be finite, that each primitive step be relatively simple (enough to understand), and that one cannot engage in supertasks. That is, the intuitive notion of a computation exactly corresponds to those computations a finite agent can compute. We are inclined to say that a function f from the natural numbers to the natural numbers is intuitively computable if, for each natural number n, $f(n)$ is intuitively computable.

The Turing machine formalism provides an abstract, precise notion of what a computation is and leads to a particular set of functions on the natural numbers as the set of Turing-computable functions. Any computation a modern computer can do, a Turing machine can, in principle, do also; and vice versa. There is the well known hypothesis that the set of functions that are intuitively computable just is the set of Turing-computable functions; this hypothesis is referred to as *Church's Thesis* (or the Church-Turing Thesis).

The hypothesis is not a mathematical assertion; it refers to our intuitions and it does not make sense to ask whether it has been mathematically proven. Nearly everyone believes in Church's Thesis, though, as do I. One reason for this is that no one has yet provided a convincing case of an intuitively computable function that is not Turing-computable; the longer we go without such a case being found, the higher our inductive probability goes toward one that the sets are identical. A second, more slippery, reason nearly everyone believes in Church's Thesis is that half a dozen very different formalizations of computation have been concocted by different people and each leads to precisely the same set of computable functions.

If a finite agent can compute a function on the natural numbers, then the function must be intuitively computable. But then by Church's Thesis that function must be Turing-computable. Therefore, the only functions on the natural numbers a finite agent can possibly compute are those that are Turing-computable.

But any finite agent worth considering carries out computations on objects besides the natural numbers. What constraints are these computations under?

Although there are objects besides natural numbers that are objects of such computations (i.e., such an agent computes functions over objects besides the natural numbers), we can encode all of the objects the finite agent can grasp—including natural numbers—onto the natural numbers. Supposing each different possible state of the finite agent's mind is finitely describable, the set of all such finite descriptions can be bijectively encoded onto the natural numbers (hopefully in an intuitively computable fashion). (Such an encoding is *bijective* if and only if each object gets assigned to a unique natural number and each natural number is used in the encoding.) '4' may now be the code for mental state p_1 which holds the information of the finite agent's mother, '37' the code for mental state p_2 which holds the information of a particular fist fight the finite agent once witnessed, '18' the code for mental state p_3 which holds the information of the feeling of love-at-first-sight the finite agent felt upon meeting its spouse, '103' the code for the mental state p_4 which holds the information of the natural number 5, '1000' for the mental state p_5 which holds the information of the finite agent's favorite shade of blue, etc. Intuitively, every possible dog, every possible shade of color, every possible action, etc., is given its own natural number. With such an encoding, all of the finite agent's computations may be interpreted as computations on the natural numbers, and *the finite agent's computational power is constrained in such a way that it can compute only the Turing-computable functions on this set of codings.*

One should not find this too fantastic, given that the same sort of thing is true about every computer. In a physical computer, as opposed to an abstract model, there are no numbers actually input into the machine nor output from the machine; numbers are abstract objects. Rather, an input or output is some physical state and it encodes certain information. Each physical state is finitely describable and can be coded onto the natural numbers. '4' may be the code for physical state p_1 which holds the information of a black and white picture of a rooster, '37' may be the code for physical state p_2 which holds the information of natural number 5, '18' may be the code for physical state p_3 which holds the information of the sentence "Press any key to continue," etc. It is only through such means that one can meaningfully say that computers are subject to the same ultimate computational constraints as Turing machines, and it is also only through such means that one can meaningfully say that a finite agent is subject to the same ultimate computational constraints as Turing machines.

Worries over which coding is being employed for the finite agent are sometimes raised. For example, what if the coding makes intuitively uncomputable problems computable by having a non-intuitively computable coding? Or, is there a privileged coding and, if so, what determines it? I wish to sidestep all such issues. To whatever extent these are legitimate worries, they are worries for anyone claiming that even computers are bound by Church's Thesis. This latter claim is uncontroversial, however, and so I am under no special obligation to explain or address issues of coding with respect to finite agents.

One might complain that the universe has uncountably many possible objects, and so no bijection is possible onto the natural numbers. Supposing for the moment that there are indeed uncountably many possible objects, I only care about what possible objects the finite agent can hold before its mind. Since it is finite, it can only entertain countably many possible objects. Its universe is countable, regardless of the cardinality of the real universe. This brings in its own trouble: if the universe is uncountable and the finite agent's universe countable, is not it going to have a false model of the world? The Downward Lowenheim-Skolem Theorem can help to alleviate this worry to an extent: as long as the finite agent notices only first-order properties of the universe, it is possible for its model to be such that the set of all truths is the same as God's (whose model is the true uncountable one). Should we, however, believe that it is confined to first-order properties? Perhaps, perhaps not; there are many things that can be said on this issue, but I have no need to pursue them here since nothing hinges on the agent's model being true.

Thus, I am confining discussion to finite agents, which means that I am confining discussion to agents capable of computing *only* the Turing-computable.

By "sufficiently powerful" I mean that the finite agent is capable of computing *at least* the Turing-computable.

Together, "finite" and "sufficiently powerful" imply that the computational powers of the agents I wish to discuss are bound by Church's Thesis and only bound by Church's Thesis. I sometimes say "Church-bound" instead of "finite and sufficiently powerful." I record this as the Church-Bound Constraint.

Church-Bound Constraint: *The agent can compute any function (over natural numbers coding mental states, which in turn represent objects in the world) so long as it is Turing-computable.*

Related to this constraint is the Church-Bound Hypothesis, which states that you are under the Church-Bound Constraint. The Church-Bound Hypothesis is, by assumption, true of the fantasy you. Is it true of the real you? Yes, and here is why. It is plausible that you are finite in the four senses discussed above (although see Penrose, 1994) and so cannot compute the Turing-uncomputable. Furthermore, you are, in principle, able (given enough time and scratch paper) to compute any Turing-computable function. We know this because any of us can easily mimic the simple actions of a Turing machine as long as we please.

4.2.4   Programs in the head

Suppose that you, in the fantasy, are exposed to enough examples of things you have reason to believe are in the true extension of 'bald' (BALD) and others that are not that you acquire an educated guess as to what BALD is. Your guess determines some set as your "shot" at BALD, and this is your interpretation of 'bald'. In what can such a guess consist? There are infinitely many (possible) objects in your universe, infinitely many of them are bald and infinitely many are not. You are not, then, able to simply guess what the extension is, for you cannot store the extension since you are finite.

You must employ some sort of intension. You need to find some finite description of the set that determines your interpretation of 'bald' and your educated guess at BALD, and some finite description of 'not bald'. Recalling that these sets may be considered to be sets of natural numbers, one may wonder whether your interpretation of 'bald' can be described "in your head" as, say, a first-order sentence in the language of arithmetic (such sets are called *arithmetically definable*). For example, you may interpret 'bald' to be the set $\{n \mid \exists x \forall y \ R(n, x, y)\}$, where $R(n, x, y)$ is some recursive formula without quantifiers. The problem with this is that although it is indeed a finite description, the set is not recursively enumerable and since you are Church-bound it is generally too difficult for you to handle. (A set is *recursive* if and only if it is decidable by some program. A formula is recursive if and only if the set of objects satisfying it is recursive. A set is *recursively enumerable* if and only if it is semidecidable by some program.)

The same is true for any arithmetically definable set... except those that are recursively enumerable. For a recursively enumerable set it is possible for you to have a program in the head that says YES when and only when presented with objects in the set (although the program may never halt at all when presented with objects not in the set), but sets any more computationally difficult than recursively enumerable are beyond your reach. A program in your head, then, is what you must be employing to determine your interpretation of 'bald' if you wish to have an interpretation that is accessible to you. Your interpretation would then be the set of objects for which the program for 'bald' outputs YES, and this is recursively enumerable. This motivates the first rationality principle.

Principle of Program-Favoring: *Without good reason to the contrary, you should assume that the extension of natural language predicate P and its natural language negation 'not P' are capable of being correctly determined using programs in the head.*

This does seem to be a compelling principle of rationality: why choose interpretations that are not generally possible for you to actually use unless you have a good reason?

Supposing we believe that the Principle of Program-Favoring really is a constraint on rationality, is there good reason for believing that programs will not suffice to correctly interpret natural language predicates and their natural language negations? For example, in mathematics there *is* good reason for believing that programs do not suffice for certain predicates because there are predicates with interpretations that you know are not recursively enumerable. Consider the predicate 'not a theorem of Peano Arithmetic', for example. You know its extension is not recursively enumerable (since its complement is known to be recursively enumerable but not recursive). Your interpretation of 'not a theorem of PA' is set to its extension, regardless of the fact that you are incapable of generally recognizing things that are not theorems of PA. "To me, something is not a theorem of PA exactly if it does not follow from Peano's Axioms; I have no program for it, though." You might acquire a program in the head as a heuristic device aimed to approximately semidecide your interpretation of 'not a theorem of PA', but you are not confused into conflating your heuristic with your interpretation; you know that no such heuristic can possibly be the extension. Thus, you as a mathematician do have predicates for which you have good reason to believe the extension is not determinable via a program in the head, and your interpretations are, accordingly, not determined using programs in the head. (This is, in passing, why mathematical predicates such as 'not a theorem of PA' are not vague.) Given that you can acquire good reasons to believe programs are inadequate and can have interpretations that are not recursively enumerable, what reason is there for you not to do the same for natural language predicates?

The answer is that in the case of such a mathematical predicate you know what the definition of the extension is, and so you set your interpretation accordingly. For a natural language predicate, however, you have no God's eye view of its extension. The extension of 'bald' is learned via induction; you infer your interpretation of 'bald' from seeing objects you have reason to believe (somehow) are in BALD or its complement. You cannot easily acquire the definition for BALD, and as many examples of BALDness and its complement you might confidently find, you still will not have access to its definition in the way you have access to that of 'not a theorem of PA', for you have no luxury of setting your interpretation to that determined by the definition written on paper before you as you do for mathematical predicates (and this has nothing to do with the fact that you are Church-bound). Given that you cannot have access to BALD in the way you have access to the extension of 'not a theorem of PA', it is also reasonable to suppose that you cannot learn that BALD is not recursive enumerable (supposing this were indeed true) in the way you learn that the extension of 'not a theorem of PA' is not recursively enumerable. I cannot discount the logical possibility of you, a Church-bound agent, learning (in the fantasy) through time that no recursively enumerable interpretation of 'bald' seems to fit the examples of BALD and its complement, and in this way assigning high probability to the hypothesis that BALD is not recursively enumerable, and therefore no program in the head is sufficient. The reasonable hypothesis, though, seems to be that for most (if not all) natural language predicates you have no good reason for believing that programs will not work. I record this as the following hypothesis.

No-Good-Reason-for-Non-Programs Hypothesis: *For most natural language predicates P and their natural language negation 'not P' you have no good reason to believe that programs in the head are inadequate for correctly determining their interpretation.*

The No-Good-Reason-for-Non-Programs Hypothesis together with the Principle of Program-Favoring imply the following hypothesis.

Programs-in-Head Hypothesis: *For most natural language predicates P and their natural language negation 'not P', their interpretations are determined by you using programs in the head.*

You in the fantasy scenario are, then, very likely to fall under the Programs-in-Head Hypothesis. "Very likely" because it is very likely that the No-Good-Reason-for-Non-Programs Hypothesis is true, and given that you are rational you will follow the Principle of Program-Favoring and thus fall under the Programs-in-Head Hypothesis.

Does the Programs-in-Head Hypothesis apply to the real you? Here is an intuitive reason to think so. For most natural language predicates P you are capable of recognizing, given enough time, any cases of P and 'not P'. E.g., given enough time you are capable of recognizing, for any bald-to-you person, that he is bald to you; and, for any not-bald-to-you person, that he is not bald to you. To suppose otherwise would imply, implausibly, that there is a person that is bald (not bald) to you, but you are utterly incapable of recognizing him as such. The only way for you, who are Church-bound, to have this recognition capability is to have programs in the head doing the work.

4.2.5   Any algorithm

By the Programs-in-Head Hypothesis you have for most natural language predicates a program in the head as the intension determining your recursively enumerable interpretation of the predicate, and this interpretation is your attempt to fit the extension of the predicate. You would like to have a single program in the head capable of determining your interpretation of both 'bald' and 'not bald'; that is, a program that not only says YES exactly when an object is in your interpretation of 'bald', but says NO exactly when an object is not in your interpretation of 'bald'. This is just to say that you would like to have a program to decide the interpretation of 'bald', not just semidecide it. Such a program would be an algorithm since it would halt on every input, and the corresponding recursively enumerable interpretation of 'bald' would be recursive.

But alas, you are Church-bound, and a well-known undecidability result says that there is no algorithm for algorithmhood; there is no general procedure by which either you or a Turing machine can always choose programs (from the set of all possible programs) that are algorithms. It is not, then, generally the case that your programs in the head are algorithms, and your corresponding interpretations for natural language predicates and their natural language negations may generally be only semidecided by the programs for them. (And in fact things are even worse than this, for a related undecidability result says that the corresponding interpretations are not generally even recursive; semidecide is all that any possible program can do in these cases.) If the interpretation of 'bald' ('not bald') is determined by a program in the head that semidecides but not decides it, then supposing that 'bald' ('not bald') is one of the predicates covered by the Programs-in-Head Hypothesis, that program cannot be what is determining the interpretation of 'not bald' ('bald'). This is because the Programs-in-Head Hypothesis states that 'not bald' ('bald') must have a program semideciding its interpretation, and the program for 'bald' ('not bald') cannot possibly be that program. Thus, 'not bald' ('bald') must have its own program in the head. I have now shown 1 and 2 of the Thesis.

How about 3 from the Thesis? It is possible for the interpretation of 'bald' and that of 'not bald' to cover every object, but by the Church-Bound Hypothesis this is not generally possible for you to accomplish. If it were generally possible, then the two programs semideciding each interpretation could serve as a single algorithm (run both programs simultaneously until one halts), and you could therefore always acquire algorithms. But this is impossible. Thus, it is not generally the case that your interpretation of 'bald' and that of 'not bald' cover every object.

Notice that none of this hinges on either interpretation being non-recursive; what matters is the program for the interpretation semideciding but not deciding it. Predicates with finite interpretations (arguably 'small natural number') are therefore subject to the same conclusion just made concerning 'bald'.

Except for the use of "most" in the statement of the Thesis, I now seem to have shown that you are subject to the Thesis. Concerning "most," it is easier to acquire non-algorithms than algorithms, since in order to achieve algorithmic status the program must halt on every input, whereas to achieve non-algorithmic status there needs to be only one input on which the program does not halt.[3] This observation makes it convenient and informally true to say that for "most" natural language predicates your corresponding programs are not algorithms. This is really just elliptical for the proposition that you are not generally able to acquire algorithms and that it is more difficult to acquire algorithms than non-algorithms. To differentiate this use of 'most' (or 'usually') with genuine uses of it, I always put scare quotes around it.

With this it appears I have now shown you are subject to the Thesis. There is just one remaining problem. I wrote above (second paragraph of this subsection) that "there is no general procedure by which either you or a Turing machine can always choose programs (from the set of all possible programs) that are algorithms." The parenthetic remark merits some examination. Why should you be required to choose from among the set of *all* possible programs? Although the set of all algorithms is not recursively enumerable, there do exist proper subsets of the set of all algorithms that are recursively enumerable, and even recursive. Could you be choosing your programs from one of these subsets? For example, the set of primitive recursive programs is recursive, and perhaps you are choosing from this. If so, you can be sure that every program you choose is an algorithm, and thus that every one of your interpretations for natural language predicates is decidable by the program responsible for it (and is therefore recursive). The Thesis would, then, not follow after all.

---

[3] More formally and in recursion theoretic terminology, this is captured by the fact that the set of algorithms is Π_2, and the set of non-algorithms Σ_2; the relative difficulty of acquiring algorithms versus non-algorithms is analogous to the relative difficulty of determining cases where a program does not halt versus when it does.

There is a good reason for the fantasy you not to confine yourself in such a fashion. Your interpretations of natural language predicates are a result of a learning process of some sort. You see cases you have reason to believe are in the extension of 'bald' (i.e., you are guided by the true semantics somehow), and you make an educated guess at the extension with your interpretation. *A priori*, you have no reason to believe that all concepts of the world can be correctly determined (or even adequately approximated) with algorithms from some recursively enumerable subset of the set of algorithms. Why should you believe that all extensions may be correctly determined with, say, primitive recursive interpretations? This motivates the following rationality principle.

Principle of No-R.E.-Subsets-of-Algorithms: *Without good reason to the contrary, you should not presume that there is a recursively enumerable subset of the set of all algorithms such that for all natural language predicates P (or 'not P'), algorithms from this subset supply the best interpretation for P ('not P').*

This is a compelling principle: why purposely choose a language with less rich interpretations without good reason? In fact, any recursively enumerable subset of the set of all algorithms is, in a certain real mathematical sense, infinitely less rich than the set of all algorithms.

Supposing we believe that the Principle of No-R.E.-Subsets-of-Algorithms is a constraint on rationality, is there good reason to believe that there are recursively enumerable subsets of the set of all algorithms sufficiently rich for natural language predicate interpretations? Although I am willing to suppose that it may be logically possible for you to acquire high probability in such a hypothesis (after, say, many years of searching for uses of algorithms outside of this recursively enumerable subset and not finding one), there would not appear to be actual evidence for such a supposition. This goes to support the following hypothesis.

No-Good-Reason-for-R.E.-Subsets-of-Algorithms Hypothesis: *There is no good reason for you to presume that there is a recursively enumerable subset of the set of all algorithms such that for all natural language predicates P ('not P'), algorithms from this subset supply the best interpretation for P ('not P').*

One might wonder whether there is nevertheless the following good pragmatic reason for confining algorithm choice to a recursively enumerable subset of the set of all algorithms: by so confining oneself one does indeed avoid the Thesis (and vagueness). The solution comes with a painful price, though. For all you know there are algorithms that can provide the correct interpretation. Yes, not confining yourself to a recursively enumerable subset of the set of all algorithms brings with it the cost of there being objects in neither the interpretation of P nor the interpretation of 'not P'. However, it is possible for the interpretations to be only "finitely mistaken," where by this I mean that they are complements save for finitely many objects in neither interpretation. Constraining yourself to a recursively enumerable subset of the set of algorithms only for the pragmatic reason of avoiding the Thesis runs the risk that there are predicates, perhaps many, that are not only not correctly interpretable using algorithms from that subset, but will be infinitely mistaken. For example, if one constrains oneself to the set of primitive recursive functions without reason to believe that no predicates should best be interpreted using non-primitive recursive algorithms, then in all those cases where a predicate should be best interpreted using a non-primitive recursive algorithm you are guaranteed to incorrectly classify the objects on infinitely many occasions. Worse than this, it may be that no primitive recursive algorithm even "comes close" to the best algorithm for the predicate. It might be like using the set of odd numbers as an approximation to the prime numbers.

The No-Good-Reason-for-R.E.-Subsets-of-Algorithms Hypothesis conjoined with the Principle of No-R.E.-Subsets-of-Algorithms imply the following hypothesis.

No-R.E.-Subsets-of-Algorithms Hypothesis: *You do not confine your choice of programs to a recursively enumerable subset of the set of all algorithms when interpreting natural language predicates and their natural language negations.*

You in the fantasy scenario are, then, very likely to fall under the No-R.E.-Subsets-of-Algorithms Hypothesis. "Very likely" because it is very likely that the No-Good-Reason-for-R.E.-Subsets-of-Algorithms Hypothesis is true, and given that you are rational you will follow the Principle of No-R.E.-Subsets-of-Algorithms and thus fall under the No-R.E.-Subsets-of-Algorithms Hypothesis.

Does the No-R.E.-Subsets-of-Algorithms Hypothesis apply to the real you? There are reasons to think so. In fact, there is reason to think that the real you is subject to the following hypothesis.

Any-Algorithm Hypothesis: *You are free to choose from the set of all algorithms when interpreting natural language predicates or their natural language negations.*

If the Any-Algorithm Hypothesis is true of you, then so is the No-R.E.-

Subsets-of-Algorithms Hypothesis. This is because any set containing the set of all algorithms is not a recursively enumerable subset of the set of all algorithms.

What reasons are there to think that the Any-Algorithm Hypothesis is true of the real you? It is difficult to tell a plausible story about how (the real) you could have come to restrict program choice to exclude some algorithms, especially since by the Church-Bound Hypothesis you are capable of computing any algorithm. The man on the street does not know recursion theory, and even if he does, as I do, I cannot imagine attempting to restrict myself to, say, primitive recursive intensions for every new interpretation I acquire. Nor does it seem plausible to suppose that we humans might have evolved to exclude certain algorithms. It is, in fact, very difficult to avoid allowing yourself the choice of any algorithm since once you allow yourself the use of *'while' loops*—i.e., the ability to implement programs including statements like "while such and such is true, continue doing blah"—you are able to build, in principle, any algorithm (presuming you can also carry out some trivial basic operations).

To avoid this conclusion you would have to ban the use of 'while' loops, using only *'for' loops*—i.e., the ability to implement programs including statements like "for i becomes equal to 1 to n do blah", or "do blah n times"— which is very restrictive. One could argue that your 'while' loops are in reality bounded since you do not (and cannot) let them run forever; thus, it is not the case that every algorithm can be implemented. But this does not mean that the proper representation of your program does not use a 'while' loop. No real computer, after all, can actually implement unbounded 'while' loops, but it would be a mistake to say they cannot run unbounded 'while' loops and any algorithm.

It can be noted that the idea of animals employing 'while' loops has some empirical support, namely in the *Sphex ichneumoneus* wasp, which has been observed to enter into what is plausibly represented as an infinite loop. Consider the following often quoted excerpt from Woolridge (1963, p. 82).

> When the time comes for egg laying, the wasp *Sphex* builds a burrow for the purpose and seeks out a cricket which she stings in such a way as to paralyze but not kill it. She drags the cricket into the burrow, lays her eggs alongside, closes the burrow, then flies away, never to return. In due course, the eggs hatch and the wasp grubs feed off the paralyzed cricket, which has not decayed, having been kept in the wasp equivalent of deep freeze. To the human mind, such an elaborately organized and seemingly purposeful routine conveys a convincing flavor of logic

and thoughtfulness—until more details are examined. For example, the Wasp's routine is to bring the paralyzed cricket to the burrow, leave it on the threshold, go inside to see that all is well, emerge, and then drag the cricket in. If the cricket is moved a few inches away while the wasp is inside making her preliminary inspection, the wasp, on emerging from the burrow, will bring the cricket back to the threshold, but not inside, and will then repeat the preparatory procedure of entering the burrow to see that everything is all right. If again the cricket is removed a few inches while the wasp is inside, once again she will move the cricket up to the threshold and re-enter the burrow for a final check. The wasp never thinks of pulling the cricket straight in. On one occasion this procedure was repeated forty times, always with the same result.

I am not suggesting that you are possibly subject to such infinite loops. I am only suggesting that 'while' loops are plausibly part of your computational grammar. In fact, one might say that the wasp has a "concept" of 'readied burrow' which is determined by the following program:

```
WHILE burrow not ready do
  IF burrow clear & cricket not moved when I emerge
  THEN burrow is ready;
```

As an example showing that you regularly engage in 'while' loops (or an equivalent) as well, in order to determine if the bath temperature is good, you may well keep increasing the hot until it is comfortable or too hot; if the latter then you keep decreasing until comfortable; and so on. That is, you implement the following program:

```
WHILE temperature not comfortable do
  IF temperature too cold
  THEN increase hot water;
  ELSE decrease hot water;
```

'while' loops seem to be an integral part of your (and my) computational grammar. And if this is true, the Any-Algorithm Hypothesis is sure to apply to the real you. Thus, the No-R.E.-Subsets-of-Algorithms Hypothesis also applies to the real you.

What does the No-R.E.-Subsets-of-Algorithms Hypothesis tell us? There is a set of all possible programs you can attain (and this is recursively enumerable since you are bound by Church's Thesis). This set is not a subset of the set of all algorithms, as the No-R.E.-Subsets-of-Algorithms Hypothesis requires. This means you are not generally capable of choosing algorithms. In particular, if the Any-Algorithm Hypothesis is true, then since the set of all algorithms is undecidable, you are not generally capable of choosing algorithms. We saw above that the Church-Bound Hypothesis and the Programs-in-Head Hypothesis "almost" imply the Thesis. What was missing was some reason to believe that the set of algorithms from which you determine your interpretations is not recursively enumerable. The No-R.E.-Subsets-of-Algorithms Hypothesis finishes the argument, and these three hypotheses together entail the Thesis.

4.2.6 Thesis

In this subsection I bring together the previous three subsections. Here is the Thesis again.

For "most" natural language predicates P

1. your interpretation of P is determined by a program in your head that is capable of semideciding but not deciding it,

2. your interpretation of 'not P' is determined by another program in your head that is capable of semideciding but not deciding it, and

3. there are objects neither in your interpretation of P nor in your interpretation of 'not P'.

There are two ways of arguing toward the Thesis. The first is via the fantasy scenario and is largely but not entirely prescriptive, concluding that the Thesis, and thus vagueness, follows largely but not entirely from rationality considerations alone (and the Church-Bound Constraint). The second is via the real you scenario and is descriptive, concluding that the Thesis follows from hypotheses that are true about us. The first would be rather worthless without the second, because a theory that claims that vagueness would exist in the fantasy scenario but says nothing about the real us would be incomplete at best, since it is us who experience vagueness, not some idealized, rational fantasy agents. The second, however, is made more interesting by the first. The Thesis, and thus vagueness, does not follow from some human irrationality or quirk, but is, on the contrary, something to which nearly any rational, sufficiently powerful, finite agent will converge.

I finish this section by (i) cataloguing the premises of both the prescriptive (fantasy you) and the descriptive (real you) argument, (ii) reminding us that the premises of the prescriptive argument entail those of the descriptive argument, and (iii) summarizing how the Thesis follows from the descriptive argument (and thus by (ii) also from the prescriptive argument). Below are the two arguments.

Descriptive Argument (Real you)	Prescriptive Argument (Fantasy you)
Church-Bound Hypothesis.	Church-Bound Constraint.
Programs-in-Head Hypothesis.	Principle of Program-Favoring.
	No-Good-Reason-for-Non-Programs Hypothesis.
No-R.E.-Subsets-of-Algorithms Hypothesis.	Principle of No-R.E.-Subsets-of-Algorithms.
	No-Good-Reason-for-R.E.-Subsets-of-Alg Hyp

The prescriptive argument says that any rational (Principles of Program-Favoring and No-R.E.-Subsets-of-Algorithms), Church-bound (Church-Bound Constraint) agent is subject to the Thesis so long as (a) he has no good reason to believe that the extensions of most natural language predicates are not recursively enumerable (No-Good-Reason-for-Non-Programs Hypothesis), and (b) he has no good reason to presume that there is a recursively enumerable subset of the set of all algorithms that suffices for adequate interpretations of natural language predicates (No-Good-Reason-for-R.E.-Subsets-of-Algorithms Hypothesis). Because (a) and (b) are very difficult to imagine being false, the Thesis follows "largely" from the Church-Bound Constraint and the two principles of rationality. Supposing (a) and (b) are true, the Thesis (and thus vagueness) is good for you, fantasy and real.

The descriptive argument says that (α) we humans are Church-bound (Church-Bound Hypothesis), (β) for most natural language predicates and their natural language negations we use programs in the head to determine our interpretations of them (Programs-in-Head Hypothesis), and (γ) any algorithm may possibly be used by us as a determiner of the interpretations of natural language predicates or their natural language negations (Any-Algorithm Hypothesis), and thus we do not confine ourselves to a recursively enumerable subset of the set of all algorithms for interpreting natural language predicates or their natural language negations (No-R.E.-Subsets-of-Algorithms Hypothesis).

The prescriptive premises (for the fantasy scenario) imply the descriptive premises in the following sense. If you satisfy the Church-Bound Constraint then the Church-Bound Hypothesis is true. If you follow the Principle of Program-Favoring and the No-Good-Reason-for-Non-Programs Hypothesis is true, then the Programs-in-Head Hypothesis is true; the converse is not true. If you follow the Principle of No-R.E-Subsets-of-Algorithms and the No-Good-Reason-for-R.E.-Subsets-of-Algorithms Hypothesis is true, then the No-R.E.-Subsets-of-Algorithms Hypothesis is true; the converse is not true.

The descriptive premises entail the Thesis as follows. The Programs-in-Head Hypothesis states that you use programs in the head to determine most of your interpretations of natural language predicates and their natural language negations. Most of your interpretations are therefore semidecidable by the programs responsible for them. The No-R.E.-Subsets-of-Algorithms Hypothesis states that the set of programs at your disposal for natural language predicate interpretations is not a recursively enumerable subset of the set of all algorithms. This entails that the set of algorithms from which you can possibly choose is not recursively enumerable. The Church-Bound Hypothesis states that you can only compute the Turing-computable, and thus you cannot generally choose programs for your interpretations of natural language that are algorithms (even if your interpretations are recursive, or even finite). For "most" of your interpretations of natural language predicates P (or 'not P') your program for it will be able to semidecide but not decide it. Since "most" predicates can only semidecide their interpretation, this means that for "most" predicates there must be one program for P, and another program for 'not P', and each can only semidecide the interpretation for which it is responsible. We have so far concluded that "most" natural language predicates satisfy 1 and 2 of the Thesis. Your interpretations of P and 'not P' cannot cover every object, because if they could be then it would be possible to take the two programs and use them as one algorithm to simultaneously decide the interpretations, and this would contradict the impossibility of generally acquiring algorithms. Thus, "most" of the time your interpretations of predicates and their natural language negations do not cover all objects; there are objects in neither interpretation. We now have that "most" predicates satisfy 1, 2 and 3; the Thesis follows from the three hypotheses.

It is important to note that the Thesis is an important claim about natural language whether or not one believes that the Thesis has anything to do with vagueness. If it is true, then, informally, our concepts have "unseeable holes" in them. As for vagueness, my theory's characterization of vagueness is that a predicate is *vague* if and only if it satisfies 1, 2 and 3 from the Thesis; the truth of the Thesis implies that "most" natural language predicates are indeed vague, as we know they are.

4.3 From theory to vagueness

In this section I demonstrate how the Undecidability Theory of Vagueness explains vagueness. Recall that the theory's characterization of vagueness from Subsection 4.2.6 is as follows: Predicate P is vague to you if and only if

1. your interpretation of P is determined by a program in your head that is capable of semideciding but not deciding it,

2. your interpretation of 'not P' is determined by a program in your head that is capable of semideciding but not deciding it, and

3. there are objects in neither your interpretation of P nor your interpretation of 'not P'.

The Thesis stated that "most" natural language predicates satisfy 1, 2 and 3, i.e., "most" natural language predicates are vague. In terms of a single program $C_{P/nonP}$ that outputs YES whenever an object is P and outputs NO whenever an object is 'not P', the characterization is that a predicate P is vague to you if and only if there are objects on which your program $C_{P/nonP}$ does not halt. The corresponding Thesis is that "most" natural language predicates have a region of objects for which the program does not halt. In what follows the Thesis is assumed to be true.

Please notice that in my theory's characterization of vagueness, vague predicates are not in any way required to be computationally complex. I have a great deal of difficulty with people thinking that my theory somehow equates non-recursiveness with vagueness. The only appeal to non-recursiveness has been to the non-recursiveness of the set of algorithms and the halting set (the set of all pairs of programs and inputs such that the program halts on that input), not to the non-recursiveness of natural language predicate interpretations. The interpretations of vague predicates may well be recursive, and even finite, and vagueness is unscathed. And even if a predicate's interpretation is not recursive, the vagueness comes *not* from this but, as we will see, from the facts that the interpretations do not cover all objects and that the programs are not algorithms.

4.3.1 Borderline region

Your interpretations of P and 'not P' do not cover all objects; there are objects c such that the natural language sentences 'c is P' and 'c is not P' are both false. These objects comprise the borderline region. This fits well with the datum of a borderline region: that there are objects which do not seem to fit neatly into just one category. The development in Section 4.2 served in part to show that (i) any rational, Church-bound agent in the fantasy is very likely to have a borderline region for "most" natural language predicates, and (ii) you do, in fact, have such a borderline region for "most" natural language predicates.

In epistemic theories of vagueness the borderline region is characterized differently than merely "not fitting neatly into just one category." Rather, for epistemicists the borderline region is comprised of those objects for which knowledge of membership is unattainable, where "membership" refers to membership in the true extension. The Undecidability Theory explains this sort of borderline region as well. Suppose that BALD is the true extension of 'bald'. You are not generally capable of acquiring a program in the head that decides BALD, even if BALD is decidable, because you are not generally capable of acquiring algorithms. Your interpretation of 'bald' is semidecidable but not generally decidable by the program responsible for it, and even if you are so lucky to correctly interpret it (i.e., your interpretation is equal to the extension BALD), if you want to be able to respond to queries about 'not bald' you must acquire a second program in the head, and this program will not generally correctly interpret 'not bald' (i.e., the 'not bald' program will not semidecide the complement of BALD). Your interpretations of 'bald' and 'not bald' do not cover every object, and the programs for each only semidecide them. There are therefore objects for which you are incapable of determining or even knowing, using your programs in your head, whether or not it is a member of BALD.

4.3.2 Higher-order vagueness

Although you cannot draw a sharp line between 'bald' and 'not bald', can you draw a sharp line between 'bald' and 'borderline bald'? There is, in fact, a sharp line here posited by my theory, but are you generally capable of drawing it? No. The two programs in the head for baldness (one for 'bald' and one for 'not bald') are not powerful enough to determine the lines. To see this intuitively, imagine starting in the 'bald' region and moving toward the borderline bald region. While in the 'bald' region your program for 'bald' halts and says YES and the program for 'not bald' never halts. When you move into the borderline region the program for 'not bald' still does not halt, but the program for 'bald' suddenly now never halts as well. You are not, though, generally able to know that the program for 'bald' will never halt—you cannot generally know when you have crossed the line. This seems to be consistent with our observations of higher-order vagueness, and it solves the problem without having to postulate semantically distinct higher-order borderline regions. This latter aspect is good since it puts a stop to the regress of higher and higher order semantically distinct borderline regions, all which amount to nothing if when one is finished there is still a knowable sharp line between the definite region and the non-definite region.

Can this phenomenon really be the phenomenon of higher-order vagueness? In my theory what does it "feel like" to not be capable of determining the boundaries of the borderline region? Well it feels like whatever it feels like to attempt to decide a set using a program that only semidecides it. One might try to make the following criticism: Let us take the set of even numbers and supply you with a program that only says YES exactly when a number is even, and is otherwise silent. Do the evens now seem vague through the lens of this program? There are a number of problems with such a criticism as stated. First, it is not enough that the program simply says YES when an input is even and is silent otherwise. When we say that the program semidecides but does not decide the set of evens we mean that if the program is silent we are not sure whether it will at any moment converge and say YES. The program's silence is not translatable to NO. Second, it is difficult to judge our intuitions with a predicate like 'even' for which we already have a program in the head for deciding it. We should imagine instead that it is some new predicate P for which we have no intuitions. The third problem is that even with these fixes the question the critic needs to ask is not whether P-ness seems vague, but whether P-ness seems to have whatever feel higher-order vagueness has. This is because P is not vague according to my theory since it does not satisfy part 2 of the characterization of vagueness (i.e., we are not given a program for semideciding 'not P'). On this modified question it is unclear that we have any compelling intuitions that the answer is NO. When using the given program to attempt to decide the extension of P, you will be incapable of seeing where exactly the boundary is, and therefore you will be unable to classify many objects. These objects plausibly are just like the borderline borderline objects (i.e., second-order borderline objects).

Another critic may ask the following: Let us suppose that you have two programs that only semidecide their respective interpretations, and let us also suppose that the interpretations do not cover every object. If these programs are for some predicate P and 'not P' then is P-ness necessarily vague? For example, let us take the predicate 'theorem of arithmetic', whose extension is not even recursively enumerable. You are surely capable of determining some theorems and some non-theorems, and you must therefore utilize a program in the head for 'theorem' and another for 'not theorem'. But surely 'theorem' is not now vague! There is a difference between this case and vague natural language predicates. You as a mathematician are conscious that you are not actually deciding theoremhood with your programs. You understand that they are only heuristics, and it is possible that each might even occasionally be incorrect, e.g., saying that a theorem is not a theorem. That is, your programs for theoremhood do not determine what you mean by 'theorem'. You mean by 'theorem of arithmetic' whatever follows from its definition. 1 and 2 from the characterization of vagueness are not satisfied.

4.3.3 The sorites paradox

Finally we arrive at the sorites paradox, which I give here in the following form: (i) 0 hairs is bald, (ii) for all n, if n hairs is bald, so is $n + 1$, (iii) therefore you are bald no matter how many hairs you have. Notice that I have stated the argument in natural language; many researchers on vagueness state the paradox in some logical language, which is strange since the paradox is one in natural language. Presenting it in a logical language inevitably makes certain biased presumptions; for example that 'not' is to be translated to the classical negation '$\neg$'.

What is usually dangerous about rejecting premise (ii) is that it implies there is an n_0 such that n_0 hairs is bald but $n_0 + 1$ hairs is not; i.e., it usually leads to there being no borderline region. This is bad because borderline regions surely exist. In my theory's case, though, what happens? A sorites series moves along a "path" that is most gradual from P to 'not P'; it must therefore cross the borderline region lest it not be "most gradual." Imagine starting in the 'bald' region and moving toward the borderline region. Eventually there will be a number n_0 such that n_0 hairs is bald but it is not case that $n_0 + 1$ is bald, and you cannot in general determine where this occurs. However, this in no way prevents $n_0 + 1$ from being borderline bald, i.e., being neither bald nor not bald. Eventually the denial of (ii) will occur—and you will not know when—but it does not imply the lack of a borderline region. The sorites paradox is thus prevented without losing vagueness.

4.3.4 Essentialness

There is a widespread feeling (since Wittgenstein, it seems) that vagueness is an essential part of natural language. That is, even if it were eliminable (see

Subsection 4.3.5 to see why it is not), we would not want to eliminate it since it serves an essential role.

My Undecidability Theory of Vagueness has its own explanation. Recall from Section 4.2 that the Undecidability Theory in largely prescriptive dress rested upon one constraint, two weak (weak relative to the three hypotheses in the descriptive argument) hypotheses, and two principles of rationality. The constraint was the Church-Bound Constraint, which states that I am concentrating only on agents that are bound by Church's Thesis and able to compute any computable function. The first of the two weak hypotheses is the No-Good-Reason-for-Non-Programs Hypothesis which says that we have no good reason to believe that programs are not sufficient to describe the world. The second of the two weak hypotheses is the No-Good-Reason-to-Exclude-Algorithms Hypothesis which says that we have no good reason to believe that some algorithms may not be useful in describing the world. Supposing the truth of these two weak hypotheses the truth of the two principles of rationality suffices to secure the Two-Programs Thesis and the resulting vagueness (as seen in Subsections 4.3.1 and 4.3.2). Principles of rationality claim that one ought to do something, where there is some implication that not doing that something would be very bad, whatever that might mean. The essentialness of vagueness is bound up with the rationality principles: vagueness is essential because the only way to avoid it is through irrationality, which would be bad. Avoid badness... get vagueness. Let us examine the two principles of rationality in turn.

The Principle of Program-Favoring says that without good reason to the contrary, you should assume that the extension of natural language predicate P and its natural language negation 'not P' are capable of being correctly determined using programs in the head. Recall that this helps lead to the Two-Programs Thesis and thus vagueness because 'not P' is required to be semidecidable by the program for it as well as P, and it is this dual requirement that is difficult to satisfy. How "essential" is this rationality principle; i.e., how "bad" would it be to act in non-accordance with it? You could, after all, avoid the vagueness of 'bald' if you were only willing to live with just one program in the head—the one for 'bald', say. However, this benefit would come at great cost since you would be generally able to identify bald things but not generally things that are not bald. Is seeing the other half of a concept really that essential? Alternatively, is not being able to see the other half of a concept so bad? Yes, it is so bad; I take this to be obvious. The utility gained by bringing in the program for 'not bald' is that it helps you see the "other half" of the concept.

Since it cannot do this job perfectly, vagueness is the result. [Or in "single program" form (see the discussion near the start of Subsection 4.2.5), as soon as you allow your single program to say NO and make your interpretation of 'not P' be the set of objects on which the program says NO rather than simply the complement of the interpretation of P, vagueness is the result since you cannot put in the NOs perfectly.]

Now let us look at the second principle of rationality, the Principle of Any-Algorithm, which says that without good reason to the contrary, you should not presume that there are particular algorithms such that for all natural language predicates P (or 'not P') the algorithm does not supply the best interpretation for P ('not P'). You could avoid the vagueness of 'bald' if you were willing to confine your choice of programs to some recursive subset of the set of algorithms. I spent a little time near the end of Subsection 4.2.5 defending why it is bad not to act in accordance with this principle, and I largely refer you to that discussion. The short of it is that violating the Principle of Any-Algorithm would be very costly since you would thereby confine yourself to much less rich interpretations for natural language predicates and you would not be as capable—possibly even incapable—of adequately classifying an in principle, classifiable world. Vagueness is essential, in addition to the earlier reason, because it is essential that we be capable of classifying our world.

4.3.5 Ineliminability

Vagueness is not to be easily circumvented, or so it is usually thought, and my theory of vagueness leads to several ways in which it may be said that vagueness is ineliminable. One major notion of ineliminability emanates from the fact that there is nothing particular to us humans assumed in the theory; ideal computing devices such as HAL from *2001 Space Odyssey* and Data from *Star Trek* are subject to vagueness as well. Greater powers of discrimination and computation cannot overcome the dilemma of a borderline region and higher-order vagueness. Why, though, is vagueness ineliminable for them?

Let us consider whether the borderline region may be completely eliminated. Once the two programs exist for, say, baldness, perhaps it is possible to find a single new algorithm for baldness that divvies up the universe into YES and NO in such a way that anything that is definitely bald (with respect to the two programs) falls into YES, and anything that is definitely not bald falls into NO (i.e., it respects the definite cases). This algorithm would act by classifying each member of the borderline region as either bald or not, and would serve

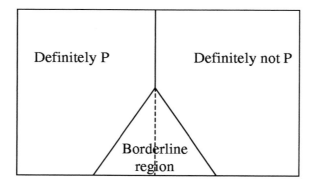

Figure 4.2: A successful precisification would recursively cut through the borderline region as shown by the dotted line, leaving the definite regions untouched. This is, however, not generally possible.

to redefine baldness so as to be non-vague all the while preserving the definite cases (see Figure 4.2). The algorithm would amount to a precisification of baldness. But if it were generally possible to precisify the borderline region and obtain an algorithm for a precisification of baldness, then it would have been generally possible to find such an algorithm in the first place (i.e., pick two programs and then precisify them), contradicting the non-recursiveness of the set of algorithms. Therefore it is not generally possible to eliminate the borderline region, and I call this sort of ineliminability *non-precisifiability*. [If, under supervaluationism, (super)truth is meant to be determined by running through all precisifications and checking to see if the sentence is true in each, then (super)truth of natural language utterances is not generally possible to determine since it is not generally possible to precisify at all.]

May you carefully restrict yourself to certain well-defined contexts, and within these contexts might vagueness be eliminated? We usually do not think so. For example, we do not seem to find ourselves able to identify a group of people (say, infants) such that baldness is no longer vague amongst that group. My theory explains this sort of ineliminability. What you would like to find is a subset of the universe of objects such that there is no longer a borderline region for baldness; you would like to eliminate vagueness by restricting the context

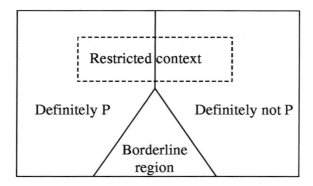

Figure 4.3: A successful restriction would consist of a recursive subset (a context) consisting of no borderline region as shown by the dotted box. This is, however, not generally possible.

to one where there is no borderline region (see Figure 4.3). Not just any subset (or context) will do—you need to be able to recognize (via a program in the head) when something is or is not in that subset, and this implies that you need an algorithm. But now you are back to the same old problem yet again: you cannot generally acquire algorithms. Your contexts are not generally decidable by the programs for them. You may then acquire a context which does not include any of the borderline region but be presented with objects for which you are incapable of knowing whether it is in the context. The objects may in actuality not be in the context, but you may then judge them to be borderline cases and thereby see vagueness. One might respond in two ways here. First, perhaps you only judge an object to be part of the context if the program for the context actually says YES that it is part of the context; if this were so then a single program only semideciding the context is sufficient. The difficulty with this response is that you may well be asked about some object whether it is part of the context and what its categorization is with respect to the vague predicate, and you cannot just refuse to answer. A second response is that we have no reason to believe that contexts require an Any-Algorithm hypothesis; perhaps the allowable programs for contexts are confined to a recursive subset of the set of algorithms. The difficulty with this is that contexts very often are natural language concepts; e.g., attractiveness among bald people, or quickness among cats. Therefore, (a) the arguments from Section 4.2.5 toward allowing any algorithm apply here, and therefore (b) the context itself is vague. Even if you manage to secure a context that is decided by the program for it, because you cannot generally determine where the borderlines are it is not generally possible for you to be assured that the context does not include some of the borderline region. It is not, then, generally possible for you to restrict yourself to a context wherein the borderline region is eliminated, and I call this sort of ineliminability *non-restrictability*.

I have discussed two sorts of ineliminabilities concerning the borderline phenomenon. Higher-order vagueness is also ineliminable. To begin with, it is not generally possible to correct the two programs so that they decide their respect interpretations. If it were possible, the program determining the interpretation of P could have been "corrected" to an algorithm in the first place, and there would be no need for a second program in the head at all. But it is not possible to generally acquire an algorithm, and thus it is not possible to so correct the programs.

Although the two programs for baldness cannot determine the boundaries of the borderline region, it is possible for the borderline region to be recursive and thus it is in principle possible for you to have an algorithm deciding it. If you had such an algorithm there would be no more higher-order vagueness since you could determine the boundaries of the borderline region. However, it is not generally possible to find the algorithm. For one, it is not generally possible to pick an algorithm rather than a non-algorithm for the job of attempting to decided the borderline region. And two, you cannot be sure, even given that you have an oracle handing you any desired algorithm, whether what it is deciding is the borderline region since you cannot generally know what things are in the borderline region.

4.3.6   Degrees of vagueness

The previous section concludes the sections on the phenomena of vagueness. I want to discuss "degrees of membership" in this section. Membership seeming to come in degrees is not a phenomenon unique to vagueness; non-vague predicates such as 'even' have been shown (Armstrong et al., 1983) to come in degrees to subjects. Also, consider the set $HALT$ of ordered pairs of programs C and inputs x such that the program halts on that input. In a natural sense we are inclined to say that $\langle C_1, x_1 \rangle$ is more $HALT$-like than $\langle C_2, x_2 \rangle$ if $C_1(x_1)$ halts after fewer steps than does $C_2(x_2)$. Such judgements may depend on a number of factors, such as typicality, probability, degree of difficulty in determining membership, and so on.

Prior to noticing such examples of precise sets that nevertheless display to us a phenomenon of degrees, one might be worried about the fact that since in my theory concepts are pairs of precise sets, the phenomenon of seeming degrees of membership might be precluded. If single precise sets are not inconsistent with the phenomenon, then it seems there is no prima facie inconsistency to two pragmatically related precise sets displaying the phenomenon—i.e., vague predicates within my theory—and it seems there is no particular responsibility for me to explain seeming degrees of membership. I do not know the explanation for degrees, and I do not care; I defer to whatever is the best theory.

4.3.7   Summing up

I have argued that it is very likely to be in a Church-bound agent's (i.e., finite and sufficiently powerful) best interest to have accessible (i.e., interpreted via programs in the head), maximally accurate (i.e., any algorithm a possible meaning determiner) interpretations of natural language predicates. I have also argued that such an agent having such interpretations experiences vagueness. Vagueness, then, is in such an agent's best interest. If we, too, are Church-bound, then vagueness is very likely *good* for us; the only possible ways for us to avoid vagueness are either to lose the accessibility of our natural language meanings or to confine our meanings to an "infinitely less rich" choice, each very likely to be more costly than the costs of vagueness.

4.4   Discussion

The Undecidability Theory of Vagueness is now out on the table. We have seen how it is motivated in Section 4.2 and how it explains the phenomena linked to vagueness in Section 4.3. There are number of issues to be discussed concerning it that I take up in this section.

Nonstandard concepts

Section 4.2 concluded with the Two-Programs Thesis, which says that "most" of your interpretations of natural language predicates P and 'not P' are semidecidable but not decidable by the programs for them ((i) and (ii)), and are not

complements ((iii)). This is a stunning conclusion irrespective of the vagueness to which we saw in Section 4.3 it leads. 'not P' cannot ("usually") be
translated into logic as '$\neg P$' as it is usually thought. Rather, 'not P' should be
represented as a distinct predicate of its own, $nonP$, the *dual* to P. $nonP$ has
no logical connection to P (other than non-overlap), although, informally, they
should be thought of as rough approximations of the complement of the other.

My theory leads to a nonstandard notion of what is a concept, where by
'concept' I mean your or my concept, extensionally construed. There is no single set of objects which fall under a given concept; the interpretation of 'bald' is
not the concept baldness, and neither is the interpretation of 'non-bald'. Rather,
a concept is comprised of two sets with only a pragmatic relation; the concept
baldness is comprised of the interpretation of 'bald' and that of 'non-bald'.
Your concept, or your semantics of, P-ness is the ordered pair $(A, nonA)$,
where A and $nonA$ are the interpretations of P and 'not P', respectively. If
a single two-sided coin represents the usual view of a concept—i.e., 'bald'
on one side and '$\neg$bald' on the other—my theory leads to a 'two single-sided
coins' view of what is a concept: you have access to only one side of each
coin, and the coins are independent of one another (although they are disjoint,
and are likely to be rough approximations of the complement of the other's
interpretation).

There is an intuitive sense in which this notion of a concept is incoherent.
By incoherent I do not just mean that it is non-classical; there are certainly
many other non-classical notions of what it is to be a concept which are not
incoherent in the way that I mean. For example, in fuzzy logic a concept is a
fuzzy set, and in some connectionist models of mind a concept is a vector of the
weights of the connections. In each of these cases there is a *single* intuitively
natural object representing a concept—a single fuzzy set and a single weights
vector. In my case though the single object representing a concept is an ordered
pair of sets and this complex object is entirely unnatural. Rather, my notion of
a concept consists of *two* intuitively natural objects, namely sets, unnaturally
paired together. Whether one should share this intuition of incoherence is not
particularly crucial, but to the extent that there is incoherence it helps to explain
one of the aspects typically thought to pertain to vagueness. Since Frege (1970)
(see also Dummett (1975) and Wright (1975)) there has been the thought that
vague predicates are incoherent, and we can see from where the feeling of
incoherence springs. This "incoherent" notion of a concept is, according to
my theory, an *essential* part of language for any rational, sufficiently powerful,
finite agent. Our concepts are essentially incoherent, and this is just the sort of

intuitive feeling people have had about vagueness since Wittgenstein.

The notion of "your concept" just discussed is really "your concept competently employed." In your actual performance you will sometimes incorrectly conclude, for example, that an object is borderline since you cannot afford to let your programs run forever. There are thus three levels of concepts that may be distinguished. First, there is the (fantasy) level of true concepts "out there in the world." I have been presuming they are determinate. Second, there is the level of your concepts "in the head" determined by the pair of programs for P and 'not P', respectively. Third, there is the level of your concepts as you actually perform using them; these will be, at best, approximations of the second-level concepts.

Associated with these distinctions is the following criticism. One might complain that my explanation of vagueness is too abstract to possibly be correct. The theory depends crucially on the notion that we run programs in our head that do not halt on some inputs. This, one could charge, cannot be the explanation for vagueness since our programs never actually do diverge forever. The critic can even admit that perhaps the methods in the head are indeed best represented as non-algorithmic programs in the head, but deny that this can be the explanation for vagueness since the programs are never actually allowed to run forever.

It is true that programs in your head certainly do simply give up after some point; you don't run into infinite loops. When presented with something in the borderline region, where both programs diverge, after some period of introspection you will inductively conclude that the object is borderline. You could have been too hasty, for on the very next step it could be that one program would have halted. All this is no difficulty for my theory. In fact I need it for my theory, for it is just these difficulties the actual agent runs into in dealing with his programs in the head that accounts for his inability to determine the boundaries of the borderline region, or higher-order vagueness. It is this third level mentioned above whose non-identity with the second level helps to explain higher-order vagueness.

Universality of vagueness

I now touch on three issues related to the universality of vagueness.

The first concerns whether all natural language predicates are vague. By the Thesis and the characterization of vagueness, "most" of your natural language predicates are vague. "most," however, does not mean all, and according to my theory there may be some non-vague predicates. But are not all natural language (nonrelational) predicates vague? It is not clear that the answer to this question is 'Yes'. For example, Sorensen (1988, p. 201) cites 'flat', 'clean' and 'empty' as example non-vague predicates. These predicates are often applied in "restricted domains of discourse; not all bumps, dirt, and contents are relevant" (ibid.), but if I am asked if, strictly speaking, some surface is flat, I am sure that my answer is either YES or NO (i.e., not neither). "Strictly speaking," surfaces are either flat or not, whereas for 'heap' there is no such "strictly speaking" analog. I also believe 'mortal' and 'everlasting', for example, to be non-vague. There are explanations consistent with my theory for why non-vague predicates are rare at best. The first emanates from the observation made in Subsection 4.2.5 that the set of algorithms is much more difficult than its complement, and this is what motivated the scare quotes around 'most' in the first place. The second explanation helps to explain why there are few to no non-vague "complex" natural language predicates. By complex predicates I informally mean those predicates like 'dog', 'bald', 'people', 'chair', etc., that depend on a number of more "primitive" predicates like 'red', 'circular', etc., for their application. Most of our every day predicates—the ones we use to carve up the world—are complex. In order for one of these predicates to be non-vague, *every* more primitive concept it employs must be non-vague—although see Sorensen (1988, pp. 228–229) for some nice and unusual counterexamples—and this is probably never the case, given that "most" (primitive) concepts are, according to my theory, vague.

The second universality issue is that given that some predicates might be non-vague, we do not find in our experiences cases where, say, 'dog' is vague but 'cat' is not; similar sorts of predicates should either both be vague or neither. The observation just mentioned concerning complex versus primitive concepts explains this datum. Similar predicates make use of similar more primitive concepts, and thus inherit the vagueness (or lack thereof) of the more primitive concepts.

On the third issue of universality, the Thesis is about your interpretations, stating that "most" of the time your natural language predicates are vague. The Thesis obviously also applies to any of us individually. One datum of vagueness seems to be that we don't find ourselves disagreeing about the vagueness of predicates. What reason have we to believe, in my theory's sights, that you and I have the same vague predicates? Why should your "most" and my "most" coincide? The answer to this query is as follows: If you believe 'bald' is vague and I believe it is non-vague, then it is not the case that we have the same concept of baldness save that one is vague and the other not. Your concept of baldness consists of two interpretations which do not cover every object. My concept of baldness, on the other hand, consists of just a single classical interpretation; I have no "hole" in my concept. We disagree about more than just baldness's vagueness since our concepts are genuinely different. Therefore, in order to explain why we all agree on which predicates are vague, it suffices to explain why we all tend to have the same concepts for predicates. Explaining *this*, however, is not something my theory is subject to any more than any other theory; any adequate account of our shared concepts suffices to explain why we agree about the vagueness of predicates.

Non-vague metalanguage

I do not equate precision and non-vagueness; you can be precise and vague since vague concepts in my theory are, after all, two precise sets of a particular sort. For convenience here is the Undecidability Theory of Vagueness's characterization of vagueness again: Predicate P is vague if and only if

1. your interpretation of P is determined by a program in your head that is capable of semideciding but not deciding it,

2. your interpretation of 'not P' is determined by a program in your head that is capable of semideciding but not deciding it, and

3. your interpretation of P is not the complement of your interpretation of 'not P'.

The metalanguage used to state the characterization is precise. Furthermore it is non-vague. Do not be confused into thinking my use of "your head" in the metalanguage brings in vagueness; it is no more vague within my model than is "the computer I am typing this on," and we may suppose that it is implicitly "your head right now" to allay worries about the identity of your head through time. Predicates are possibly vague, names (or individual constants) are not; or, at least, they are different issues.

Non-vague metalanguage does not necessarily imply that the characterization of vagueness, and thus 'vague', is non-vague, although it does imply that it is precise. Is 'vague' vague according to my theory? This question can be read in two ways: (a) is *your* concept of vagueness vague?, and (b) is *the* concept emanating from my theory's characterization vague. Let me answer (b) first. The concept of vagueness emanating from the characterization is the set of all predicates P satisfying the characterization; i.e., it is just a set. The extension of 'vague' is that set, and the extension of 'not vague' is the complement of that set. Part (iii) of the characterization is thus violated and so *the* concept of vagueness is not vague.

What about question (a)? Although the true extension of 'vague' is not vague, might your concept of it possibly be vague? According to the characterization, 'vague' is vague to you if and only if

1. your interpretation of 'vague' is determined by a program in your head that is capable of semideciding but not deciding it,

2. your interpretation of 'not vague' is determined by a program in your head that is capable of semideciding but not deciding it, and

3. your interpretation of 'vague' is not the complement of your interpretation of 'not vague'.

The characterization is not just a characterization—it is also the explanation for why some predicates are vague. It says that vague P seems vague to you because your programs in the head for P and 'not P' semidecide but do not decide their respective interpretations and these interpretations are not complements. When you are presented with P and are asked whether it is vague, my theory's claim is that you do some introspection—running both programs on various inputs—and see if things "feel" like whatever things feel like when (i), (ii) and (iii) of the characterization obtain. Therefore, your interpretation of 'vague' according to my account might seem to be the same as the true extension. The problem with this suggestion is that it is not generally possible for you to successfully do such introspection. Your interpretations might be complements yet you not be able to know this, or not be complements and you not be able to know this. Also, your programs might not decide their interpretations but you may not be capable of verifying this, or vice versa. Thus, it is doubtful that your interpretation of 'vague' actually is the true extension. The question of whether 'vague' is vague to you is still open, and it comes down to a factual matter. As far as my characterization goes, your concept of vagueness may be vague or not.

Sorensen (1985) has argued that 'vague' is vague. His argument tactic is to show that one can build a sorites argument using 'vague'. He proposes the disjunctive predicates 'n-small' for each natural number n, each which applies to those natural numbers that are either small or less than n. The sorites argument is as follows:

(1) '1-small' is vague.

(2) For all n, if 'n-small' is vague, then '$n + 1$-small' is vague.

(3) 'One-billion-small' is vague.

'1-small' is obviously vague, so (1) is true. (2) seems compelling, but (1) and (2) imply (3) which, supposing that our interpretation of 'small' is such that one billion is definitely and clearly not small, is false since it is now equivalent to the non-vague predicate 'less than one billion.' 'vague' is vague because "it is unclear as to where along the sequence the predicates with borderline cases end and the ones without borderline cases begin" (Sorensen, 1985, p. 155).

Such unclarity, or even unknowability, of the line is not sufficient for vagueness. The sorites argument for 'vague' is not paradoxical, since it is not fantastic to deny the induction step; such a denial means asserting that there is an n such that 'n-small' is vague but '$n + 1$-small' is not. Is this difficult to believe? I do not see why. For 'bald', on the other hand, the proposition that n hairs is bald but $n + 1$ hairs is not strains credulity, and we are accordingly unhappy to deny the induction step. It strains credulity because we feel, right or wrong, that there are borderline cases of baldness. Is there any such intuition for vague predicates? Are there cases of predicates we find are borderline vague? I do not know of any such examples; any natural language predicates I have encountered either have borderline cases or do not. (See also Deas (1989) for one who agrees.)

'vague' is not vague, then; or at least if it is vague it is not so merely by being able to be put it into what seems to be a sorites series. A sorites series is only genuine when it is paradoxical, and it is only paradoxical when the denial of the induction step seems counter-intuitive. Since the denial of the induction step for the sorites series above is not counter-intuitive, this suggests that any sorites series with 'vague' will also not be paradoxical. In fact, since there are no clear cases of borderline vague predicates, there is not even any prima facie reason to believe 'vague' is vague.

Hyde (1994) utilizes Sorensen's argument for the vagueness of 'vague' to argue, in turn, that higher-order vagueness is a pseudo-problem. If 'vague' is vague and vagueness is defined as the phenomenon that there are borderline cases, then since the existential 'there are' is not vague 'borderline case' must be vague because otherwise 'vague' would not be vague. I.e., if 'vague' is vague then 'borderline case' is vague. But higher-order vagueness is the phenomenon that there are borderline cases of borderline cases—it is that 'borderline case' is vague—and this is already built into the original concept of vagueness. Higher-order vagueness comes for free from the vagueness of 'vague', and thus one need not tell any special story concerning higher-order vagueness. Hyde's argument fails without the vagueness of 'vague', though, and with reason now to deny the vagueness of 'vague', there is reason to believe that higher-order vagueness is a genuine problem needing possibly a separate explanation. (See also Tye (1994) for other criticisms of Hyde's argument.)

What kind of theory is this?

Where does the Undecidability Theory of Vagueness fit in amongst the spectrum of theories of vagueness?

(1) It is an epistemic theory. Vagueness exists in part because of your inadequacies: you are finite. Furthermore, no semantic indeterminacy concerning the true concepts is required. In this sense it is a full-fledged epistemic theory. (2) However, despite the consistency with a determinist account of true semantics, my theory has an indeterminist aspect in that the semantics for the natural language user's concept P-ness consists of two distinct interpretations, one for P and another for 'not P'. The borderline region is semantically distinct from the definite regions. The account of natural language semantics for your concepts is, then, indeterminist. If one holds that (true) meaning is competent use (of a community even), then the semantics at this level *is* the true semantics, and one would have to hold semantic indeterminism. (3) Finally, the underlying logic of your concepts is determinist in that P and 'not P' become P and $nonP$, each which gets its own determinist classical interpretation.

How does the Undecidability Theory compare to other explanatory theories of vagueness? Is there anything about the Undecidability Theory that is favorable? I think so: its paucity of assumptions. It rests on the three weak descriptive hypotheses from Section 4.2: the Church-Bound, Programs-in-Head, and Any-Algorithm Hypotheses. Each is, independent of vagueness, quite plausible. The Church-Bound Hypothesis (see Subsection 4.2.3) says that you are finite, bound by Church's Thesis, and capable of, in principle, computing any Turing-computable function. The Programs-in-Head Hypothesis (see Subsection 4.2.4) says that your interpretation of a predicate is determined by a program in your head for it. That is, natural language predicates are not like 'theorem of Peano Arithmetic' for which your interpretation is set to that given by its (not recursively enumerable) arithmetic definition, not the set determined by whatever program you might use as a heuristic for responding to queries about it. The Any-Algorithm Hypothesis (see Subsection 4.2.5) says that you allow yourself the use of any algorithm for your interpretations of natural language predicates. We saw that if you allow yourself 'while' loops in the building of programs, then it is difficult to reject this hypothesis. Each is compelling, and vagueness follows. Alternative theories of vagueness must either deny one of these, or argue that the phenomenon my theory explains is not vagueness.

Bibliography

Abeles M (1991) *Corticonics: Neural Circuits of the Cerebral Cortex.* Cambridge University Press, Cambridge.

Adams AM and Gathercole SE (2000) Limitations in working memory: Implications for language development. *Int. J. Lang. Comm. Dis.* 35: 95–116.

Adams DR (1986) *Canine Anatomy.* The Iowa State University Press, Ames.

Agur AMR and Lee MJ (1991) *Grant's Atlas of Anatomy.* Williams and Wilkins, Baltimore.

Aitkin M (1991) Posterior Bayes factors. *Journal of the Royal Statistical Society, B* 53:111–142.

Allman JM (1999) *Evolving Brains.* Scientific American Library, New York.

Allott R (1992) The motor theory of language: Origin and function. In Wind J, Bichakjian BH, Nocentini A and Chiarelli B (eds.) *Language Origin: A Multidisciplinary Approach.* Kluwer, Dordrecht.

Alpert CJ, Hu TC, Huang JH, Kahng AB and Karger D (1995) Prim-Dijkstra Tradeoffs for Improved Performance-Driven Global Routing. *IEEE Trans. on CAD* 14: 890–896.

Altmann S (1965) Sociobiology of rhesus monkeys. II: Stochastics of social communication. *J. Theor. Biol.* 8: 490–522.

Anderson BL (1999) Stereoscopic surface perception. *Neuron* 24: 919-928.

Anson BJ (1966) *Morris' Human Anatomy.* McGraw-Hill, New York.

Anstis S (1989) Kinetic edges become displaced, segregated, and invisible. In Lam DM-K and Gilbert CD, *Neural Mechanisms of Visual Perception*, Portfolio, The Woodlands, TX, pp. 247–260.

Arend LE and Goldstein R (1990) Lightness and brightness over spatial illumination gradients. *J. Opt. Soc. Am. A* 7: 1929–1936.

Arend LE and Reeves A (1986) Simultaneous color constancy. *J. Opt. Soc. Am. A* 3: 1743–1751.

Arend LE, Reeves A, Schirillo J and Goldstein R (1991) Simultaneous color constancy: papers with diverse Munsell values. *J. Opt. Soc. Am. A* 8: 661–672.

Arend LE and Spehar B (1993) Lightness, brightness, and brightness contrast: 1. Illumination variation. *Perc. and Psychophys.* 54: 446–456.

Ariew R and Garber D (1989) *G. W. Leibniz. Philosophical Essays.* Hackett Publishing Company, Indianapolis.

Armstrong SL, Gleitman LR, and Gleitman H (1983) What some concepts might not be. *Cognition* 13:263–308.

Ashdown RR and Done S (1984) *Color Atlas of Veterinary Anatomy: The Ruminants.* University Park Press, Baltimore.

Baerends GP (1976) The functional organization of behaviour. *Anim. Behav.* 24: 726–738.

Baldo MVC and Klein SA (1995) Extrapolation or attention shift? *Nature* 378: 565–566.

Barnes RD (1963) *Invertebrate zoology.* W. B. Saunders, Philadelphia

Bast TH, Christensen K, Cummins H, Geist FD, Hartman CG, Hines M, Howell AB, Huber E, Kuntz A, Leonard SL, Lineback P, Marshall JA, Miller GS Jr, Miller RA, Schultz AH, Stewart TD, Straus WL Jr, Sullivan WE and Wislocki GB (1933) *The Anatomy of the Rhesus Monkey.* Williams and Wilkins, Baltimore.

Bastock M and Blest AD (1958) An analysis of behaviour sequences in Automeris aurantiaca Weym (Lepidoptera). *Behaviour* 12: 243–284.

Bell G and Mooers AO (1997) Size and complexity among multicellular organisms. *Biol. J. Linnean Soc.* 60: 345–363.

Berg CJ (1974) A comparative ethological study of strombid gastropods. *Behaviour* 51: 274–322.

Berkinblit MB, Feldman AG and Fukson OI (1986) Adaptability of innate motor patterns and motor control mechanisms. *Behav. Brain Sci.* 9: 585–638.

Bernstein NP and Maxson SJ (1982) Behaviour of the antarctic blue-eyed shag, Phalacrocorax atriceps bransfieldensis. *Nortonis* 29: 197–207.

Berry MJ II, Brivanlou IH, Jordan TA and Melster M (1999) Anticipation of moving stimuli by the retina. *Nature* 398: 334–338.

Bertrand J (1889) *Calcul des probabilites.* Gauthier-Villars, Paris.

Betz O (1999) A behavioural inventory of adult Stenus species (Coleoptera: Staphylinidae). *J. Natural Hist.* 33: 1691–1712.

Bishop SC (1943) *Handbook of salamanders.* Comstock, Ithaca.

Bizzi E and Mussa-Ivaldi FA (1998) Neural basis of motor control and its cognitive implications. *Trends Cogn. Sci.* 2: 97–102.

Bollabás B (1985) *Random Graphs.* Academic Press, London.

Bolles RC and Woods PJ (1964) The ontogeny of behaviour in the albino rat. *Anim. Behav.* 12: 427–441.

Bolwig N (1959) A study of the behaviour of the Chacma baboon, *Papio ursinus. Behaviour* 14: 136–163.

Bonin G von (1937) Brain-weight and body-weight of mammals. *J. Gen. Psych.* 16: 379–389.

Bouma H and Andriessen JJ (1970) Induced changes in the perceived orientation of line segments. *Vision Res.* 10: 333–349.

Boyd JS, Paterson C and May AH (1991) *Clinical Anatomy of the Dog and Cat.* Mosby, St. Louis.

Bradley OC and Grahame T (1959) *Topographical Anatomy of the Dog.* Macmillan, New York.

Brainard DH and Freeman WT (1997) Bayesian color constancy. *J. Opt. Soc. Am. A* 14: 1393–1411.

Braitenberg V (1978) Cortical architectonics: General and areal. In Brazier MAB, Petsche H (eds) *Architectonics of the Cerebral Cortex.* Raven Press, New York, pp. 443–465.

Brenner E and Smeets JBJ (2000) Motion extrapolation is not responsible for the flash-lag effect. *Vision Res.* 40: 1645–1648.

Brewer DW and Sites RW (1994) Behavioral inventory of Pelocoris femoratus (Hemiptera: Naucoridae). *J. Kansas Entomol. Soc.* 67: 193–198.

Broad CD (1918) On the relation between induction and probability. *Mind* 108:389–404.

Brockway BF (1964a) Ethological studies of the Budgerigar (*Melopsittacus undulatus*): Non-reproductive behavior. *Behaviour* 22: 193–222.

Brockway BF (1964b) Ethological studies of the Budgerigar: Reproductive behavior. *Behaviour* 23: 294–324.

Brown ED, Farabaugh SM and Veltman CJ (1988) Song sharing in a group-living songbird, the Australian Magpie, Gymnorhina tibicen. Part I. Vocal sharing within and among social groups. *Behavior* 104: 1–28.

Brusca RC, Brusca CJ (1990) *Invertebrates.* Sinauer Associates, Sunderland.

Buchsbaum R (1956) *Animals without backbones. An introduction to the invertebrates.* The University of Chicago Press, Chicago.

Buchsbaum R, Buchsbaum M, Pearse J and Pearse V (1987) *Animals without backbones.* The University of Chicago Press, Chicago.

Budras K-D and Sack WO (1994) *Anatomy of the Horse: An Illustrated Text.* Mosby-Wolfe, London.

Burnie D, Elphick J, Greenaway T, Taylor B, Walisiewicz M and Walker R (1998) *Nature encyclopedia.* DK Publishing, London

Busam JF (1937) *A Laboratory Guide on the Anatomy of the Rabbit.* Spaulding-Moss, Boston.

Calder, WA III (1996) *Size, Function, and Life History.* Dover, Mineola.

Campbell R (1974) The sorites paradox. *Phil. Studies* 26: 175–191.

Cargile J (1969) The sorites paradox. *Brit. J. Phil. Sci.* 20: 193–202.

Carnap R (1950) *Logical Foundations of Probability.* The University of Chicago Press, Chicago.

Carnap R (1952) *The Continuum of Inductive Methods.* The University of Chicago Press, Chicago.

Carnap R (1989) Statistical and inductive probability. In Brody BA and Grandy RE (eds) *Readings in the Philosophy of Science.* Prentice Hall, Eaglewood Cliffs, pp. 279–287.

Carpenter RHS and Blakemore C (1973) Interactions between orientations in human vision. *Exp. Brain Res.* 18: 287–303.

Case R, Kurland DM and Goldberg J (1982) Operational efficiency and the growth of short-term memory span. *J. Exp. Child Psychol.* 33: 386–404.

Chamberlain FW (1943) *Atlas of Avian Anatomy.* Hallenbeck, East Lansing.

Changizi MA (1995) Fuzziness in classical two-valued logic. *Proc. of the Third Int. Symp. on Uncertainty Modeling and Analysis, and the Annual Conf. of the North American Fuzzy Information Processing Soc..* IEEE, pp 483–488.

Changizi MA (1999a) Vagueness and computation. *Acta Analytica* 14: 39–45.

Changizi MA (1999b) Motivation for a new semantics for vagueness. *EWIC (Irish Workshop on Formal Methods).*

Changizi MA (1999c) Vagueness, rationality and undecidability: A theory of why there is vagueness. *Synthese* 117: 345–374.

Changizi MA (2001a) The economy of the shape of limbed animals. *Biol. Cyb.* 84: 23–29.

Changizi MA (2001b) Principles underlying mammalian neocortical scaling. *Biol. Cybern.* 84: 207–215.

Changizi MA (2001c) 'Perceiving the present' as a framework for ecological explanations for the misperception of projected angle and angular size. *Perception* 30: 195–208.

Changizi MA (2001d) Universal scaling laws for hierarchical complexity in languages, organisms, behaviors and other combinatorial systems. *J. Theor. Biol.* 211: 277–295.

Changizi MA (2001e) Universal laws for hierarchical systems. *Comments Theor. Biol.* 6: 25–75.

Changizi MA (2002) The relationship between number of muscles, behavioral repertoire size, and encephalization in mammals. *J. Theor. Biol.*, in press.

Changizi MA and Barber TP (1998) A paradigm-based solution to the riddle of induction. *Synthese* 117: 419–484.

Changizi MA and Cherniak C (2000) Modeling large-scale geometry of human coronary arteries with principles of global volume and power minimization. *Can. J. Physiol. Pharmacol.* 78: 603–611.

Changizi MA and Hall WG (2001) Thirst modulates a perception. *Perception* 30: 1489–1497.

Changizi MA, McDannald M and Widders D (2002a) Scaling of differentiation in networks: Nervous systems, organisms, ant colonies, ecosystems, businesses, universities, cities, electronic circuits and Legos. *J. Theor. Biol.*, in press.

Changizi MA, McGehee RMS and Hall WG (2002b) Evidence that appetitive responses for dehydration and food-deprivation are learned. *Physiol. Behav.*, in press.

Changizi MA and Widders D (2002) Latency-correction explains the classical geometrical illusions. *Perception*, in press.

Cherniak C (1992) Local optimization of neuron arbors. *Biol. Cybern.* 66: 503–510.

Cherniak C (1994) Component placement optimization in the brain. *J. Neurosci.* 14: 2418–2427.

Cherniak C (1995) Neural component placement. *Trends Neurosci.* 18: 522–527.

Cherniak C, Changizi MA, Kang DW (1999) Large-scale optimization of neuron arbors. *Physical Review E* 59: 6001–6009.

Chierchia G and McConnell-Ginet S (1990) *Meaning and Grammar: An Introduction to Semantics*. MIT Press.

Chin E Jr (1957) *The Rabbit. An Illustrated Anatomical Guide.* Master's Thesis, College of the Pacific.

Chklovskii DB (2000) Binocular disparity can explain the orientation of ocular dominance stripes in primate V1. *Vision Res.* 40: 1765–1773.

Chklovskii DB and Koulakov AA (2000) A wire length minimization approach to ocular dominance patterns in mammalian visual cortex. *Physica A* 284: 318–334.

Chomsky N (1972) *Language and Mind.* Harcourt Brace Jovanovich, New York.

Clark EV (1993) *The Lexicon in Acquisition.* Cambridge University Press, Cambridge.

Clark HH and Wasow T (1998) Repeating words in spontaneous speech. *Cog. Psychol.* 37: 201–242.

Coelho AM Jr and Bramblett CA (1981) Interobserver agreement on a molecular ethogram of the genus *Papio. Anim. Behav.* 29: 443–448.

Cole BJ (1985) Size and behavior in ants: Constraints on complexity. *Proc. Natl. Acad. Sci. USA* 82: 8548–8551.

Cooper G and Schiller AL (1975) *Anatomy of the Guinea Pig.* Harvard University, Cambridge.

Copeland BJ (1998) Turing's O-machines, Searle, Penrose and the brain. *Analysis* 58: 128–138.

Coren S and Girgus JS (1978) *Seeing is deceiving: The psychology of visual illusions.* Lawrence Erlbaum, Hillsdale.

Cormen TH, Leiserson CE and Rivest RL (1990) *Introduction to Algorithms.* MIT Press, Cambridge.

Corrigan R (1983) The development of representational skills. In Fischer K (ed) *Levels and Transitions in Children's Development.* Jossey-Bass, San Francisco, pp. 51–64.

Courts SE (1996) An ethogram of captive Livingstone's fruit bats Pteropus livingstonii in a new enclosure at Jersey wildlife preservation trust. *Dodo J. Wildl. Preserv. Trusts* 32: 15–37.

Coxeter HSM (1962) The problem of packing a number of equal nonoverlapping circles on a sphere. *Trans. N.Y. Acad. Sci. Series II* 24: 320–331.

Craigie EH (1966) *A Laboratory Guide to the Anatomy of the Rabbit.* University of Toronto Press, Toronto.

Crile G and Quiring DP (1940) A record of the body weight and certain organ and gland weights of 3690 animals. *Ohio J. Sci.* 40: 219–259.

Danoff-Burg JA (1996) An ethogram of the ant-guest bettle trive sceptobiini (Coleoptera: Staphylinidae; Formicidae). *Sociobiol.* 27: 287–328.

Dawkins R and Dawkins M (1976) Hierarchical organization and postural facilitation: Rules for grooming in flies. *Anim. Behav.* 24: 739–755.

de Finetti B (1974) *Theory of Probability* John Wiley and Sons, New York.

Deacon T (1990) Rethinking mammalian brain evolution. *Amer. Zool.* 30: 629–705.

Deas R (1989) Sorensen's sorites. *Analysis* 49: 26–31.

De Valois RL and De Valois KK (1991) Vernier acuity with stationary moving Gabors. *Vision Res.* 31: 1619–1626.

DeVito S (1997) A gruesome problem. *British Journal for the Philosophy of Science* 48: 391–396.

Devoogd TJ, Krebs JR, Healy SD and Purvis A (1993) Relations between song repertoire size and the volume of brain nuclei related to song: comparative evolutionary analyses amongst ocine birds. *Proc. R. Soc. Lond. B* 254: 75–82.

Douglas JM and Tweed RL (1979) Analysing the patterning of a sequence of discrete behavioural events. *Anim. Behav.* 27: 1236–1252.

Downey ME (1973) *Starfishes from the Caribbean and the Gulf of Mexico.* Smithsonian Institution, Washington, D.C.

Draper WA (1967) A behavioural study of the home-cage activity of the white rat. *Behaviour* 28: 280–306.

Dummett M (1975) Wang's paradox. *Synthese* 30: 301–324.

Durbin R and Mitchison G (1990) A dimension reduction framework for understanding cortical maps. *Nature* 343: 644–647.

Eagleman DM and Sejnowski TJ (2000) Motion integration and postdiction in visual awareness. *Science* 287: 2036–2038.

Earman J (1992) *Bayes or Bust? A Critical Examination of Bayesian Confirmation Theory.* MIT Press, Cambridge .

Earman J and Norton JD (1993) Forever is a day: Supertasks in Pitowsky and Malament-Hogarth spacetimes. *Philosophy of Science* 60: 22–42.

Earman J and Norton JD (1996) Infinite pains: The trouble with supertasks. In Morton A and Stich SP (eds) *Benacerraf and his Critics*, Blackwell, pp. 65–124.

Edgington D (1992) Validity, uncertainty and vagueness. *Analysis* 52: 193–204.

Edgington D (1993) Wright and Sainsbury on higher-order vagueness. *Analysis* 53: 193–200.

Ehrlich A (1977) Social and individual behaviors in captive Greater Galagos. *Behaviour* 63: 192–214.

Ehrlich A and Musicant A (1977) Social and individual behaviors in captive Slow Lorises. *Behaviour* 60: 195–220.

Eisenberg JF (1962) Studies on the behavior of Peromyscus maniculatus gambelii and Peromyscus californicus parasiticus. *Behaviour* 29: 177–207.

Ellsworth AF (1976) *The North American Opossum: An Anatomical Atlas.* Robert E. Krieger Publishing, Huntington.

Evans HE (1993) *Miller's Anatomy of the Dog.* W.B. Saunders, Philadelphia.

Everett RA, Ostfeld RS and Davis WJ (1982) The behavioral hierarchy of the garden snail Helix aspersa. *J. Compar. Ethology* 59: 109–126.

Fagen RM and Goldman RN (1977) Behavioural catalogue analysis methods. *Anim. Behav.* 25: 261–274.

Fentress JC (1983) Ethological models of hierarchy and patterning in species-specific behavior. In Satinoff E and Teitelbaum P (eds) *Handbook of Behavioral Neurobiology.* Plenum Press, New York, pp. 185–234.

Fentress JC and Stilwell FP (1973) Grammar of a movement sequence in inbred mice. *Nature* 244: 52–53.

Fernald RD and Hirata NR (1977) Field study of Haplochromis burtoni: Quantitative behavioural observations. *Anim. Behav.* 25: 964–975.

Ferron J (1981) Comparative ontogeny of behaviour in four species of squirrels (Sciuridae). *J. Compar. Ethol.* 55: 193–216.

Ferron J and Ouellet J-P (1991) Physical and behavioral postnatal development of woodchucks (*Marmota monax*). *Can. J. Zool.* 69: 1040–1047.

Figueredo AJ, Ross DM and Petrinovich L (1992) The quantitative ethology of the zebra finch: A study in comparative psychometrics. *Multivariate Behav. Res.* 27: 435–458.

Fine K (1975) Vagueness, truth, and logic. *Synthese* 30: 265–300.

Fisher GH (1969) An experimental study of angular subtension. *Q. J. Exp. Psych.* 21: 356–366.

Forster MR and Sober E (1994) How to tell when simpler, more unified, or less *ad hoc* theories will provide more accurate predictions. *British Journal for the Philosophy of Science* 45: 1–35.

Foster C and Altschuler EL (2001) The bulging grid. *Perception* 30: 393–395.

Frahm HD, Stephan H, Stephan M (1982) Comparison of brain structure volumes in Insectivora and Primates. I. Neocortex. *J. Hirnforschung* 23: 375–389.

Freeman WT (1994) The generic viewpoint assumption in a framework for visual perception. *Nature* 368: 542–545.

Freyd JJ (1983a) Representing the dynamics of a static form. *Memory and Cognition* 11: 342–346.

Freyd JJ (1983b) The mental representation of movement when static stimuli are viewed. *Perception and Psychophysics* 33: 575–581.

Freyd JJ and Finke RA (1984) Representational momentum. *Journal of Experimental Psychology: Learning, Memory and Cognition* 10: 126–132.

Gadagkar R and Joshi NV (1983) Quantitative ethology of social wasps: Time-activity budgets and caste differentiation in Ropalidia marginata (Lep.) (Hymenoptera: Vespidae). *Anim. Behav.* 31: 26–31.

Gallistel CR (1980) *The Organization of Action: A New Synthesis.* Lawrence Erlbaum, Hillsdale.

Ganslosser U and Wehnelt S (1997) Juvenile development as part of the extraordinary social system of the Mara Dolichotis patagonum (Rodentia: Caviidae). *Mammalia* 61: 3–15.

Gärdenfors P (1990) Induction, conceptual spaces, and AI. *Philosophy of Science* 57: 78–95.

Geist V (1963) On the behaviour of the North American moose (Alces alces andersoni Peterson 1950) in British Columbia. *Behaviour* 20: 377–416.

Gilbert EN and Pollak HO (1968) Steiner minimal trees. *SIAM J. Appl. Math.* 16: 1–29.

Gibson JJ (1986) *The Ecological Approach to Visual Perception*. Lawrence Erlbaum, Hillsdale.

Gibson RN (1980) A quantitative description of the behaviour of wild juvenile plaice. *Anim. Behav.* 28: 1202–1216.

Gillam BJ (1980) Geometrical illusions. *Sci. Am.* 242: 102–111.

Gillam BJ (1998) Illusions at century's end. In Hochberg J (ed) *Perception and Cognition at Century's End*. Academic Press, San Diego, pp. 98–137.

Goodhill GJ, Bates KR and Montague PR (1997) Influences on the global structure of cortical maps. *Proc. R. Soc. Lond. B* 264: 649–655.

Greene E (1988) The corner Poggendorff. *Perception* 17: 65–70.

Greene EC (1935) *Anatomy of the Rat*. The American Philosophical Society, Philadelphia.

Greenfield PM (1991) Language, tools and brain: The ontogeny and phylogeny of hierarchically organized sequential behavior. *Behav. Brain Sci.* 14: 531–595.

Gregory RL (1963) Distortion of visual space as inappropriate constancy scaling. *Nature* 199: 678–680.

Gregory RL (1997) *Eye and Brain*. Princeton University Press, New Jersey, Fifth edition.

Gunn D and Morton DB (1995) Inventory of the behaviour of New Zealand White rabbits in laboratory cages. *Appl. Anim. Behav. Sci.* 45: 277–292.

Hailman JP (1989) The organization of major vocalizations in the paradae. *Wilson Bull.* 101: 305–343.

Hanlon RT, Maxwell MR, Shashar N, Loew ER and Boyle K-L (1999) An ethogram of body patterning behavior in the biomedically and commercially valuable squid Loligo pealei off Cape Cod, Massachusetts. *Biol. Bull.* 197: 49–62.

Harsanyi JC (1983) Bayesian decision theory. Subjective and objective probabilities, and acceptance of empirical hypotheses. *Synthese* 57: 341–365.

Haug H (1987) Brain sizes, surfaces and neuronal sizes of the cortex cerebri: A stereological investigation of man and his variability and a comparison with some mammals (primates, whales, marsupials, insectivores, and one elephant). *Am. J. Anatomy* 180: 126–142.

Hebel R and Stromberg MW (1976) *Anatomy of the Laboratory Rat*. Williams and Wilkins, Baltimore.

Hegner RW (1933) *Invertebrate Zoology*. Macmillan, New York.

Helmholtz H von (1962) *Helmholtz's Treatise on Physiological Optics*. (trans. ed. Southall JPC) Optical Society of America, New York.

Hintikka J (1966) A two-dimensional continuum of inductive methods. In Hintikka J and Suppes P (eds) *Aspects of Inductive Logic*. North-Holland, Amsterdam, pp. 113–132.

Hof PR, Glezer II, Condé F, Flagg RA, Rubin MB, Nimchinsky EA, Weisenhorn, DMV. (1999) Cellular distribution of the calcium-binding proteins parvalbumin, calbindin, and calretinin in the neocortex of mammals: phylogenetic and developmental patterns. *J. Chem. Neuroanat.* 16: 77–116.

Hofman MA (1982a) Encephalization in mammals in relation to the size of the cerebral cortex. *Brain Behav. Evol.* 20: 84–96.

Hofman MA (1982b) A two-component theory of encephalization in mammals. *J. Theor. Biol.* 99: 571–584.

Hofman MA (1989) On the evolution and geometry of the brain in mammals. *Progress in Neurobiology* 32: 137–158.

Hofman MA (1991) The fractal geometry of convoluted brains. *J. Hirnforschung* 32: 103–111.

Hogarth M (1994) Non-Turing computers and non-Turing computability. *Proceedings of the Philosophy of Science Association* 1: 126–138.

Horgan T (1994) Transvaluationism: A Dionysian approach to vagueness. *The Southern Journal of Philosophy* 33 Supplement: 97–126.

Howell AB (1926) *Anatomy of the Wood Rat.* Williams and Wilkins, Baltimore.

Howson C (1987) Popper, prior probabilities, and inductive inference. *The British Journal for the Philosophy of Science* 38: 207–224.

Howson C and Urbach P (1989) *Scientific Reasoning: The Bayesian Approach.* Open Court, Chicago.

Hrdlicka A (1907) Brain weight in vertebrates. Washington, D. C.: Smithsonian Miscellaneous Collections, pp 89–112.

Hubbard TL and Ruppel SE (1999) Representational momentum and the landmark attraction effect. *Canadian Journal of Experimental Psychology* 43: 242–255.

Hudson LC and Hamilton WP (1993) *Atlas of Feline Anatomy for Veterinarians.* W. B. Saunders, Philadelphia.

Hunt KW (1965) A synopsis of clause-to-sentence length factors. *English J.* 54: 300–309.

Hutt SJ and Hutt C (1971) *Direct Observation and Measurement of Behaviour.* C. C. Thomas, Springfield, Ill.

Hyde D (1994) Why higher-order vagueness is a pseudo-problem. *Mind* 103: 35–41.

Hyde D (1997) From heaps and gaps to heaps of gluts. *Mind* 106: 641–660.

Ince SA and Slater PJB (1985) Versatility and continuity in the songs of thrushes Turdus spp. *Ibis* 127: 355–364.

Jacobs RA and Jordan MI (1992) Computational consequences of a bias toward short connections. *J. Cogn. Neurosci.* 4: 323–336.

Jaffe K (1987) Evolution of territoriality and nestmate recognition systems in ants. In Pasteels JM and Deneubourg JL (eds), *From Individual to Collective Behavior in Social Insects.* Birkhäuser Verlag, Basel, pp. 295–311.

Jaynes ET (1973) The well-posed problem. *Foundations of Physics* 3: 477–493.

Jeffreys H (1948) *Theory of Probability.* Clarendon Press, Oxford.

Jeffreys H (1955) The present position in probability theory. *The British Journal for the Philosophy of Science* 5: 275–289.

Jerison H (1973) *The Evolution of the Brain and Intelligence.* Academic Press, New York.

Jerison HJ (1982) Allometry, brain size, cortical surface, and convolutedness. In Armstrong E and Falk F, *Primate Brain Evolution*. Plenum Press, New York, pp. 77–84.

Johnson WE (1924) *Logic, Part III: The Logical Foundations of Science*. Cambridge University Press, Cambridge.

Kamiya A and Togawa T (1972) Optimal branching structure of the vascular tree. *Bull. Math. Biophys.* 34: 431–438.

Kamp JAW (1975) Two theories about adjectives. In Keenan EL (ed) *Formal Semantics of Natural Language*. Cambridge University Press, Cambridge, pp. 123–155.

Kaufman C and Rosenblum LA (1966) A behavioral taxonomy for Macaca nemestrina and Macaca radiata: Based on longitudinal observation of family groups in the laboratory. *Primates* 7: 205–258.

Kaupp BF (1918) *The Anatomy of the Domestic Fowl*. W. B. Saunders, Philadelphia.

Kemeny JG (1953) The use of simplicity in induction. *The Philosophical Review* 62: 391–408.

Keynes JM (1921) *A Treatise on Probability*. Macmillan, New York.

Khuller S, Raghavachari B and Young N (1995) Balancing Minimum Spanning and Shortest Path Trees. *Algorithmica* 14: 305–321.

Khurana B and Nijhawan R (1995) Extrapolation or attention shift? *Nature* 378: 565–566.

Khurana B, Watanabe R and Nijhawan R (2000) The role of attention in motion extrapolation: Are moving objects 'corrected' or flashed objects attentionally delayed? *Perception* 29: 675–692.

Kingdom FAA, Blakeslee B and McCourt ME (1997) Brightness with and without perceived transparency: When does it make a difference? *Perception* 26: 493–506.

Kitazaki M and Shimojo S (1996) 'Generic-view principle' for three-dimensional-motion perception: Optics and inverse optics of a moving straight bar. *Perception* 25: 797–814.

Kneale W (1949) *Probability and Induction*. Clarendon Press, Oxford.

Knill DC and Kersten D (1991) Apparent surface curvature affects lightness perception. *Nature* 351: 228–230.

Knill D and Richards W (eds) (1996) *Perception as Bayesian Inference*. Cambridge University Press, Cambridge.

Kolmes SA (1985) An information-theory analysis of task specialization among worker honey bees performing hive duties. *Anim. Behav.* 33: 181–187.

Koons RC (1994) A new solution to the sorites problem. *Mind* 103: 439–449.

Koulakov AA and Chklovskii DB (2001) Orientation preference patterns in mammalian visual cortex: A wire length minimization approach. *Neuron* 29: 519–527.

Krekelberg B and Lappe M (1999) Temporal recruitment along the trajectory of moving objects and the perception of position. *Vis. Res.* 39: 2669–2679.

Kroodsma DE (1977) Correlates of song organization among North American wrens. *Am. Naturalist* 111: 995–1008.

Kroodsma DE (1984) Songs of the Alder Flycatcher (Empidonax alnorum) and Willow Flycatcher (Empidonax Traillii) are innate. *Auk* 101: 13–24.

Kuhn TS (1977) Objectivity, value judgments, and theory choice. In Kuhn TS (ed) *The Essential Tension*. University of Chicago Press, Chicago.

Laplace PS (1995) *Philosophical Essays on Probability*. (Edited by Toomer GJ, translated by Dale AI, first published 1820.) Springer-Verlag, New York.

Lappe M and Krekelberg B (1998) The position of moving objects. *Perception* 27: 1437–1449.

Lasek R (1988) Studying the intrinsic determinants of neuronal form and function, In Lasek R, Black M (eds) *Intrinsic Determinants of Neuronal Form and Function*, Liss, New York.

Lefebvre L (1981) Grooming in crickets: Timing and hierarchical organization. *Anim Behav.* 29: 973–984.

Lennie P (1981) The physiological basis of variations in visual latency. *Vision Res.* 21: 815–824.

Leonard JL and Lukowiak K (1984) An ethogram of the sea slug, Navanax inermis (Gastropoda, Opisthobranchia). *J. Compar. Ethol.* 65: 327–345.

Leonard JL and Lukowiak K (1986) The behavior of Aplysia californica cooper (Gastropoda; Opisthobranchia): I. Ethogram. *Behaviour* 98: 320–360.

Levine R, Chein I and Murphy G (1942) The relation of the intensity of a need to the amount of perceptual distortion: A preliminary report. *J. Psychol.* 13: 283–293.

Lewis D (1984) Putnam's paradox. *Australasian Journal of Philosophy* 62: 221–236.

Lewis JJ (1985) The ethology of captive juvenile Caiman sclerops: Predation, growth and development, and sociality (crocodilians, life history, behavior, dominance). Dissertation, Northwestern University, DAI, 46, no. 11B: 3751.

Mach E (1976) *Knowledge and Error. Sketches on the Psychology of Enquiry*. Translated by McCormack TJ and Folks P. D. Reidel, Dordrecht.

MacKay DM (1958) Perceptual stability of a stroboscopically lit visual field containing self-luminous objects. *Nature* 181: 507–508.

Di Maio MC (1994) Inductive logic: Aims and procedures. *Theoria* 60: 129–153.

Manger P, Sum M, Szymanski M, Ridgway S and Krubitzer L (1998) Modular subdivisions of Dolphin insular cortex: Does evolutionary history repeat itself? *J. Cogn. Neurosci.* 10: 153–166.

Mariappa D (1986) *Anatomy and Histology of the Indian Elephant.* Indira Publishing House, Oak Park, Michigan.

Marinoff L (1994) A resolution of Bertrand's Paradox. *Philosophy of Science* 61: 1–24.

Markus EJ and Petit TL (1987) Neocortical synaptogenesis, aging, and behavior: Lifespan development in the motor-sensory system of the rat. *Exp. Neurol.* 96: 262–278.

Martin IG (1980) An ethogram of captive Blarina brevicauda. *Am. Midland Naturalist* 104: 290–294.

Masin SC (1997) The luminance conditions of transparency. *Perception* 26: 39–50.

Mather JA (1985) Behavioural interactions and activity of captive Eledone moschata: laboratory investigations of a 'social' octopus. *Anim. Behav.* 33: 1138–1144.

Maunsell JHR and Gibson JR (1992) Visual response latencies in striate cortex of the macaque monkey. *J. Neurophysiol.* 68: 1332–1344.

McClure RC, Dallman MJ and Garrett PD (1973) *Cat Anatomy: An Atlas, Text and Dissection Guide.* Lea and Febiger, Philadelphia.

McLaughlin CA and Chiasson RB (1990) *Laboratory Anatomy of the Rabbit.* Wm. C. Brown Publishers, Dubuque.

Mellor DH (1971) *The Matter of Chance.* University Press, Cambridge.

Miller GA (1956) The magical number seven, plus or minus two: Some limits on our capacity for processing information. *Psychol. Rev.* 63: 81–97.

Miller BJ (1988) Conservation and behavior of the endangered Black-footed ferret (Mustela nigripes) with a comparative analysis of reproductive behavior between the Black-footed ferret and the congeneric domestic ferret (Mustela putorius furo). Dissertation, University of Wyoming, DAI, 50, no. 03B: 08309.

Miller HC (1963) The behavior of the Pumpkinseed sunfish, *Lepomis gibbosus* (Linneaus), with notes on the behavior of other species of Lepomis and the Pigmy sunfish, Elassoma evergladei. *Behaviour* 22: 88–151.

Miller RJ and Jearld A (1983) Behavior and phylogeny of fishes of the genus Colisa and the family Belontiidae. *Behaviour* 83: 155–185.

Mitchison G (1991) Neuronal branching patterns and the economy of cortical wiring. *Proc. R. Soc. Lond. B* 245: 151–158.

Mitchison G (1992) Axonal trees and cortical architecture. *Trends Neurosci.* 15: 122–126.

Mountcastle VB (1957) Modality and topographic properties of single neurons of cat's somatic sensory cortex. *J. Neurophysiol.* 20: 408–434.

Mountcastle VB (1997) The columnar organization of the neocortex. *Brain* 120: 701–722.

Müller M, Boutiére H, Weaver ACF and Candelon N (1998) Ethogram of the bottlenose dolphin, with special reference to solitary and sociable dolphins. *Vie et Milieu* 48: 89–104.

Mundinger PC (1999) Genetics of canary song learning: Innate mechanisms and other neurobiological considerations. In Hauser MD and Konishi M (eds) *The Design of Animal Communication* MIT Press, Cambridge, pp. 369–390.

Murray CD (1926a) The physiological principle of minimum work. I. The vascular system and the cost of blood volume. *Proc. Natl. Acad. Sci. USA* 12: 207–214.

Murray CD (1926b) The physiological principle of minimum work applied to the angle of branching of arteries. *J. Gen. Physiol.* 9: 835–841.

Murray CD (1927) A relationship between circumference and weight in trees and its bearing on branching angles. *J. Gen. Physiol.* 10: 725–729.

Nakayama K and Shimojo S (1992) Experiencing and perceiving visual surfaces. *Science* 257: 1357–1363.

Netter FH (1997) *Atlas of Human Anatomy.* East Hanover, New Jersey.

Nickel R, Schummer A, Seiferle E, Siller WG and Wight PAL (1977) *Anatomy of the Domestic Birds.* Springer-Verlag, New York.

Nijhawan R (1994) Motion extrapolation in catching. *Nature* 370: 256–257.

Nijhawan R (1997) Visual decomposition of colour through motion extrapolation. *Nature* 386: 66–69.

Nijhawan R (2001) The flash-lag phenomenon: object motion and eye movements. *Perception* 30: 263–282.

Nimchinsky EA, Gilissen E, Allman JM, Perl DP, Erwin JM (1999) A neuronal morphologic type unique to humans and great apes. *Proc. Natl. Acad. Sci. USA* 96: 5268–5273.

Nishida S and Johnston A (1999) Influence of motion signals on the perceived position of spatial pattern. *Nature* 397: 610–612.

Nowak RM (1999) *Walker's Mammals of the World.* The Johns Hopkins University Press, Baltimore.

Nundy S, Lotto B, Coppola D, Shimpi A and Purves D (2000) Why are angles misperceived? *Proc. Nat. Acad. Sci.* 97: 5592–5597.

Orbison WD (1939) Shape as a function of the vector-field. *American Journal of Psychology* 52: 31–45.

Oswald M and Lockard JS (1980) Ethogram of the De Brazza's guenon (Cercopithecus neglectus) in captivity. *Appl. Anim. Ethol.* 6: 285–296.

Owings DH, Borchert M and Virginia R (1977) The behaviour of California Ground squirrels. *Anim. Behav.* 25: 221–230.

Palmer SE (1999) *Vision Science: Photons to Phenomenology.* MIT Press, Cambridge.

Parker SP (ed) (1982) *Synopsis and classification of living organisms.* McGraw-Hill, New York.

Pascual-Leone J (1970) A mathematical model for the transition rule in Piaget's developmental stages. *Acta Psychologica* 32: 301–345.

Pasquini C, Reddy VK and Ratzlaff MH (1983) *Atlas of Equine Anatomy.* Sudz, Eureka.

Passingham RE (1973) Anatomical differences between the neocortex of man and other primates. *Brain Behav. Evol.* 7: 337–359.

Patenaude F (1984) The ontogeny of behavior of free-living beavers (*Castor canadensis*). *J. Compar. Ethol.* 66: 33–44.

Pearse V, Pearse J, Buchsbaum M, Buchsbaum R (1987) *Living invertebrates.* Blackwell Scientific, Palo Alto.

Pearson K (1992) *The Grammar of Science.* J. M. Dent and Sons, London.

Penrose R (1994) *Shadows of the Mind.* Oxford University Press, Oxford.

Pickwell G (1947) *Amphibians and Reptiles of the Pacific States.* Stanford University Press, Stanford University.

Pinker S (1994) *The Language Instinct.* HarperPerennial, New York.

Pinna B and Brelstaff GJ (2000) A new visual illusion of relative motion. *Vision Research* 40: 2091–2096.

Popesko P, Rajtov V and Horák J (1990) *A Colour Atlas of the Anatomy of Small Laboratory Animals.* Wolfe Publishing, Bratislava.

Popper KR (1959) *The Logic of Scientific Discovery*. Hutchinson and Company, London.

Prothero J (1997a) Cortical scaling in mammals: A repeating units model. *J. Brain Res.* 38: 195–207.

Prothero J (1997b) Scaling of cortical neuron density and white matter volume in mammals. *J. Brain Res.* 38: 513–524.

Prothero JW and Sundsten JW (1985) Folding of the cerebral cortex in mammals. *Brain Behav. Evol.* 24: 152–167.

Purushothaman G, Patel SS, Bedell HE and Ogmen H (1998) Moving ahead through differential visual latency. *Nature* 396: 424.

Putnam H (1980) Models and reality. *Journal of Symbolic Logic* 45: 464–482.

Putnam H (1981) *Reason, Truth and History*. Cambridge University Press, Cambridge.

Putnam H (1983) Vagueness and alternative logic. *Erkenntnis* 19: 297–314.

Quine WVO (1960) *Word and Object*. MIT Press, Cambridge.

Quine WVO (1963) On simple theories in a complex world. *Synthese* 15: 103–106.

Raghavan D (1964) *Anatomy of the Ox*. Indian Council of Agricultural Research, Calcutta.

Ramachandran NK (1998) Activity patterns and time budgets of the pheasant-tailed (Hydrophasianus chirurgus) and bronzewinged (Metopidius indicus) jacanas. *J. Bombay Nat. History Soc.* 95: 234–245.

Ramsey FP (1931) Truth and probability. *Foundations of Mathematics and Other Essays*. Routledge and Kegan Paul, London.

Read AF and Weary DM (1992) The evolution of bird song: Comparative analyses. *Phil. Trans. R. Soc. Lond.* B 338: 165–187.

Reichenbach H (1938) *Experience and Prediction*. University of Chicago Press, Chicago.

Reighard J and Jennings HS (1929) *Anatomy of the Cat*. Henry Holt and Company, New York.

Ringo JL (1991) Neuronal interconnection as a function of brain size. *Brain Behav. Evol.* 38: 1–6.

Ringo JL, Doty RW, Demeter S, Simard PY (1994) Time is of the essence: A conjecture that hemispheric specialization arises from interhemispheric conduction delay. *Cereb. Cortex* 4: 331–343.

Robinson BF and Mervis CB (1998) Disentangling early language development: Modeling lexical and grammatical acquisition using an extension of case-study methodology. *Devel. Psychol.* 34: 363–375.

Rock I (1975) *An Introduction to Perception*. Macmillan, New York.

Rock I (1983) *The Logic of Perception*. MIT Press, Cambridge.

Rock I (1984) *Perception*. Scientific American Library, New York.

Rockel AJ, Hiorns RW and Powell TPS (1980) The basic uniformity in structure of the neocortex. *Brain* 103: 221–244.

Rodger RS and Rosebrugh RD (1979) Computing the grammar for sequences of behavioural acts. *Anim. Behav.* 27: 737–749.

Rohen JW and Yokochi C (1993) *Color Atlas of Anatomy.* Igaku-Shoin, New York.

Roper TJ and Polioudakis E (1977) The behaviour of Mongolian gerbils in a semi-natural environment, with special reference to ventral marking, dominance and sociability. *Behaviour* 61: 207–237.

Ross MH, Romrell LJ and Kaye GI (1995) *Histology: A Text and Atlas.* Williams and Wilkins, Baltimore.

Roy AG and Woldenberg MJ (1982) A generalization of the optimal models of arterial branching. *Bull. of Math. Bio.* 44: 349–360.

Ruby DE and Niblick HA (1994) A behavioral inventory of the desert tortoise: Development of an ethogram. *Herpetol. Monogr.* 8: 88–102.

Ruppin E, Schwartz EL and Yeshurun Y (1993) Examining the volume efficiency of the cortical architecture in a multi-processor network model. *Biol. Cybern.* 70: 89–94.

Russell B (1918) *Mysticism and Logic.* Longmans; now Allen and Unwin, New York.

Sainsbury M (1990) Concepts without boundaries. Inaugural lecture delivered November 6, 1990, published by King's College London Department of Philosophy, London WC2R 2LS.

Salmon WC (1966) *The Foundations of Scientific Inference.* University of Pittsburgh Press, Pittsburgh.

Salmon WC (1990) Rationality and objectivity in science, or Tom Kuhn meets Tom Bayes. In Savage CW (ed) *Scientific Theories.* University of Minnesota Press, Twin Cities, pp. 175–204.

Santos RS and Barreiros JP (1993) The ethogram of Parablennius sanguinolentus parvicornis (Valenciennes in Cuvier and Valenciennes, 1836) (Pisces: Blenniidae) from the Azores. *Arquipelago ciencias da natureza* 0: 73–90.

Scannel JW and Young MP (1993) The connectional organization of neural systems in the cat cerebral cortex. *Current Biol.* 3: 191–200.

Schlag J, Cai RH, Dorfman A, Mohempour A and Schlag-Rey M (2000) Extrapolating movement without retinal motion. *Nature* 403: 38–39.

Schleidt WM, Yakalis G, Donnelly M and McGarry J (1984) A proposal for a standard ethogram, exemplified by an ethogram of the bluebreasted quail (Coturnix chinensis). *J. Compar. Ethol.* 64: 193–220.

Schlossberg L and Zuidema GD (1997) *The Johns Hopkins Atlas of Human Functional Anatomy.* Johns Hopkins University Press, Baltimore.

Schmidt-Nielson K (1984) *Scaling: Why is Animal Size So Important?* Cambridge University Press, Cambridge.

Schmolesky MT, Wang Y, Hanes DP, Thompson KG, Leutger S, Schall JD and Leventhal AG (1998) Signal timing across the macaque visual system. *J. Neurophysiol.* 79: 3272–3278.

Schrater PR, Knill DC and Simoncelli EP (2001) Perceiving visual expansion without optic flow. *Nature* 410: 816–819.

Schreiner W and Buxbaum PF (1993) Computer-optimization of vascular trees. *IEEE Transactions on Biomedical Engineering* 40: 482–491.

Schreiner W, Neumann M, Neumann F, Roedler SM, End A, Buxbaum P, Müller MR and Spieckermann P (1994) The branching angles in computer-generated optimized models of arterial trees. *J. Gen. Physiol.* 103: 975–989.

Schreiner W, Neumann F, Neumann M, End A, and Müller MR (1996) Structural quantification and bifurcation symmetry in arterial tree models generated by constrained constructive optimization. *J. Theor. Biol.* 180: 161–174.

Schwarz G (1978) Estimating the dimension of a model. *The Annals of Statistics* 6: 461–464.

Schüz A (1998) Neuroanatomy in a computational perspective. In Arbib MA (ed) *The Handbook of Brain Theory and Neural Networks.* MIT Press, Cambridge, pp. 622–626.

Scudder HH (1923) Sentence length. *English J.* 12: 617–620.

Segall MH, Campbell DT and Herskovits MJ (1966) *The Influence of Culture on Visual Perception.* The Bobbs-Merill Co., New York.

Sherwani N (1995) *Algorithms for VLSI physical design automation.* Kluwer Academic, Boston.

Sheth BR, Nijhawan R and Shimojo S (2000) Changing objects lead briefly flashed ones. *Nature Neurosci.* 3: 489–495.

Shultz JR and Wang SS-H (2001) How the cortex got its folds: Selection constraints due to preservation of cross-brain conduction time. Proceedings of Neuroscience Conference.

Siegel LS and Ryan EB (1989) The development of working memory in normally achieving and subtypes of learning disabled children. *Child Develop.* 60: 973–980.

Singh H and Roy KS (1997) *Atlas of the Buffalo Anatomy.* Indian Council of Agricultural Research, Pusa, New Delhi.

Sisson S and Grossman JD (1953) *The Anatomy of the Domestic Animals.* W. B. Saunders, Philadelphia.

Slater PJB (1973) Describing sequences of behavior. In Bateson PPG and Klopfer PH (eds) *Perspectives in Ethology.* Plenum Press, New York, pp. 131–153.

Smith AFM and Spiegelhalter DJ (1980) Bayes factors and choice criteria for linear models. *Journal of the Royal Statistical Society B* 42: 213–220.

Sober E (1975) *Simplicity.* Oxford University Press, Oxford.

Sorensen RA (1985) An argument for the vagueness of 'vague'. *Analysis* 45: 154–157.

Sorensen RA (1988) *Blindspots.* Clarendon Press, Oxford.

Sorensen RA (1994) A thousand clones. *Mind* 103: 47–54.

Stalnaker R (1979) Anti-essentialism. *Midwest Studies in Philosophy* 4: 343–355.

Stamhuis EJ, Reede-Dekker T, van Etten Y, de Wilges JJ and Videler JJ (1996) Behaviour and time allocation of the burrowing shrimp Callianassa subterranea (Decapoda, Thalassinidea). *J. Exp. Marine Biol. Ecol.* 204: 225–239.

Stebbins RC (1954) *Amphibians and reptiles of western North America.* McGraw-Hill, New York.

Stevens CE (1989) How cortical interconnectedness varies with network size. *Neural Computation* 1: 473–479.

Stevenson MF and Poole TB (1976) An ethogram of the Common Marmoset (Calithrix jacchus jacchus): General behavioural repertoire. *Anim. Behav.* 24: 428–451.

Stone RJ and Stone JA (1990) *Atlas of the Skeletal Muscles.* Wm. C. Brown, Dubuque.

Streeter V and Wylie E (1985) *Fluid Mechanics.* McGraw-Hill, New York, 8th edition.

Suppe F (1989) *The Semantic Conception of Theories and Scientific Realism.* University of Illinois Press, Urbana.

Thompson D (1992) *On Growth and Form.* Dover, New York, the complete revised edition.

Thorson J and Lange GD and Biederman-Thorson M (1969) Objective measure of the dynamics of a visual movement illusion. *Science* 164: 1087–1088.

Thouless RH (1931a) Phenomenal regression to the real object. I. *B. J. Psych.* 21: 339–359.

Thouless RH (1931b) Phenomenal regression to the real object. II. *B. J. Psych.* 22: 1–30.

Tinbergen N (1950) The hierarchical organization of nervous mechanisms underlying instinctive behaviour. *Symp. Soc. Exp. Biol.* 4: 305–312.

Torr GA and Shine R (1994) An ethogram for the small scincid lizard Lampropholis guichenoti. *Amphibia-Reptilia* 15: 21–34.

Tower DB (1954) Structural and functional organization of mammalian cerebral cortex: The correlation of neurone density with brain size. *J. Compar. Neurol.* 101: 9–52.

Traverso S, Morchio R and Tamone G (1992) Neuronal growth and the Steiner problem. *Rivista di Biologia-Biology Forum* 85: 405–418.

Turney P (1990) The curve fitting problem: A solution. *British Journal for the Philosophy of Science* 41: 509–530.

Tye M (1994) Why the vague need not be higher-order vague. *Mind* 103: 43–45.

Unger P (1979) There are no ordinary things. *Synthese* 41: 117–54.

Uribe F (1982) Quantitative ethogram of Ara ararauna and Ara macao (Aves, Psittacidae) in captivity. *Biology of Behaviour* 7: 309–323.

Van Essen DC (1997) A tension-based theory of morphogenesis and compact wiring in the central nervous system. *Nature* 385: 313–319.

Varignon M (1725) *Nouvelle Mecanique.* Claude Jombert, Paris, vol. 1 and 2.

Velten HV (1943) The growth of phonemic and lexical patterns in infant language. *Language* 19: 281–292.

Walker WF (1988) *Anatomy and Dissection of the Fetal Pig.* W.H. Freeman, New York.

Watts DJ and Strogatz SH (1998) Collective dynamics of 'small-world' networks. *Nature* 393: 440–442.

Way RF and Lee DG (1965) *The Anatomy of the Horse.* J. B. Lippincott, Philadelphia.

Webster AB and Hurnik JF (1990) An ethogram of white leghorn-type hens in battery cages. *Can. J. Anim. Sci.* 70: 751–760.

West GB, Brown JH and Enquist BJ (1997) A general model for the origin of allometric scaling laws in biology. *Science* 276: 122–126.

Wheeler SC (1979) On that which is not. *Synthese* 41: 155–194.

Whitney D and Cavanagh P (2000) Motion distorts visual space: shifting the perceived position of remote stationary objects. *Nature Neurosci.* 3: 954–959.

Whitney D and Murakami I (1998) Latency difference, not spatial extrapolation. *Nature Neurosci.* 1: 656–657.

Whitney D, Murakami I and Cavanagh P (2000) Illusory spatial offset of a flash relative to a moving stimulus is caused by differential latencies for moving and flashed stimuli. *Vision Res.* 40: 137–149.

Williams PL, Warwick R, Dyson M and Bannister LH (1989) *Gray's Anatomy.* Churchill Livingstone, New York.

Williamson T (1994) *Vagueness.* Routledge, London.

Wilson EO and Fagen R (1974) On the estimation of total behavioral repertoires in ants. *J. NY Entomol. Soc.* 82: 106–112.

Wingerd BD (1985) *Rabbit Dissection Manual.* The Johns Hopkins University Press, Baltimore.

Wittgenstein L (1961) *Tractatus Logico-philosophicus.* Routledge and Kegan Paul, London.

Woldenberg MJ and Horsfield K (1983) Finding the optimal lengths for three branches at a junction. *J. Theor. Biol.* 104: 301–318.

Woldenberg MJ and Horsfield K (1986) Relation of branching angles to optimality for four cost principles. *J. Theor. Biol.* 122: 187–204.

Woolridge D (1963) *The Machinery of the Brain.* McGraw-Hill, New York.

Wooton RJ (1972) The behaviour of the male three-spined stickleback in a natural situation: A quantitative description. *Behaviour* 41: 232–241.

Wright C (1975) On the coherence of vague predicates. *Synthese* 30: 325–365.

Young MP (1993) The organization of neural systems in the primate cerebral cortex. *Proc. R. Soc. Lond.* B 252: 13–18.

Zabell SL (1989) The rule of succession. *Erkenntnis* 31: 283–321.

Zadeh LA (1965) Fuzzy sets. *Information and Control* 8: 338–353.

Zamir M (1976) Optimality principles in arterial branching. *J. Theor. Biol.* 62: 227–251.

Zamir M (1978) Nonsymmetrical bifurcations in arterial branching. *J. Gen. Physiol.* 72: 837–845.

Zamir M (1986) Cost analysis of arterial branching in the cardiovascular systems of man and animals. *J. Theor. Biol.* 120: 111–123.

Zamir M, Wrigley SM, and Langille BL (1983) Arterial bifurcations in the cardiovascular system of a rat. *J. Gen. Physiol.* 81: 325–335.

Zamir M, Phipps S, Languille BL and Wonnacott TH (1984) Branching characteristics of coronary arteries in rats. *Can. J. Physiol. Pharmacol.* 62: 1453–1459.

Zamir M and Chee H (1986) Branching characteristics of human coronary arteries. *Can. J. Physiol. Pharmacol.* 64: 661–668.

Index

313

1. J. M. Bochénski, *A Precis of Mathematical Logic.* Translated from French and German by O. Bird. 1959   ISBN 90-277-0073-7
2. P. Guiraud, *Problèmes et méthodes de la statistique linguistique.* 1959   ISBN 90-277-0025-7
3. H. Freudenthal (ed.), *The Concept and the Role of the Model in Mathematics and Natural and Social Sciences.* 1961   ISBN 90-277-0017-6
4. E. W. Beth, *Formal Methods.* An Introduction to Symbolic Logic and to the Study of Effective Operations in Arithmetic and Logic. 1962   ISBN 90-277-0069-9
5. B. H. Kazemier and D. Vuysje (eds.), *Logic and Language.* Studies dedicated to Professor Rudolf Carnap on the Occasion of His 70th Birthday. 1962   ISBN 90-277-0019-2
6. M. W. Wartofsky (ed.), *Proceedings of the Boston Colloquium for the Philosophy of Science, 1961–1962.* [Boston Studies in the Philosophy of Science, Vol. I] 1963 ISBN 90-277-0021-4
7. A. A. Zinov'ev, *Philosophical Problems of Many-valued Logic.* A revised edition, edited and translated (from Russian) by G. Küng and D.D. Comey. 1963   ISBN 90-277-0091-5
8. G. Gurvitch, *The Spectrum of Social Time.* Translated from French and edited by M. Korenbaum and P. Bosserman. 1964   ISBN 90-277-0006-0
9. P. Lorenzen, *Formal Logic.* Translated from German by F.J. Crosson. 1965
                                                                           ISBN 90-277-0080-X
10. R. S. Cohen and M. W. Wartofsky (eds.), *Proceedings of the Boston Colloquium for the Philosophy of Science, 1962–1964.* In Honor of Philipp Frank. [Boston Studies in the Philosophy of Science, Vol. II] 1965   ISBN 90-277-9004-0
11. E. W. Beth, *Mathematical Thought.* An Introduction to the Philosophy of Mathematics. 1965
                                                                           ISBN 90-277-0070-2
12. E. W. Beth and J. Piaget, *Mathematical Epistemology and Psychology.* Translated from French by W. Mays. 1966   ISBN 90-277-0071-0
13. G. Küng, *Ontology and the Logistic Analysis of Language.* An Enquiry into the Contemporary Views on Universals. Revised ed., translated from German. 1967   ISBN 90-277-0028-1
14. R. S. Cohen and M. W. Wartofsky (eds.), *Proceedings of the Boston Colloquium for the Philosophy of Sciences, 1964–1966.* In Memory of Norwood Russell Hanson. [Boston Studies in the Philosophy of Science, Vol. III] 1967   ISBN 90-277-0013-3
15. C. D. Broad, *Induction, Probability, and Causation.* Selected Papers. 1968
                                                                           ISBN 90-277-0012-5
16. G. Patzig, *Aristotle's Theory of the Syllogism.* A Logical-philosophical Study of *Book A* of the *Prior Analytics.* Translated from German by J. Barnes. 1968   ISBN 90-277-0030-3
17. N. Rescher, *Topics in Philosophical Logic.* 1968   ISBN 90-277-0084-2
18. R. S. Cohen and M. W. Wartofsky (eds.), *Proceedings of the Boston Colloquium for the Philosophy of Science, 1966–1968, Part I.* [Boston Studies in the Philosophy of Science, Vol. IV] 1969   ISBN 90-277-0014-1
19. R. S. Cohen and M. W. Wartofsky (eds.), *Proceedings of the Boston Colloquium for the Philosophy of Science, 1966–1968, Part II.* [Boston Studies in the Philosophy of Science, Vol. V] 1969   ISBN 90-277-0015-X
20. J. W. Davis, D. J. Hockney and W. K. Wilson (eds.), *Philosophical Logic.* 1969
                                                                           ISBN 90-277-0075-3
21. D. Davidson and J. Hintikka (eds.), *Words and Objections.* Essays on the Work of W. V. Quine. 1969, rev. ed. 1975   ISBN 90-277-0074-5; Pb 90-277-0602-6
22. P. Suppes, *Studies in the Methodology and Foundations of Science. Selected Papers from 1951 to 1969.* 1969   ISBN 90-277-0020-6
23. J. Hintikka, *Models for Modalities.* Selected Essays. 1969
                                                        ISBN 90-277-0078-8; Pb 90-277-0598-4

24. N. Rescher *et al.* (eds.), *Essays in Honor of Carl G. Hempel.* A Tribute on the Occasion of His 65th Birthday. 1969 ISBN 90-277-0085-0
25. P. V. Tavanec (ed.), *Problems of the Logic of Scientific Knowledge.* Translated from Russian. 1970 ISBN 90-277-0087-7
26. M. Swain (ed.), *Induction, Acceptance, and Rational Belief.* 1970 ISBN 90-277-0086-9
27. R. S. Cohen and R. J. Seeger (eds.), *Ernst Mach: Physicist and Philosopher.* [Boston Studies in the Philosophy of Science, Vol. VI]. 1970 ISBN 90-277-0016-8
28. J. Hintikka and P. Suppes, *Information and Inference.* 1970 ISBN 90-277-0155-5
29. K. Lambert, *Philosophical Problems in Logic.* Some Recent Developments. 1970
ISBN 90-277-0079-6
30. R. A. Eberle, *Nominalistic Systems.* 1970 ISBN 90-277-0161-X
31. P. Weingartner and G. Zecha (eds.), *Induction, Physics, and Ethics.* 1970 ISBN 90-277-0158-X
32. E. W. Beth, *Aspects of Modern Logic.* Translated from Dutch. 1970 ISBN 90-277-0173-3
33. R. Hilpinen (ed.), *Deontic Logic.* Introductory and Systematic Readings. 1971
See also No. 152. ISBN Pb (1981 rev.) 90-277-1302-2
34. J.-L. Krivine, *Introduction to Axiomatic Set Theory.* Translated from French. 1971
ISBN 90-277-0169-5; Pb 90-277-0411-2
35. J. D. Sneed, *The Logical Structure of Mathematical Physics.* 2nd rev. ed., 1979
ISBN 90-277-1056-2; Pb 90-277-1059-7
36. C. R. Kordig, *The Justification of Scientific Change.* 1971
ISBN 90-277-0181-4; Pb 90-277-0475-9
37. M. Čapek, *Bergson and Modern Physics.* A Reinterpretation and Re-evaluation. [Boston Studies in the Philosophy of Science, Vol. VII] 1971 ISBN 90-277-0186-5
38. N. R. Hanson, *What I Do Not Believe, and Other Essays.* Ed. by S. Toulmin and H. Woolf. 1971 ISBN 90-277-0191-1
39. R. C. Buck and R. S. Cohen (eds.), *PSA 1970.* Proceedings of the Second Biennial Meeting of the Philosophy of Science Association, Boston, Fall 1970. In Memory of Rudolf Carnap. [Boston Studies in the Philosophy of Science, Vol. VIII] 1971
ISBN 90-277-0187-3; Pb 90-277-0309-4
40. D. Davidson and G. Harman (eds.), *Semantics of Natural Language.* 1972
ISBN 90-277-0304-3; Pb 90-277-0310-8
41. Y. Bar-Hillel (ed.), *Pragmatics of Natural Languages.* 1971
ISBN 90-277-0194-6; Pb 90-277-0599-2
42. S. Stenlund, *Combinators, γ Terms and Proof Theory.* 1972 ISBN 90-277-0305-1
43. M. Strauss, *Modern Physics and Its Philosophy.* Selected Paper in the Logic, History, and Philosophy of Science. 1972 ISBN 90-277-0230-6
44. M. Bunge, *Method, Model and Matter.* 1973 ISBN 90-277-0252-7
45. M. Bunge, *Philosophy of Physics.* 1973 ISBN 90-277-0253-5
46. A. A. Zinov'ev, *Foundations of the Logical Theory of Scientific Knowledge (Complex Logic).* Revised and enlarged English edition with an appendix by G. A. Smirnov, E. A. Sidorenka, A. M. Fedina and L. A. Bobrova. [Boston Studies in the Philosophy of Science, Vol. IX] 1973
ISBN 90-277-0193-8; Pb 90-277-0324-8
47. L. Tondl, *Scientific Procedures.* A Contribution concerning the Methodological Problems of Scientific Concepts and Scientific Explanation. Translated from Czech by D. Short. Edited by R.S. Cohen and M.W. Wartofsky. [Boston Studies in the Philosophy of Science, Vol. X] 1973
ISBN 90-277-0147-4; Pb 90-277-0323-X
48. N. R. Hanson, *Constellations and Conjectures.* 1973 ISBN 90-277-0192-X

49. K. J. J. Hintikka, J. M. E. Moravcsik and P. Suppes (eds.), *Approaches to Natural Language.*
1973 ISBN 90-277-0220-9; Pb 90-277-0233-0
50. M. Bunge (ed.), *Exact Philosophy.* Problems, Tools and Goals. 1973 ISBN 90-277-0251-9
51. R. J. Bogdan and I. Niiniluoto (eds.), *Logic, Language and Probability.* 1973
ISBN 90-277-0312-4
52. G. Pearce and P. Maynard (eds.), *Conceptual Change.* 1973
ISBN 90-277-0287-X; Pb 90-277-0339-6
53. I. Niiniluoto and R. Tuomela, *Theoretical Concepts and Hypothetico-inductive Inference.* 1973
ISBN 90-277-0343-4
54. R. Fraissé, *Course of Mathematical Logic* – Volume 1: *Relation and Logical Formula.* Translated from French. 1973 ISBN 90-277-0268-3; Pb 90-277-0403-1
(For *Volume 2* see under No. 69).
55. A. Grünbaum, *Philosophical Problems of Space and Time.* Edited by R.S. Cohen and M.W.
Wartofsky. 2nd enlarged ed. [Boston Studies in the Philosophy of Science, Vol. XII] 1973
ISBN 90-277-0357-4; Pb 90-277-0358-2
56. P. Suppes (ed.), *Space, Time and Geometry.* 1973 ISBN 90-277-0386-8; Pb 90-277-0442-2
57. H. Kelsen, *Essays in Legal and Moral Philosophy.* Selected and introduced by O. Weinberger.
Translated from German by P. Heath. 1973 ISBN 90-277-0388-4
58. R. J. Seeger and R. S. Cohen (eds.), *Philosophical Foundations of Science.* [Boston Studies in
the Philosophy of Science, Vol. XI] 1974 ISBN 90-277-0390-6; Pb 90-277-0376-0
59. R. S. Cohen and M. W. Wartofsky (eds.), *Logical and Epistemological Studies in Contemporary
Physics.* [Boston Studies in the Philosophy of Science, Vol. XIII] 1973
ISBN 90-277-0391-4; Pb 90-277-0377-9
60. R. S. Cohen and M. W. Wartofsky (eds.), *Methodological and Historical Essays in the Natural
and Social Sciences. Proceedings of the Boston Colloquium for the Philosophy of Science,
1969–1972.* [Boston Studies in the Philosophy of Science, Vol. XIV] 1974
ISBN 90-277-0392-2; Pb 90-277-0378-7
61. R. S. Cohen, J. J. Stachel and M. W. Wartofsky (eds.), *For Dirk Struik. Scientific, Historical
and Political Essays.* [Boston Studies in the Philosophy of Science, Vol. XV] 1974
ISBN 90-277-0393-0; Pb 90-277-0379-5
62. K. Ajdukiewicz, *Pragmatic Logic.* Translated from Polish by O. Wojtasiewicz. 1974
ISBN 90-277-0326-4
63. S. Stenlund (ed.), *Logical Theory and Semantic Analysis.* Essays dedicated to Stig Kanger on
His 50th Birthday. 1974 ISBN 90-277-0438-4
64. K. F. Schaffner and R. S. Cohen (eds.), *PSA 1972. Proceedings of the Third Biennial Meeting of
the Philosophy of Science Association.* [Boston Studies in the Philosophy of Science, Vol. XX]
1974 ISBN 90-277-0408-2; Pb 90-277-0409-0
65. H. E. Kyburg, Jr., *The Logical Foundations of Statistical Inference.* 1974
ISBN 90-277-0330-2; Pb 90-277-0430-9
66. M. Grene, *The Understanding of Nature.* Essays in the Philosophy of Biology. [Boston Studies
in the Philosophy of Science, Vol. XXIII] 1974 ISBN 90-277-0462-7; Pb 90-277-0463-5
67. J. M. Broekman, *Structuralism: Moscow, Prague, Paris.* Translated from German. 1974
ISBN 90-277-0478-3
68. N. Geschwind, *Selected Papers on Language and the Brain.* [Boston Studies in the Philosophy
of Science, Vol. XVI] 1974 ISBN 90-277-0262-4; Pb 90-277-0263-2
69. R. Fraissé, *Course of Mathematical Logic* – Volume 2: *Model Theory.* Translated from French.
1974 ISBN 90-277-0269-1; Pb 90-277-0510-0
(For *Volume 1* see under No. 54)

70. A. Grzegorczyk, *An Outline of Mathematical Logic*. Fundamental Results and Notions explained with all Details. Translated from Polish. 1974   ISBN 90-277-0359-0; Pb 90-277-0447-3

71. F. von Kutschera, *Philosophy of Language*. 1975   ISBN 90-277-0591-7

72. J. Manninen and R. Tuomela (eds.), *Essays on Explanation and Understanding*. Studies in the Foundations of Humanities and Social Sciences. 1976   ISBN 90-277-0592-5

73. J. Hintikka (ed.), *Rudolf Carnap, Logical Empiricist*. Materials and Perspectives. 1975
    ISBN 90-277-0583-6

74. M. Čapek (ed.), *The Concepts of Space and Time*. Their Structure and Their Development. [Boston Studies in the Philosophy of Science, Vol. XXII] 1976
    ISBN 90-277-0355-8; Pb 90-277-0375-2

75. J. Hintikka and U. Remes, *The Method of Analysis*. Its Geometrical Origin and Its General Significance. [Boston Studies in the Philosophy of Science, Vol. XXV] 1974
    ISBN 90-277-0532-1; Pb 90-277-0543-7

76. J. E. Murdoch and E. D. Sylla (eds.), *The Cultural Context of Medieval Learning*. [Boston Studies in the Philosophy of Science, Vol. XXVI] 1975
    ISBN 90-277-0560-7; Pb 90-277-0587-9

77. S. Amsterdamski, *Between Experience and Metaphysics*. Philosophical Problems of the Evolution of Science. [Boston Studies in the Philosophy of Science, Vol. XXXV] 1975
    ISBN 90-277-0568-2; Pb 90-277-0580-1

78. P. Suppes (ed.), *Logic and Probability in Quantum Mechanics*. 1976
    ISBN 90-277-0570-4; Pb 90-277-1200-X

79. H. von Helmholtz: *Epistemological Writings. The Paul Hertz / Moritz Schlick Centenary Edition of 1921 with Notes and Commentary by the Editors*. Newly translated from German by M. F. Lowe. Edited, with an Introduction and Bibliography, by R. S. Cohen and Y. Elkana. [Boston Studies in the Philosophy of Science, Vol. XXXVII] 1975
    ISBN 90-277-0290-X; Pb 90-277-0582-8

80. J. Agassi, *Science in Flux*. [Boston Studies in the Philosophy of Science, Vol. XXVIII] 1975
    ISBN 90-277-0584-4; Pb 90-277-0612-2

81. S. G. Harding (ed.), *Can Theories Be Refuted?* Essays on the Duhem-Quine Thesis. 1976
    ISBN 90-277-0629-8; Pb 90-277-0630-1

82. S. Nowak, *Methodology of Sociological Research*. General Problems. 1977
    ISBN 90-277-0486-4

83. J. Piaget, J.-B. Grize, A. Szemińsska and V. Bang, *Epistemology and Psychology of Functions*. Translated from French. 1977   ISBN 90-277-0804-5

84. M. Grene and E. Mendelsohn (eds.), *Topics in the Philosophy of Biology*. [Boston Studies in the Philosophy of Science, Vol. XXVII] 1976   ISBN 90-277-0595-X; Pb 90-277-0596-8

85. E. Fischbein, *The Intuitive Sources of Probabilistic Thinking in Children*. 1975
    ISBN 90-277-0626-3; Pb 90-277-1190-9

86. E. W. Adams, *The Logic of Conditionals*. An Application of Probability to Deductive Logic. 1975   ISBN 90-277-0631-X

87. M. Przełęcki and R. Wójcicki (eds.), *Twenty-Five Years of Logical Methodology in Poland*. Translated from Polish. 1976   ISBN 90-277-0601-8

88. J. Topolski, *The Methodology of History*. Translated from Polish by O. Wojtasiewicz. 1976
    ISBN 90-277-0550-X

89. A. Kasher (ed.), *Language in Focus: Foundations, Methods and Systems*. Essays dedicated to Yehoshua Bar-Hillel. [Boston Studies in the Philosophy of Science, Vol. XLIII] 1976
    ISBN 90-277-0644-1; Pb 90-277-0645-X

90. J. Hintikka, *The Intentions of Intentionality and Other New Models for Modalities*. 1975
ISBN 90-277-0633-6; Pb 90-277-0634-4

91. W. Stegmüller, *Collected Papers on Epistemology, Philosophy of Science and History of Philosophy*. 2 Volumes. 1977 Set ISBN 90-277-0767-7

92. D. M. Gabbay, *Investigations in Modal and Tense Logics with Applications to Problems in Philosophy and Linguistics*. 1976 ISBN 90-277-0656-5

93. R. J. Bogdan, *Local Induction*. 1976 ISBN 90-277-0649-2

94. S. Nowak, *Understanding and Prediction*. Essays in the Methodology of Social and Behavioral Theories. 1976 ISBN 90-277-0558-5; Pb 90-277-1199-2

95. P. Mittelstaedt, *Philosophical Problems of Modern Physics*. [Boston Studies in the Philosophy of Science, Vol. XVIII] 1976 ISBN 90-277-0285-3; Pb 90-277-0506-2

96. G. Holton and W. A. Blanpied (eds.), *Science and Its Public: The Changing Relationship*. [Boston Studies in the Philosophy of Science, Vol. XXXIII] 1976
ISBN 90-277-0657-3; Pb 90-277-0658-1

97. M. Brand and D. Walton (eds.), *Action Theory*. 1976 ISBN 90-277-0671-9

98. P. Gochet, *Outline of a Nominalist Theory of Propositions*. An Essay in the Theory of Meaning and in the Philosophy of Logic. 1980 ISBN 90-277-1031-7

99. R. S. Cohen, P. K. Feyerabend, and M. W. Wartofsky (eds.), *Essays in Memory of Imre Lakatos*. [Boston Studies in the Philosophy of Science, Vol. XXXIX] 1976
ISBN 90-277-0654-9; Pb 90-277-0655-7

100. R. S. Cohen and J. J. Stachel (eds.), *Selected Papers of Léon Rosenfield*. [Boston Studies in the Philosophy of Science, Vol. XXI] 1979 ISBN 90-277-0651-4; Pb 90-277-0652-2

101. R. S. Cohen, C. A. Hooker, A. C. Michalos and J. W. van Evra (eds.), *PSA 1974*. *Proceedings of the 1974 Biennial Meeting of the Philosophy of Science Association*. [Boston Studies in the Philosophy of Science, Vol. XXXII] 1976 ISBN 90-277-0647-6; Pb 90-277-0648-4

102. Y. Fried and J. Agassi, *Paranoia*. A Study in Diagnosis. [Boston Studies in the Philosophy of Science, Vol. L] 1976 ISBN 90-277-0704-9; Pb 90-277-0705-7

103. M. Przełęçki, K. Szaniawski and R. Wójcicki (eds.), *Formal Methods in the Methodology of Empirical Sciences*. 1976 ISBN 90-277-0698-0

104. J. M. Vickers, *Belief and Probability*. 1976 ISBN 90-277-0744-8

105. K. H. Wolff, *Surrender and Catch*. Experience and Inquiry Today. [Boston Studies in the Philosophy of Science, Vol. LI] 1976 ISBN 90-277-0758-8; Pb 90-277-0765-0

106. K. Kosík, *Dialectics of the Concrete*. A Study on Problems of Man and World. [Boston Studies in the Philosophy of Science, Vol. LII] 1976 ISBN 90-277-0761-8; Pb 90-277-0764-2

107. N. Goodman, *The Structure of Appearance*. 3rd ed. with an Introduction by G. Hellman. [Boston Studies in the Philosophy of Science, Vol. LIII] 1977
ISBN 90-277-0773-1; Pb 90-277-0774-X

108. K. Ajdukiewicz, *The Scientific World-Perspective and Other Essays, 1931-1963*. Translated from Polish. Edited and with an Introduction by J. Giedymin. 1978 ISBN 90-277-0527-5

109. R. L. Causey, *Unity of Science*. 1977 ISBN 90-277-0779-0

110. R. E. Grandy, *Advanced Logic for Applications*. 1977 ISBN 90-277-0781-2

111. R. P. McArthur, *Tense Logic*. 1976 ISBN 90-277-0697-2

112. L. Lindahl, *Position and Change*. A Study in Law and Logic. Translated from Swedish by P. Needham. 1977 ISBN 90-277-0787-1

113. R. Tuomela, *Dispositions*. 1978 ISBN 90-277-0810-X

114. H. A. Simon, *Models of Discovery and Other Topics in the Methods of Science*. [Boston Studies in the Philosophy of Science, Vol. LIV] 1977 ISBN 90-277-0812-6; Pb 90-277-0858-4

115. R. D. Rosenkrantz, *Inference, Method and Decision.* Towards a Bayesian Philosophy of Science. 1977   ISBN 90-277-0817-7; Pb 90-277-0818-5

116. R. Tuomela, *Human Action and Its Explanation.* A Study on the Philosophical Foundations of Psychology. 1977   ISBN 90-277-0824-X

117. M. Lazerowitz, *The Language of Philosophy.* Freud and Wittgenstein. [Boston Studies in the Philosophy of Science, Vol. LV] 1977   ISBN 90-277-0826-6; Pb 90-277-0862-2

118. Not published 119.   J. Pelc (ed.), *Semiotics in Poland, 1894–1969.* Translated from Polish. 1979   ISBN 90-277-0811-8

120. I. Pörn, *Action Theory and Social Science.* Some Formal Models. 1977   ISBN 90-277-0846-0

121. J. Margolis, *Persons and Mind.* The Prospects of Nonreductive Materialism. [Boston Studies in the Philosophy of Science, Vol. LVII] 1977   ISBN 90-277-0854-1; Pb 90-277-0863-0

122. J. Hintikka, I. Niiniluoto, and E. Saarinen (eds.), *Essays on Mathematical and Philosophical Logic.* 1979   ISBN 90-277-0879-7

123. T. A. F. Kuipers, *Studies in Inductive Probability and Rational Expectation.* 1978   ISBN 90-277-0882-7

124. E. Saarinen, R. Hilpinen, I. Niiniluoto and M. P. Hintikka (eds.), *Essays in Honour of Jaakko Hintikka on the Occasion of His 50th Birthday.* 1979   ISBN 90-277-0916-5

125. G. Radnitzky and G. Andersson (eds.), *Progress and Rationality in Science.* [Boston Studies in the Philosophy of Science, Vol. LVIII] 1978   ISBN 90-277-0921-1; Pb 90-277-0922-X

126. P. Mittelstaedt, *Quantum Logic.* 1978   ISBN 90-277-0925-4

127. K. A. Bowen, *Model Theory for Modal Logic.* Kripke Models for Modal Predicate Calculi. 1979   ISBN 90-277-0929-7

128. H. A. Bursen, *Dismantling the Memory Machine.* A Philosophical Investigation of Machine Theories of Memory. 1978   ISBN 90-277-0933-5

129. M. W. Wartofsky, *Models.* Representation and the Scientific Understanding. [Boston Studies in the Philosophy of Science, Vol. XLVIII] 1979   ISBN 90-277-0736-7; Pb 90-277-0947-5

130. D. Ihde, *Technics and Praxis.* A Philosophy of Technology. [Boston Studies in the Philosophy of Science, Vol. XXIV] 1979   ISBN 90-277-0953-X; Pb 90-277-0954-8

131. J. J. Wiatr (ed.), *Polish Essays in the Methodology of the Social Sciences.* [Boston Studies in the Philosophy of Science, Vol. XXIX] 1979   ISBN 90-277-0723-5; Pb 90-277-0956-4

132. W. C. Salmon (ed.), *Hans Reichenbach: Logical Empiricist.* 1979   ISBN 90-277-0958-0

133. P. Bieri, R.-P. Horstmann and L. Krüger (eds.), *Transcendental Arguments in Science.* Essays in Epistemology. 1979   ISBN 90-277-0963-7; Pb 90-277-0964-5

134. M. Marković and G. Petrović (eds.), *Praxis.* Yugoslav Essays in the Philosophy and Methodology of the Social Sciences. [Boston Studies in the Philosophy of Science, Vol. XXXVI] 1979   ISBN 90-277-0727-8; Pb 90-277-0968-8

135. R. Wójcicki, *Topics in the Formal Methodology of Empirical Sciences.* Translated from Polish. 1979   ISBN 90-277-1004-X

136. G. Radnitzky and G. Andersson (eds.), *The Structure and Development of Science.* [Boston Studies in the Philosophy of Science, Vol. LIX] 1979   ISBN 90-277-0994-7; Pb 90-277-0995-5

137. J. C. Webb, *Mechanism, Mentalism and Metamathematics.* An Essay on Finitism. 1980   ISBN 90-277-1046-5

138. D. F. Gustafson and B. L. Tapscott (eds.), *Body, Mind and Method.* Essays in Honor of Virgil C. Aldrich. 1979   ISBN 90-277-1013-9

139. L. Nowak, *The Structure of Idealization.* Towards a Systematic Interpretation of the Marxian Idea of Science. 1980   ISBN 90-277-1014-7

140. C. Perelman, *The New Rhetoric and the Humanities.* Essays on Rhetoric and Its Applications. Translated from French and German. With an Introduction by H. Zyskind. 1979
ISBN 90-277-1018-X; Pb 90-277-1019-8

141. W. Rabinowicz, *Universalizability.* A Study in Morals and Metaphysics. 1979
ISBN 90-277-1020-2

142. C. Perelman, *Justice, Law and Argument.* Essays on Moral and Legal Reasoning. Translated from French and German. With an Introduction by H.J. Berman. 1980
ISBN 90-277-1089-9; Pb 90-277-1090-2

143. S. Kanger and S. Öhman (eds.), *Philosophy and Grammar.* Papers on the Occasion of the Quincentennial of Uppsala University. 1981   ISBN 90-277-1091-0

144. T. Pawlowski, *Concept Formation in the Humanities and the Social Sciences.* 1980
ISBN 90-277-1096-1

145. J. Hintikka, D. Gruender and E. Agazzi (eds.), *Theory Change, Ancient Axiomatics and Galileo's Methodology.* Proceedings of the 1978 Pisa Conference on the History and Philosophy of Science, Volume I. 1981   ISBN 90-277-1126-7

146. J. Hintikka, D. Gruender and E. Agazzi (eds.), *Probabilistic Thinking, Thermodynamics, and the Interaction of the History and Philosophy of Science.* Proceedings of the 1978 Pisa Conference on the History and Philosophy of Science, Volume II. 1981 ISBN 90-277-1127-5

147. U. Mönnich (ed.), *Aspects of Philosophical Logic.* Some Logical Forays into Central Notions of Linguistics and Philosophy. 1981   ISBN 90-277-1201-8

148. D. M. Gabbay, *Semantical Investigations in Heyting's Intuitionistic Logic.* 1981
ISBN 90-277-1202-6

149. E. Agazzi (ed.), *Modern Logic – A Survey.* Historical, Philosophical, and Mathematical Aspects of Modern Logic and Its Applications. 1981   ISBN 90-277-1137-2

150. A. F. Parker-Rhodes, *The Theory of Indistinguishables.* A Search for Explanatory Principles below the Level of Physics. 1981   ISBN 90-277-1214-X

151. J. C. Pitt, *Pictures, Images, and Conceptual Change.* An Analysis of Wilfrid Sellars' Philosophy of Science. 1981   ISBN 90-277-1276-X; Pb 90-277-1277-8

152. R. Hilpinen (ed.), *New Studies in Deontic Logic.* Norms, Actions, and the Foundations of Ethics. 1981   ISBN 90-277-1278-6; Pb 90-277-1346-4

153. C. Dilworth, *Scientific Progress.* A Study Concerning the Nature of the Relation between Successive Scientific Theories. 3rd rev. ed., 1994   ISBN 0-7923-2487-0; Pb 0-7923-2488-9

154. D. Woodruff Smith and R. McIntyre, *Husserl and Intentionality.* A Study of Mind, Meaning, and Language. 1982   ISBN 90-277-1392-8; Pb 90-277-1730-3

155. R. J. Nelson, *The Logic of Mind.* 2nd. ed., 1989   ISBN 90-277-2819-4; Pb 90-277-2822-4

156. J. F. A. K. van Benthem, *The Logic of Time.* A Model-Theoretic Investigation into the Varieties of Temporal Ontology, and Temporal Discourse. 1983; 2nd ed., 1991   ISBN 0-7923-1081-0

157. R. Swinburne (ed.), *Space, Time and Causality.* 1983   ISBN 90-277-1437-1

158. E. T. Jaynes, *Papers on Probability, Statistics and Statistical Physics.* Ed. by R. D. Rozenkrantz. 1983   ISBN 90-277-1448-7; Pb (1989) 0-7923-0213-3

159. T. Chapman, *Time: A Philosophical Analysis.* 1982   ISBN 90-277-1465-7

160. E. N. Zalta, *Abstract Objects.* An Introduction to Axiomatic Metaphysics. 1983
ISBN 90-277-1474-6

161. S. Harding and M. B. Hintikka (eds.), *Discovering Reality.* Feminist Perspectives on Epistemology, Metaphysics, Methodology, and Philosophy of Science. 1983
ISBN 90-277-1496-7; Pb 90-277-1538-6

162. M. A. Stewart (ed.), *Law, Morality and Rights.* 1983   ISBN 90-277-1519-X

163. D. Mayr and G. Süssmann (eds.), *Space, Time, and Mechanics*. Basic Structures of a Physical Theory. 1983 ISBN 90-277-1525-4
164. D. Gabbay and F. Guenthner (eds.), *Handbook of Philosophical Logic*. Vol. I: Elements of Classical Logic. 1983 ISBN 90-277-1542-4
165. D. Gabbay and F. Guenthner (eds.), *Handbook of Philosophical Logic*. Vol. II: Extensions of Classical Logic. 1984 ISBN 90-277-1604-8
166. D. Gabbay and F. Guenthner (eds.), *Handbook of Philosophical Logic*. Vol. III: Alternative to Classical Logic. 1986 ISBN 90-277-1605-6
167. D. Gabbay and F. Guenthner (eds.), *Handbook of Philosophical Logic*. Vol. IV: Topics in the Philosophy of Language. 1989 ISBN 90-277-1606-4
168. A. J. I. Jones, *Communication and Meaning*. An Essay in Applied Modal Logic. 1983
ISBN 90-277-1543-2
169. M. Fitting, *Proof Methods for Modal and Intuitionistic Logics*. 1983 ISBN 90-277-1573-4
170. J. Margolis, *Culture and Cultural Entities*. Toward a New Unity of Science. 1984
ISBN 90-277-1574-2
171. R. Tuomela, *A Theory of Social Action*. 1984 ISBN 90-277-1703-6
172. J. J. E. Gracia, E. Rabossi, E. Villanueva and M. Dascal (eds.), *Philosophical Analysis in Latin America*. 1984 ISBN 90-277-1749-4
173. P. Ziff, *Epistemic Analysis*. A Coherence Theory of Knowledge. 1984
ISBN 90-277-1751-7
174. P. Ziff, *Antiaesthetics*. An Appreciation of the Cow with the Subtile Nose. 1984
ISBN 90-277-1773-7
175. W. Balzer, D. A. Pearce, and H.-J. Schmidt (eds.), *Reduction in Science*. Structure, Examples, Philosophical Problems. 1984 ISBN 90-277-1811-3
176. A. Peczenik, L. Lindahl and B. van Roermund (eds.), *Theory of Legal Science*. Proceedings of the Conference on Legal Theory and Philosophy of Science (Lund, Sweden, December 1983). 1984 ISBN 90-277-1834-2
177. I. Niiniluoto, *Is Science Progressive?* 1984 ISBN 90-277-1835-0
178. B. K. Matilal and J. L. Shaw (eds.), *Analytical Philosophy in Comparative Perspective*. Exploratory Essays in Current Theories and Classical Indian Theories of Meaning and Reference. 1985 ISBN 90-277-1870-9
179. P. Kroes, *Time: Its Structure and Role in Physical Theories*. 1985 ISBN 90-277-1894-6
180. J. H. Fetzer, *Sociobiology and Epistemology*. 1985 ISBN 90-277-2005-3; Pb 90-277-2006-1
181. L. Haaparanta and J. Hintikka (eds.), *Frege Synthesized*. Essays on the Philosophical and Foundational Work of Gottlob Frege. 1986 ISBN 90-277-2126-2
182. M. Detlefsen, *Hilbert's Program*. An Essay on Mathematical Instrumentalism. 1986
ISBN 90-277-2151-3
183. J. L. Golden and J. J. Pilotta (eds.), *Practical Reasoning in Human Affairs*. Studies in Honor of Chaim Perelman. 1986 ISBN 90-277-2255-2
184. H. Zandvoort, *Models of Scientific Development and the Case of Nuclear Magnetic Resonance*. 1986 ISBN 90-277-2351-6
185. I. Niiniluoto, *Truthlikeness*. 1987 ISBN 90-277-2354-0
186. W. Balzer, C. U. Moulines and J. D. Sneed, *An Architectonic for Science*. The Structuralist Program. 1987 ISBN 90-277-2403-2
187. D. Pearce, *Roads to Commensurability*. 1987 ISBN 90-277-2414-8
188. L. M. Vaina (ed.), *Matters of Intelligence*. Conceptual Structures in Cognitive Neuroscience. 1987 ISBN 90-277-2460-1

189. H. Siegel, *Relativism Refuted*. A Critique of Contemporary Epistemological Relativism. 1987
ISBN 90-277-2469-5
190. W. Callebaut and R. Pinxten, *Evolutionary Epistemology*. A Multiparadigm Program, with a Complete Evolutionary Epistemology Bibliograph. 1987 ISBN 90-277-2582-9
191. J. Kmita, *Problems in Historical Epistemology*. 1988 ISBN 90-277-2199-8
192. J. H. Fetzer (ed.), *Probability and Causality*. Essays in Honor of Wesley C. Salmon, with an Annotated Bibliography. 1988 ISBN 90-277-2607-8; Pb 1-5560-8052-2
193. A. Donovan, L. Laudan and R. Laudan (eds.), *Scrutinizing Science*. Empirical Studies of Scientific Change. 1988 ISBN 90-277-2608-6
194. H.R. Otto and J.A. Tuedio (eds.), *Perspectives on Mind*. 1988 ISBN 90-277-2640-X
195. D. Batens and J.P. van Bendegem (eds.), *Theory and Experiment*. Recent Insights and New Perspectives on Their Relation. 1988 ISBN 90-277-2645-0
196. J. Österberg, *Self and Others*. A Study of Ethical Egoism. 1988 ISBN 90-277-2648-5
197. D.H. Helman (ed.), *Analogical Reasoning*. Perspectives of Artificial Intelligence, Cognitive Science, and Philosophy. 1988 ISBN 90-277-2711-2
198. J. Woleński, *Logic and Philosophy in the Lvov-Warsaw School*. 1989 ISBN 90-277-2749-X
199. R. Wójcicki, *Theory of Logical Calculi*. Basic Theory of Consequence Operations. 1988
ISBN 90-277-2785-6
200. J. Hintikka and M.B. Hintikka, *The Logic of Epistemology and the Epistemology of Logic*. Selected Essays. 1989 ISBN 0-7923-0040-8; Pb 0-7923-0041-6
201. E. Agazzi (ed.), *Probability in the Sciences*. 1988 ISBN 90-277-2808-9
202. M. Meyer (ed.), *From Metaphysics to Rhetoric*. 1989 ISBN 90-277-2814-3
203. R.L. Tieszen, *Mathematical Intuition*. Phenomenology and Mathematical Knowledge. 1989
ISBN 0-7923-0131-5
204. A. Melnick, *Space, Time, and Thought in Kant*. 1989 ISBN 0-7923-0135-8
205. D.W. Smith, *The Circle of Acquaintance*. Perception, Consciousness, and Empathy. 1989
ISBN 0-7923-0252-4
206. M.H. Salmon (ed.), *The Philosophy of Logical Mechanism*. Essays in Honor of Arthur W. Burks. With his Responses, and with a Bibliography of Burk's Work. 1990
ISBN 0-7923-0325-3
207. M. Kusch, *Language as Calculus vs. Language as Universal Medium*. A Study in Husserl, Heidegger, and Gadamer. 1989 ISBN 0-7923-0333-4
208. T.C. Meyering, *Historical Roots of Cognitive Science*. The Rise of a Cognitive Theory of Perception from Antiquity to the Nineteenth Century. 1989 ISBN 0-7923-0349-0
209. P. Kosso, *Observability and Observation in Physical Science*. 1989 ISBN 0-7923-0389-X
210. J. Kmita, *Essays on the Theory of Scientific Cognition*. 1990 ISBN 0-7923-0441-1
211. W. Sieg (ed.), *Acting and Reflecting*. The Interdisciplinary Turn in Philosophy. 1990
ISBN 0-7923-0512-4
212. J. Karpiński, *Causality in Sociological Research*. 1990 ISBN 0-7923-0546-9
213. H.A. Lewis (ed.), *Peter Geach: Philosophical Encounters*. 1991 ISBN 0-7923-0823-9
214. M. Ter Hark, *Beyond the Inner and the Outer*. Wittgenstein's Philosophy of Psychology. 1990
ISBN 0-7923-0850-6
215. M. Gosselin, *Nominalism and Contemporary Nominalism*. Ontological and Epistemological Implications of the Work of W.V.O. Quine and of N. Goodman. 1990 ISBN 0-7923-0904-9
216. J.H. Fetzer, D. Shatz and G. Schlesinger (eds.), *Definitions and Definability*. Philosophical Perspectives. 1991 ISBN 0-7923-1046-2
217. E. Agazzi and A. Cordero (eds.), *Philosophy and the Origin and Evolution of the Universe*. 1991 ISBN 0-7923-1322-4

218. M. Kusch, *Foucault's Strata and Fields*. An Investigation into Archaeological and Genealogical Science Studies. 1991   ISBN 0-7923-1462-X
219. C.J. Posy, *Kant's Philosophy of Mathematics*. Modern Essays. 1992   ISBN 0-7923-1495-6
220. G. Van de Vijver, *New Perspectives on Cybernetics*. Self-Organization, Autonomy and Connectionism. 1992   ISBN 0-7923-1519-7
221. J.C. Nyíri, *Tradition and Individuality*. Essays. 1992   ISBN 0-7923-1566-9
222. R. Howell, *Kant's Transcendental Deduction*. An Analysis of Main Themes in His Critical Philosophy. 1992   ISBN 0-7923-1571-5
223. A. García de la Sienra, *The Logical Foundations of the Marxian Theory of Value*. 1992
                                                                      ISBN 0-7923-1778-5
224. D.S. Shwayder, *Statement and Referent*. An Inquiry into the Foundations of Our Conceptual Order. 1992   ISBN 0-7923-1803-X
225. M. Rosen, *Problems of the Hegelian Dialectic*. Dialectic Reconstructed as a Logic of Human Reality. 1993   ISBN 0-7923-2047-6
226. P. Suppes, *Models and Methods in the Philosophy of Science: Selected Essays*. 1993
                                                                      ISBN 0-7923-2211-8
227. R. M. Dancy (ed.), *Kant and Critique: New Essays in Honor of W. H. Werkmeister*. 1993
                                                                      ISBN 0-7923-2244-4
228. J. Woleński (ed.), *Philosophical Logic in Poland*. 1993   ISBN 0-7923-2293-2
229. M. De Rijke (ed.), *Diamonds and Defaults*. Studies in Pure and Applied Intensional Logic. 1993   ISBN 0-7923-2342-4
230. B.K. Matilal and A. Chakrabarti (eds.), *Knowing from Words*. Western and Indian Philosophical Analysis of Understanding and Testimony. 1994   ISBN 0-7923-2345-9
231. S.A. Kleiner, *The Logic of Discovery*. A Theory of the Rationality of Scientific Research. 1993
                                                                      ISBN 0-7923-2371-8
232. R. Festa, *Optimum Inductive Methods*. A Study in Inductive Probability, Bayesian Statistics, and Verisimilitude. 1993   ISBN 0-7923-2460-9
233. P. Humphreys (ed.), *Patrick Suppes: Scientific Philosopher*. Vol. 1: Probability and Probabilistic Causality. 1994   ISBN 0-7923-2552-4
234. P. Humphreys (ed.), *Patrick Suppes: Scientific Philosopher*. Vol. 2: Philosophy of Physics, Theory Structure, and Measurement Theory. 1994   ISBN 0-7923-2553-2
235. P. Humphreys (ed.), *Patrick Suppes: Scientific Philosopher*. Vol. 3: Language, Logic, and Psychology. 1994   ISBN 0-7923-2862-0
                                                 Set ISBN (Vols 233–235) 0-7923-2554-0
236. D. Prawitz and D. Westerståhl (eds.), *Logic and Philosophy of Science in Uppsala*. Papers from the 9th International Congress of Logic, Methodology, and Philosophy of Science. 1994
                                                                      ISBN 0-7923-2702-0
237. L. Haaparanta (ed.), *Mind, Meaning and Mathematics*. Essays on the Philosophical Views of Husserl and Frege. 1994   ISBN 0-7923-2703-9
238. J. Hintikka (ed.), *Aspects of Metaphor*. 1994   ISBN 0-7923-2786-1
239. B. McGuinness and G. Oliveri (eds.), *The Philosophy of Michael Dummett*. With Replies from Michael Dummett. 1994   ISBN 0-7923-2804-3
240. D. Jamieson (ed.), *Language, Mind, and Art*. Essays in Appreciation and Analysis, In Honor of Paul Ziff. 1994   ISBN 0-7923-2810-8
241. G. Preyer, F. Siebelt and A. Ulfig (eds.), *Language, Mind and Epistemology*. On Donald Davidson's Philosophy. 1994   ISBN 0-7923-2811-6
242. P. Ehrlich (ed.), *Real Numbers, Generalizations of the Reals, and Theories of Continua*. 1994
                                                                      ISBN 0-7923-2689-X

243. G. Debrock and M. Hulswit (eds.), *Living Doubt*. Essays concerning the epistemology of Charles Sanders Peirce. 1994　　　　　ISBN 0-7923-2898-1

244. J. Srzednicki, *To Know or Not to Know*. Beyond Realism and Anti-Realism. 1994
　　　　　ISBN 0-7923-2909-0

245. R. Egidi (ed.), *Wittgenstein: Mind and Language*. 1995　　ISBN 0-7923-3171-0

246. A. Hyslop, *Other Minds*. 1995　　　　　ISBN 0-7923-3245-8

247. L. Pólos and M. Masuch (eds.), *Applied Logic: How, What and Why*. Logical Approaches to Natural Language. 1995　　　　　ISBN 0-7923-3432-9

248. M. Krynicki, M. Mostowski and L.M. Szczerba (eds.), *Quantifiers: Logics, Models and Computation*. Volume One: Surveys. 1995　　　　　ISBN 0-7923-3448-5

249. M. Krynicki, M. Mostowski and L.M. Szczerba (eds.), *Quantifiers: Logics, Models and Computation*. Volume Two: Contributions. 1995　　　　　ISBN 0-7923-3449-3
　　　　　Set ISBN (Vols 248 + 249) 0-7923-3450-7

250. R.A. Watson, *Representational Ideas from Plato to Patricia Churchland*. 1995
　　　　　ISBN 0-7923-3453-1

251. J. Hintikka (ed.), *From Dedekind to Gödel*. Essays on the Development of the Foundations of Mathematics. 1995　　　　　ISBN 0-7923-3484-1

252. A. Wiśniewski, *The Posing of Questions*. Logical Foundations of Erotetic Inferences. 1995
　　　　　ISBN 0-7923-3637-2

253. J. Peregrin, *Doing Worlds with Words*. Formal Semantics without Formal Metaphysics. 1995
　　　　　ISBN 0-7923-3742-5

254. I.A. Kieseppä, *Truthlikeness for Multidimensional, Quantitative Cognitive Problems*. 1996
　　　　　ISBN 0-7923-4005-1

255. P. Hugly and C. Sayward: *Intensionality and Truth*. An Essay on the Philosophy of A.N. Prior. 1996　　　　　ISBN 0-7923-4119-8

256. L. Hankinson Nelson and J. Nelson (eds.): *Feminism, Science, and the Philosophy of Science*. 1997　　　　　ISBN 0-7923-4162-7

257. P.I. Bystrov and V.N. Sadovsky (eds.): *Philosophical Logic and Logical Philosophy*. Essays in Honour of Vladimir A. Smirnov. 1996　　　　　ISBN 0-7923-4270-4

258. Å.E. Andersson and N-E. Sahlin (eds.): *The Complexity of Creativity*. 1996
　　　　　ISBN 0-7923-4346-8

259. M.L. Dalla Chiara, K. Doets, D. Mundici and J. van Benthem (eds.): *Logic and Scientific Methods*. Volume One of the Tenth International Congress of Logic, Methodology and Philosophy of Science, Florence, August 1995. 1997　　　　　ISBN 0-7923-4383-2

260. M.L. Dalla Chiara, K. Doets, D. Mundici and J. van Benthem (eds.): *Structures and Norms in Science*. Volume Two of the Tenth International Congress of Logic, Methodology and Philosophy of Science, Florence, August 1995. 1997　　　　　ISBN 0-7923-4384-0
　　　　　Set ISBN (Vols 259 + 260) 0-7923-4385-9

261. A. Chakrabarti: *Denying Existence*. The Logic, Epistemology and Pragmatics of Negative Existentials and Fictional Discourse. 1997　　　　　ISBN 0-7923-4388-3

262. A. Biletzki: *Talking Wolves*. Thomas Hobbes on the Language of Politics and the Politics of Language. 1997　　　　　ISBN 0-7923-4425-1

263. D. Nute (ed.): *Defeasible Deontic Logic*. 1997　　　　　ISBN 0-7923-4630-0

264. U. Meixner: *Axiomatic Formal Ontology*. 1997　　　　　ISBN 0-7923-4747-X

265. I. Brinck: *The Indexical 'I'*. The First Person in Thought and Language. 1997
　　　　　ISBN 0-7923-4741-2

266. G. Hölmström-Hintikka and R. Tuomela (eds.): *Contemporary Action Theory*. Volume 1: Individual Action. 1997　　　　　ISBN 0-7923-4753-6; Set: 0-7923-4754-4

267. G. Hölmström-Hintikka and R. Tuomela (eds.): *Contemporary Action Theory.* Volume 2: Social Action. 1997   ISBN 0-7923-4752-8; Set: 0-7923-4754-4

268. B.-C. Park: *Phenomenological Aspects of Wittgenstein's Philosophy.* 1998   ISBN 0-7923-4813-3

269. J. Paśniczek: *The Logic of Intentional Objects.* A Meinongian Version of Classical Logic. 1998   Hb ISBN 0-7923-4880-X; Pb ISBN 0-7923-5578-4

270. P.W. Humphreys and J.H. Fetzer (eds.): *The New Theory of Reference.* Kripke, Marcus, and Its Origins. 1998   ISBN 0-7923-4898-2

271. K. Szaniawski, A. Chmielewski and J. Woleński (eds.): *On Science, Inference, Information and Decision Making.* Selected Essays in the Philosophy of Science. 1998   ISBN 0-7923-4922-9

272. G.H. von Wright: *In the Shadow of Descartes.* Essays in the Philosophy of Mind. 1998   ISBN 0-7923-4992-X

273. K. Kijania-Placek and J. Woleński (eds.): *The Lvov–Warsaw School and Contemporary Philosophy.* 1998   ISBN 0-7923-5105-3

274. D. Dedrick: *Naming the Rainbow.* Colour Language, Colour Science, and Culture. 1998   ISBN 0-7923-5239-4

275. L. Albertazzi (ed.): *Shapes of Forms.* From Gestalt Psychology and Phenomenology to Ontology and Mathematics. 1999   ISBN 0-7923-5246-7

276. P. Fletcher: *Truth, Proof and Infinity.* A Theory of Constructions and Constructive Reasoning. 1998   ISBN 0-7923-5262-9

277. M. Fitting and R.L. Mendelsohn (eds.): *First-Order Modal Logic.* 1998   Hb ISBN 0-7923-5334-X; Pb ISBN 0-7923-5335-8

278. J.N. Mohanty: *Logic, Truth and the Modalities from a Phenomenological Perspective.* 1999   ISBN 0-7923-5550-4

279. T. Placek: *Mathematical Intiutionism and Intersubjectivity.* A Critical Exposition of Arguments for Intuitionism. 1999   ISBN 0-7923-5630-6

280. A. Cantini, E. Casari and P. Minari (eds.): *Logic and Foundations of Mathematics.* 1999   ISBN 0-7923-5659-4   set ISBN 0-7923-5867-8

281. M.L. Dalla Chiara, R. Giuntini and F. Laudisa (eds.): *Language, Quantum, Music.* 1999   ISBN 0-7923-5727-2; set ISBN 0-7923-5867-8

282. R. Egidi (ed.): *In Search of a New Humanism.* The Philosophy of Georg Hendrik von Wright. 1999   ISBN 0-7923-5810-4

283. F. Vollmer: *Agent Causality.* 1999   ISBN 0-7923-5848-1

284. J. Peregrin (ed.): *Truth and Its Nature (if Any).* 1999   ISBN 0-7923-5865-1

285. M. De Caro (ed.): *Interpretations and Causes.* New Perspectives on Donald Davidson's Philosophy. 1999   ISBN 0-7923-5869-4

286. R. Murawski: *Recursive Functions and Metamathematics.* Problems of Completeness and Decidability, Gödel's Theorems. 1999   ISBN 0-7923-5904-6

287. T.A.F. Kuipers: *From Instrumentalism to Constructive Realism.* On Some Relations between Confirmation, Empirical Progress, and Truth Approximation. 2000   ISBN 0-7923-6086-9

288. G. Holmström-Hintikka (ed.): *Medieval Philosophy and Modern Times.* 2000   ISBN 0-7923-6102-4

289. E. Grosholz and H. Breger (eds.): *The Growth of Mathematical Knowledge.* 2000   ISBN 0-7923-6151-2

SYNTHESE LIBRARY

290. G. Sommaruga: *History and Philosophy of Constructive Type Theory*. 2000
ISBN 0-7923-6180-6
291. J. Gasser (ed.): *A Boole Anthology*. Recent and Classical Studies in the Logic of George Boole.
2000 ISBN 0-7923-6380-9
292. V.F. Hendricks, S.A. Pedersen and K.F. Jørgensen (eds.): *Proof Theory*. History and Philosophical Significance. 2000 ISBN 0-7923-6544-5
293. W.L. Craig: *The Tensed Theory of Time*. A Critical Examination. 2000 ISBN 0-7923-6634-4
294. W.L. Craig: *The Tenseless Theory of Time*. A Critical Examination. 2000
ISBN 0-7923-6635-2
295. L. Albertazzi (ed.): *The Dawn of Cognitive Science*. Early European Contributors. 2001
ISBN 0-7923-6799-5
296. G. Forrai: *Reference, Truth and Conceptual Schemes*. A Defense of Internal Realism. 2001
ISBN 0-7923-6885-1
297. V.F. Hendricks, S.A. Pedersen and K.F. Jørgensen (eds.): *Probability Theory*. Philosophy, Recent History and Relations to Science. 2001 ISBN 0-7923-6952-1
298. M. Esfeld: *Holism in Philosophy of Mind and Philosophy of Physics*. 2001
ISBN 0-7923-7003-1
299. E.C. Steinhart: *The Logic of Metaphor*. Analogous Parts of Possible Worlds. 2001
ISBN 0-7923-7004-X
300. To be published.
301. T.A.F. Kuipers: *Structures in Science Heuristic Patterns Based on Cognitive Structures*. An Advanced Textbook in Neo-Classical Philosophy of Science. 2001 ISBN 0-7923-7117-8
302. G. Hon and S.S. Rakover (eds.): *Explanation*. Theoretical Approaches and Applications. 2001
ISBN 1-4020-0017-0
303. G. Holmström-Hintikka, S. Lindström and R. Sliwinski (eds.): *Collected Papers of Stig Kanger with Essays on his Life and Work*. Vol. I. 2001
ISBN 1-4020-0021-9; Pb ISBN 1-4020-0022-7
304. G. Holmström-Hintikka, S. Lindström and R. Sliwinski (eds.): *Collected Papers of Stig Kanger with Essays on his Life and Work*. Vol. II. 2001
ISBN 1-4020-0111-8; Pb ISBN 1-4020-0112-6
305. C.A. Anderson and M. Zelëny (eds.): *Logic, Meaning and Computation*. Essays in Memory of Alonzo Church. 2001 ISBN 1-4020-0141-X
306. P. Schuster, U. Berger and H. Osswald (eds.): *Reuniting the Antipodes – Constructive and Nonstandard Views of the Continuum*. 2001 ISBN 1-4020-0152-5
307. S.D. Zwart: *Refined Verisimilitude*. 2001 ISBN 1-4020-0268-8
308. A.-S. Maurin: *If Tropes*. 2002 ISBN 1-4020-0656-X
309. H. Eilstein (ed.): *A Collection of Polish Works on Philosophical Problems of Time and Spacetime*. 2002 ISBN 1-4020-0670-5
310. Y. Gauthier: *Internal Logic*. Foundations of Mathematics from Kronecker to Hilbert. 2002
ISBN 1-4020-0689-6
311. E. Ruttkamp: *A Model-Theoretic Realist Interpretation of Science*. 2002
ISBN 1-4020-0729-9
312. V. Rantala: *Explanatory Translation*. Beyond the Kuhnian Model of Conceptual Change. 2002
ISBN 1-4020-0827-9
313. L. Decock: *Trading Ontology for Ideology*. 2002 ISBN 1-4020-0865-1

314. O. Ezra: *The Withdrawal of Rights*. Rights from a Different Perspective. 2002
ISBN 1-4020-0886-4

315. P. Gärdenfors, J. Woleński and K. Kijania-Placek: *In the Scope of Logic, Methodology and Philosophy of Science*. Volume One of the 11th International Congress of Logic, Methodology and Philosophy of Science, Cracow, August 1999. 2002
ISBN 1-4020-0929-1; Pb 1-4020-0931-3

316. P. Gärdenfors, J. Woleński and K. Kijania-Placek: *In the Scope of Logic, Methodology and Philosophy of Science*. Volume Two of the 11th International Congress of Logic, Methodology and Philosophy of Science, Cracow, August 1999. 2002
ISBN 1-4020-0930-5; Pb 1-4020-0931-3

317. M.A. Changizi: *The Brain from 25,000 Feet*. High Level Explorations of Brain Complexity, Perception, Induction and Vagueness. 2003
ISBN 1-4020-1176-8

Previous volumes are still available.

KLUWER ACADEMIC PUBLISHERS – DORDRECHT / BOSTON / LONDON

DATE DUE